Fast Satellite Attitude Maneuver and Control

Fast Satellite Attitude Maneuver and Control

Dong Ye
School of Astronautics
Harbin Institute of Technology

You Li
School of Aerospace Science and Technology
Xidian University

Yan Xiao
School of Astronautics
Harbin Institute of Technology

Series Editor
Quan Min Zhu

ELSEVIER

ACADEMIC PRESS
An imprint of Elsevier

Academic Press is an imprint of Elsevier
125 London Wall, London EC2Y 5AS, United Kingdom
525 B Street, Suite 1650, San Diego, CA 92101, United States
50 Hampshire Street, 5th Floor, Cambridge, MA 02139, United States
The Boulevard, Langford Lane, Kidlington, Oxford OX5 1GB, United Kingdom

ISBN 978-0-323-95455-6

For information on all Academic Press publications
visit our website at https://www.elsevier.com/books-and-journals

Publisher: Mara E. Conner
Acquisitions Editor: Sonnini R. Yura
Editorial Project Manager: Sara Greco
Production Project Manager: Fizza Fathima
Cover Designer: Matthew Limbert

Typeset by STRAIVE, India

Contents

Part II
Research on attitude control methods for fast maneuver satellites

Part IV
Hybrid actuator-based fast maneuver and stabilization control for rigid satellites

Preface

Modern space missions such as staring imaging to Earth, push-broom imaging, and single-line array camera imaging require satellites to have strong abilities to maneuver quickly at a large angle. Satellites with these rapid mobility requirements for attitude or orbit control are referred to as agile satellites. This concept is the inevitable product of the diversified development of space missions, and represents higher requirements for rapid mobility of spacecraft attitude and orbit control systems. Compared with traditional spacecraft, the application scope and efficiency of agile satellites are significantly improved, which can ensure a longer transmission time of load data. Agile satellites can realize functions that were previously difficult to achieve using conventional technology, especially under the mission background of remote sensing information acquisition, on orbit service and noncooperative target approaching. Today, in the face of frequent natural disasters and changeable regional situations, agile satellite technology has become a focus issue in the field of spacecraft research. Mastering the control technology of agile satellites will help to obtain ground information in a timely and accurately manner, provide first-hand information for disaster resistance and relief and for planning and decision-making, and is an important indication of improving national strength.

The advantage of agile satellites is based on the rapid mobility of attitude or orbit control. Large-angle attitude rapid maneuver is an important and difficult problem in the design of conventional satellite attitude control systems. This problem is involved in the satellite from the separation of satellite and rocket to the start of load operation, and then to various attitude control modes in the process of data transmission to the ground. An agile satellite needs to have the ability of rapid attitude maneuvering while maintaining high-precision tracking of observation targets. Therefore, the technical core of agile satellite is how to realize the attitude control of large-angle rapid maneuvering with high precision.

Due to the limitations of aerospace technology, hardware, mission requirements, and other aspects, previous space missions did not have high requirements for the attitude rapid maneuver ability of satellites. The design of a satellite attitude controller usually takes stability as the primary goal, and focuses more on the improvement of steady-state accuracy and system robustness. Since the control torque of the system cannot be used to the greatest extent, there are some defects in the convergence rate of the system. Most of the control actuators used in the field of satellite attitude control are angular momentum exchange devices such as flywheels and moment gyros. In order to improve the convergence rate, traditionally control torque and attitude angular velocity are increased; this will have a certain adverse impact on the satellite system, such as control saturation, the problem of Kalman filter accuracy reduction, and star sensor failure in the case of large-angle rapid maneuvering. Therefore, it is necessary to discuss the constraints of control torque and attitude angular velocity when dealing with the problem of large-angle rapid maneuvering of satellites.

Considering the high launch cost and the limitation of the launch carrying capacity, it is necessary to reduce the spacecraft weight as much as possible while meeting the functional requirements of payloads on board. Therefore, many large flexible appendages made of lightweight materials have appeared in modern spacecraft, such as solar panels, flexible antennas, and robot manipulators, resulting in a class of representative spacecraft known as rigid-flexible coupling spacecraft, composed of a rigid body and flexible appendages. This kind of spacecraft experiences a serious coupling effect, which means the attitude maneuvering of the central rigid body can easily stimulate flexible deformation and vibration of large flexible appendages, and the long-term vibrations of flexible appendages will in turn reduce the overall attitude control precision of the satellite. The coupling effect is even worse when flexible satellites are in fast maneuver mode. The linear model used in most existing studies cannot reflect this coupling dynamic characteristic. The zero-order rigid-flexible coupling dynamic model established by the traditional modeling method considers the rigid-flexible coupling characteristics, but it has severe restrictions on the satellite maneuver angle and can only be used in the case of small-angle maneuver. For a large flexible satellite with flexible appendages, if the maneuvering time is long, this will inevitably cause deformation and vibration of its attached flexible appendages. The coupling characteristics of flexible vibration and attitude maneuvering will affect the satellite attitude maneuvering control accuracy, stability, and reliability, and even make the satellite unstable and affect its safety. Therefore, for rigid-flexible coupling satellites, the design of a controller to ensure that the attitude control operation can be completed in a limited time has become a new research direction in the field of aerospace engineering.

Existing satellite attitude control systems mostly use thrusters and flywheels as actuators. A thruster is a very effective attitude control mechanism, and can use stored fuel to output the control torque required in the process of large-angle rapid maneuvering. Although the output of the thruster is much larger than that of the flywheel, the output control torque is not accurate enough and does not meet the requirement of high-precision attitude control. In contrast, although the flywheel is not like a thruster and a control moment gyroscope (CMG), which can provide strong torque, it can output more accurate control torque and consume only electric energy. After comparison of these two, it is obvious that the large-angle rapid maneuvering of agile satellites cannot be realized by relying solely on thrusters and flywheels. If the advantages of the two actuators are effectively combined to avoid their respective disadvantages, the large-angle rapid maneuvering of agile satellites can be achieved with sufficient torque and high precision. In addition, because of its torque amplification characteristics, the CMG has also become one of the preferred actuators for agile satellites. However, due to its singularity, complex mechanism, and limited reduction of rotor speed, it is still difficult to apply in practice. Since a flywheel can supplement the function of a CMG, it is also possible to combine the advantages of the CMG and flywheel effectively to provide significant and high-precision control torque to accomplish large-angle rapid maneuvering of agile satellites.

Based on the above considerations, this book studies the attitude control of agile satellites in rapid maneuvering on the basis of a detailed investigation of current research and considering a variety of on-orbit conditions. Part I is the preliminary of the book, which gives a brief introduction to the attitude kinematics and dynamics for rigid and flexible satellites, as well as some related control theories to facilitate theoretical analysis in the rest of the book. Part II of this book investigates attitude control design that is robust to model uncertainty, disturbance torque, and other disturbances for rigid agile satellites, while considering the constraints of control torque and attitude angular velocity at the same time, to improve the convergence rate of existing classical controllers. In Part III, the finite time attitude control for rigid-flexible coupling agile satellites with large flexible appendages is further considered, and a first-order approximate rigid-flexible coupling dynamic model is established. At the same time, the unmeasurable flexible modal information, chattering phenomenon of sliding mode variable structure control, unknown total dynamic perturbation, and actuator faults are considered. Part IV further studies the trajectory planning of time-optimal attitude maneuver, the hybrid actuator control strategy with thruster-flywheel, and control torque gyro-flywheel of agile satellites in fast attitude maneuver mode. This book mainly provides a systematic attitude control solution for the study of agile satellites in rapid attitude maneuvering.

Ultimately, it is hoped that the research content of this book will lead to more and richer research results of attitude control for agile satellites in rapid maneuvering with advanced control theories in the future.

Dong Ye[a]
You Li[b]
Yan Xiao[a]
[a]Harbin, China
[b]Xi'an, China

Acknowledgments

This book contains the research of three authors that addresses the problem of attitude dynamics and control for spacecraft fast maneuver in the past several years. The research works involved in this book are supported by the National Natural Science Foundation of China under Grants 62073102, 61603115, and 61903289. The authors' gratitude first goes to their supervisor, Prof. Zhaowei Sun, for his enduring support in their academic careers. The authors are also grateful to all the students and friends they have studied and worked with in the past decade at the Research Center of Satellite Technology, Harbin Institute of Technology. In addition, the first author and the third author are grateful to the China Scholarship Council (CSC) for supporting their studies as visiting PhD students at the Georgia Institute of Technology, USA, and Ryerson University, Canada, respectively. Finally, the authors want to thank Fizza Fathima, Srinivasan Bhaskaran, and all other editors at Elsevier Limited who worked on this book for their professionalism and patience throughout the publication process.

Chapter 1

Overview

1.1 Introduction

With the development of space technology, space missions represented by long-time staring at specific targets and stereo imaging with single line array cameras require satellites to be agile. Attitude agility includes three abilities: rapid attitude maneuver, rapid attitude stability, and stability maintaining during maneuver. Different tasks have different requirements for satellite agility.

Compared with conventional satellites, agile satellites can provide the following functions [1]:

(1) Persistent earth gaze: it can enable satellites to have the ability of long-term observation of designated targets on the ground, near earth space and in the universe, and can also be used to maintain the communication link between satellites and ground communication stations for a long time. Especially for natural disaster monitoring, with the help of rapid attitude maneuver and fast attitude stabilization of agile satellites, real-time image information and even video can be provided by high-resolution optical loads on small satellites through real-time target tracking. At present, such satellites, represented by the tubsat series satellites developed by Germany, have been used in wartime monitoring, the development of new ocean crossing routes, disaster information acquisition, and many other aspects [2].

(2) Stereo imaging with single-line-array camera: many space missions require satellites to have the ability to obtain stereo remote sensing images in specific areas on the ground or globally. The key of stereo imaging is to ensure fast attitude stabilization of the satellite after rapid maneuver and avoid the impact of attitude jitter on the re-imaging quality of the imaging system of satellite payloads. In particular, when the satellite works in push scan imaging mode, it is more necessary to maintain attitude and angular velocity stability. The above factors have brought a wide development space for agile satellites. Countries in European and America have already started a series of earth observation plans for relevant technical verification [3].

(3) Observable strip expansion: when using agile satellites in earth observation, the increase of attitude angle can correspondingly expand the width

Fast Satellite Attitude Maneuver and Control. https://doi.org/10.1016/B978-0-323-95455-6.00011-1

of observation strip, increase the observable range and amount of observation data during satellite transit, and shorten the time of obtaining the same amount of information. With the advantages of agile satellites, the time to obtain the same amount of information when carrying the same imaging load for push scanning is much shorter than that of traditional satellites. It has achieved remarkable results in tasks such as rapid acquisition of global remote sensing information and missile early warning [4].

In addition, agile satellites provide technical support for global communication, space monitoring, micro interstellar detection, synthetic aperture radar remote sensing, land and resources observation, and rapid response in military reconnaissance, which is also of great significance for the improvement of national defense strength [4]. Therefore, agile satellite has become one of the hotspots in the field of satellite research [5].

1.2 Development status and trend of agile satellite

Compared with traditional satellites, agile satellites have great advantages. More specifically, they broaden the earth observation ability of satellites and realize satellite missions that were difficult to complete before. At present, agile satellite is a research hotspot in the current aerospace field. Research institutions from countries represented by the United States, France, Italy, and India have put forward a series of agile satellite related plans, most of which have entered or are about to enter the implementation stage. Typical agile satellite programs are as follows:

1.2.1 Pleiades project

As shown in Fig. 1.1, Pleiades project is the ultrahigh-resolution panchromatic multispectral earth observation program launched by ESA, and two satellites of the program were successfully launched in 2011 and 2012, respectively. In order to improve the maneuverability of the satellite and ensure the attitude control accuracy, the two satellites in the Pleiades program adopt the integrated structural frame design to effectively reduce the rotational inertia. In addition, the satellite attitude system adopts three-axis stable control mode. The attitude measurement unit has three star sensors and four fiber optic gyroscopes to provide measurements required for attitude determination during rapid maneuver. The attitude maneuver actuator is four control moment gyroscopes installed on the satellite, which can adjust the attitude of roll and pitch axes by 5° within 8 s, 10° within 10 s, and 60° within 25 s.

1.2.2 Worldview series satellites

Worldview series satellite system is developed by American "Digital Earth" company to replace QuickBird series satellites as a new generation of

FIG. 1.1 Pleiades satellite. © CNES/Mira Productions/Rémy PAROT, 2012.

commercial imaging satellite system. WorldView- I and WorldView-II satellites belonging to this series were launched in 2007 and 2009, respectively. Both satellites adopt a modular design scheme. Compared with WorldView- I, WorldView-II has added new remote sensing equipment and installed a vibration isolation system to suppress satellite chattering and improve imaging quality. Both two satellites adopt three-axis stabilized attitude control system, including star sensor, GPS, inertial measurement unit, CMG, etc., which have $\pm 40^\circ$ mobility of sub satellite points. WorldView-I has attitude angular acceleration of $2.5^\circ/s^2$ and attitude angular velocity of $4.5^\circ/s$. WorldView-II can provide attitude angular acceleration of $1.5^\circ/s^2$ and attitude angular velocity of $3.5^\circ/s$. The corresponding distance to the ground that the satellite can maneuver in 9s is 300km.

1.2.3 CartoSat-2

Cartosat-2 is a high-resolution imaging satellite developed by the Indian Space Research Institute. It was successfully launched in 2007. Cartosat-2 adopts lightweight and compact structure design and carries a panchromatic sensor with a resolution of 0.8m (at present, India is developing a cartosat-3 satellite with a resolution of 0.5m). The imaging width is 9.6km and the revisit cycle is 4days. The satellite also carries an experimental recovery capsule and returned to the ground 12days after launch to do preliminary experimental work for India's manned spaceflight.

Cartosat-2 adopts a three-axis stable attitude control mode. The attitude information is mainly measured by star sensor and inertial measurement unit, while supplemented by sun sensor and magnetometer. The attitude actuator adopts large torque reaction flywheel (0.3 Nm/15Nms), thruster (1 N) and magnetic torquer. The attitude determination accuracy and pointing accuracy are 0.01° and 0.05° respectively, and the three-axis attitude stability is 0.0003°/s, providing the maneuver control capability of $\pm 45^\circ$ in all directions of the imaging strip.

1.2.4 ROCSAT-2

ROCSAT-2, developed by the US orbital Science Corporation and Taiwan's "space program office," was launched on May 21, 2004, Beijing time by the Taurus XL rocket. ROCSAT-2 is the first telemetry satellite with high resolution designed by Taiwan. Their resolution targets are panchromatic 2 m and multispectral 8 m, and its imaging width is 24 km. This satellite can be used for land use, agriculture and forestry planning, environmental monitoring, natural disaster assessment, and lightning research. Make a detour over Taiwan twice a day. Each time it passes over Taiwan Island, if the weather permits, it can shoot four adjacent image strips within 8 min to cover the whole island of Taiwan. It can also change the front and rear elevation of the satellite for three-dimensional shooting. Therefore, it will play a great role in military and disaster relief.

The attitude control mode of Huawei 2 is three-axis stability control. The attitude measurement equipment includes star sensor and gyroscope, while the attitude actuators are four reaction flywheels and thrusters. The pointing accuracy of the satellite can be less than 0.7 km (0.12°). The attitude control system can provide $\pm 45^\circ$ maneuver of both the roll axis and the pitch axis, and the attitude maneuver capability is 45° within 60 s of the pitch axis, 10° within 25 s and 30°, and within 60 s of the roll axis.

Table 1.1 shows the comparison of maneuverability of several agile satellites.

At present, the research on agile satellites is mainly reflected in two aspects:

(a) Fast dynamic task planning under constraints. In 2001, Young et al. [6] studied the systematic architecture and control methods of agile control. In 2002, Lemaitre et al. [7] studied the observation scheduling problem of the French agile earth observation satellite AEOs for regional targets. First, the regional targets were decomposed into a series of adjacent and nonoverlapping strips, and then a constrained optimization model was established with the maximum coverage of regional targets as the optimization objective. Finally, the greedy algorithm, dynamic programming, and constrained programming algorithm were used to solve this optimization problem. In 2007, Li et al. [8] studied the mission planning of earth

TABLE 1.1 Comparison of several agile satellite.

Satellite	Launch time	Imaging width	Resolution	Imaging mode	actuators	Manoeuvor angle	Manoeuvor angular velocity	Orbital altitude
ROCSat-2	2004.5	24 km	2 m/8 m	Body maneuver	Flywheels + thrusters	Roll and pitch 45°	Roll: 10° in 25 s, 30° in 60 s, pitch: 60° in 45 s	891 km
OrbView-5	2008.9	15.2 km	0.41 m/1.64 m	Body maneuver	8 enhanced flywheel (low chatter) + magnetic torquer + 8 nozzle	±60°		681 km
TecSar	2008.1	100 km	1 m/8 m	Body maneuver/ electron beam rotation	High energy and light reaction wheels			450 km
Pleiades	Autumn 2010	20 km	0.7 m/2.8 m	Body maneuver	4 CMGs	Roll and pitch 60°	Roll and pitch: 5° in 8 s, 10° in 10s, 25° in 60 s	695 km
QuickBird	2001.10	16.5 km	0.62 m/2.44 m	Body maneuver	Flywheel + 4 reaction wheels + thrusters	±30°		450 km
THEOS	2008.10	22 km	2 m	Body maneuver	Magnetic torquer + 4 reaction wheels + thrusters	±45°	45° in 60 s, 10° in 25 s, 30° in 60 s	822 km

Continued

TABLE 1.1 Comparison of several agile satellite—cont'd

Satellite	Launch time	Imaging width	Resolution	Imaging mode	actuators	Manoeuvor angle	Manoeuvor angular velocity	Orbital altitude
WorldView 2	2009.10	16.4 km	0.46 m/1.8 m	Body maneuver	CMGs	$\pm 40^{\circ}$	$3.5^{\circ}/s$	770 km
NigeriaSat-2	2010.9	300 (20 km)	32 m (5 m)	Body maneuver + antenna rotation	Magnetic torquer + reaction wheels	$\pm 45^{\circ}$	Antenna $20^{\circ}/s$ (satellite $4^{\circ}/s$)	700 km
CartoSat-2	2007.1	9.6 km	0.8 m	Body maneuver + antenna rotation	Large torque flywheel (0.3 Nm) + magnetic torquer + thrusters	$\pm 45^{\circ}$		635 km

observation agile spacecraft based on genetic algorithm. In 2013, started from the characteristics of agile satellite and the actual needs of users, Hao [9] handled the task planning of agile satellite by using the task allocation method of economical contract network based on integrity mechanism.

(b) Attitude and orbit rapid maneuver. In 2000, Bailey et al. [5] studied the multitarget fast pointing control of agile spacecraft. In 2002, Lappas from Surry University conducted a lot of in-depth research on attitude control of small agile spacecraft based on control moment gyro in his doctoral thesis [10]. In 2003, Turkey participated in the design of the satellite BILSAT-1 for the first time, and the control moment gyro was used for the first time to realize rapid and agile attitude maneuver [11]. In 2007, Zhang conducted relevant research on rapid tilt maneuver control of agile spacecraft based on micro control moment gyro parameters and moment estimation [12]. In 2008, Xing et al. studied the high-stability attitude control design of agile satellite using variable speed control moment gyroscope group as attitude control actuator [13].

1.3 Research status of optimal attitude trajectory planning

The primary problem involved in satellite attitude rapid maneuver is the time optimal path planning problem. These optimal trajectories can be used not only as the input of open-loop control and feedforward in closed-loop control but also as the reference trajectories of tracking control. For a single degree of freedom system considering only one-axis maneuver, its optimal solution is the well-known “bang bang” control, which only needs to change the control direction once in the middle of the maneuver [14]. For the satellite three-axis time optimal attitude maneuver problem, although the description of this optimal control problem is relatively compact and simple, due to the strong nonlinearity of the kinematic equation, so far there is no analytical solution, but only numerical one [15].

In practical application, the amplitude limitation of actuator control torque should be considered in satellite attitude trajectory planning. When solving the optimal control problem with constraints, there are generally two methods, direct method and indirect method [15]. The indirect method is mainly based on the maximum principle, variational method, and other mathematical methods. The maximum principle is based on the derivation of the first-order necessary condition for optimal control and transforms the trajectory optimization problem into a two-point boundary value problem. When solving the two-point boundary value problem, if the nonlinear equation describing the spacecraft motion is complex, it will be difficult to solve or the accuracy of the optimal solution cannot be guaranteed. The direct method includes point matching method and shooting method

[16]. If there are too many inequality constraints, the convergence speed of the indirect method will be seriously affected.

For symmetric rigid bodies, rotating around the Euler axis is a common solution to time optimal attitude maneuver. Based on this, Li and Bainum believe that even if the rotation around the Euler axis is not time optimal, at least the time optimal solution should be near it. Based on this assumption, they adopt the quasi linearization method to optimize the solution of Euler axis rotation until it cannot converge [17].

Li takes the integral of the control input quadratic function as the objective function and uses the Quasilinear method to deduce the optimal control problem of rigid satellite to obtain the two-point boundary value problem (TPBVP) through the Pontryagin minimum principle. The time optimal maneuver control trajectory is the corresponding maneuver trajectory by continuously reducing the maneuver time until the TPBVP with Euler axis rotation as the initial value cannot converge or the obtained control is in bang-bang form. According to the simulation results, the author points out that rotating around the Euler axis is certainly not time optimal for the general three-axis maneuver problem, unless the Euler axis corresponding to the satellite maneuver is along the main axis [18]. The subsequent authors applied the proposed method to the three-axis attitude pointing equipment of the US Naval Research Laboratory and achieved good results [19]. Bilimoria and Wie, based on the three-axis decoupled spherical symmetric satellite, proposed a calculation method to modify the constraints in the optimization process by using a parameter and solve the TPBVP in combination with the shooting method to obtain the optimal solution. The numerical calculation results showed that the time was shortened by 8.5% compared with the Euler axis rotation [20], thus pointing out that the Euler axis rotation is not the time optimal solution in general. They also pointed out that when the satellite attitude maneuver Euler angle is greater than 72°, the form of 5 control switches is time optimal, while when the maneuver Euler angle is less than 72°, the form of 7 control switches is time optimal. It also can be seen from the optimization results that when the maneuver angle is less than 72°, the optimal maneuver time of 5 control switches and 7 control switches is almost the same. For the case that the control amplitude is limited, but the control direction is not limited, Bai theoretically proved that the maneuver time for rotating around the Euler axis is the least by using the Pontryagin principle [21]. When the three components of the controller are limited separately, 6 and 7 switching control results are found when the attitude maneuver angle is less than 72° by using a hybrid numerical method. At the same time, it is shown that the two switching control results are completely independent in both state space and control space. Although the maneuver time of the 6 switching controls is always shorter, the local optimal maneuver time obtained by the switching of the two controls is very close.

Scrivener et al. pointed out that the problem of large-angle attitude maneuver needs to be deeply studied from the aspects of flexible influence, satellite attitude system composition, and the design of three-axis optimal maneuver control law of rigid aircraft [22]. Then, in his doctoral thesis, Scrivener transformed the continuous time optimal control problem into a nonlinear programming problem by dispersing the maneuver time (optimization parameters) at equal distance and estimating the trajectory between time points by Hermite cubic polynomial [23–26], and then solved it by Quasi Newton method [27]. Their optimization results are consistent with those described by Bilimoria and Wie, except in the case when the attitude maneuver angle is less than 10 degrees, when the attitude maneuver angle was less than 10 degrees, they found the attitude trajectory of 11 control switches.

Gaussian pseudospectral method uses the finite basis of global interpolation polynomial, constructs interpolation polynomial at Gaussian points through the values of discrete points, and approximately represents the state variables and control variables. The right end function of the differential equation is satisfied at the collocation point, so the continuous constraints of the dynamic equation are transformed into finite dimensional algebraic constraints. The integral term in the performance index is equivalent to Gaussian integral, so the continuous optimal control problem is transformed into a parameter optimization problem with algebraic constraints, that is, nonlinear programming problem (NLP) [28,29]. Gaussian pseudospectral method can provide a particularly accurate optimal solution when the control variable is a smooth curve, but it needs a long convergence time. In order to reduce the convergence time, the number of nodes should be reduced, but it can only converge to the suboptimal solution [30,31]. Because the trajectory of the satellite's three-axis time optimal attitude maneuver is a bang-bang control form, the time optimal attitude maneuver trajectory obtained by the Gaussian pseudospectrum method will have serious control torque flutter near the control switching point, so it is difficult to obtain the accurate control switching time.

The above methods for solving the time optimal attitude maneuver trajectory need to go through long-time offline calculation and cannot realize the rapid generation of satellite trajectory. In order to solve this problem, George [15] proposed a rapid generation method of satellite attitude maneuver trajectory based on virtual time domain inverse dynamics (iDVD) [32,33]. This method is not only suitable for rest to rest attitude maneuver but also suitable for move to move attitude maneuver. They use a parameterized Bezier polynomial to represent the trajectory of quaternions [34], use polynomials to represent the angular velocity trajectory, and finally transform the satellite trajectory optimization problem into a parameter optimization problem of boundary values (attitude, angular velocity, and its high-order derivatives) based on inverse dynamic. The known maneuver trajectory boundary conditions (the attitude and angular velocity at the beginning and end of the maneuver are known) greatly reduce the parameters of the optimal design, so that the quasi time optimal attitude maneuver trajectory can be obtained quickly. However, this fast optimization algorithm still needs nonlinear programming software, and the accuracy of trajectory planning is sacrificed while reducing the optimization parameters.

Since the time optimal attitude maneuver trajectory is in bang-bang form, some researchers have proposed a method to directly optimize the control switching time (STO, switching time optimization). Meier and Bryson solved the time optimal maneuver control problem of two link manipulator by using the idea of STO and the principle of variational method under the constraint of bounded control torque [35,36]. Some other researchers obtained the optimal control trajectory of spherically symmetric satellite based on STO idea according to the weighted time/fuel optimal index and found that when the fuel penalty factor exceeds a certain value, the rotation mode of Euler axis is better (because the satellite motion time increases too fast with the increase of penalty factor) [37–39]. An obvious disadvantage of STO is that the control structure (i.e., the number of control switches and the order of switching) needs to be known in advance. Joseph [40] and Singh [41], respectively, adopted a method similar to STO to solve the time optimal attitude maneuver control problem of satellites with flexible appendages, and the results are consistent with those based on input shaping theory [42,43].

1.4 Research status of large-angle maneuver attitude control algorithm

1.4.1 Large angle maneuver attitude control for rigid spacecraft

The Clementine lunar probe plans the optimal maneuver trajectory based on Euler rotation, designs the feedback control law for attitude tracking control, and realizes the large-angle attitude control algorithm [44,45].

For large-angle attitude maneuver tasks, under the limitation of bounded flywheel output torque and rotation speed, with the help of the idea of hierarchical saturation, Wie proposed a method without prior trajectory planning, which can enable the satellite to perform attitude maneuver along an approximate optimal maneuver trajectory and rotate around the Euler axis [45]. Subsequently, some scholars proposed to improve the hierarchical saturation control law by using fuzzy reasoning and variable structure control to improve the quality of hierarchical saturation control algorithm [46–48].

Newman and Dodds and Vittek [49,50] proposed a PD like closed-loop controller based on system state, which achieves the time optimization of single-axis maneuver by using state gain feedback. Both papers consider the saturation of angular velocity and angular acceleration during maneuver, but do not consider the response delay of actuator, which leads to the jitter and tremor of control torque. Crassidis [51,52] designed the satellite attitude maneuver control algorithm based on variable structure control. Through comparison, it is concluded that the variable structure control using linear switching surface cannot achieve time optimal or approximate time optimal control, but the satellite can track the time optimal maneuver trajectory by designing variable structure tracking controller to achieve the effect of quasi optimal control [53]. Lawton and Beard [54] assumed that the angular velocity and angular acceleration of a three-axis mobile satellite are unsaturated, and then a control algorithm to stabilize the satellite maneuver in the presence of external interference and model uncertainty was proposed. In Ref. [55], according to the time scale separation principle in singular perturbation theory, the author used accurate linearization method and model reference adaptive method to design the controller. The theoretical design and engineering implementation of this method are simple and easy.

Verbin [56] proposed an integrated state feedback control law for attitude maneuver of rigid satellite based on flywheel control. The control law comprehensively considered torque saturation, angular velocity, angular acceleration saturation, and actuator response delay and can achieve the control goal for general three-axis maneuver or two-point boundary value maneuver. Because the controller can make up for the time delay of the actuator and the uncertainty of the controller model, the robustness of the control algorithm is improved. In Ref. [57], aiming at the problems of external interference and uncertainty of inertia matrix in attitude tracking control of agile satellites, an adaptive control law was designed to suppress these adverse factors. Liu and Li [58] proposed a ∞ norm-based fast attitude maneuver method for single frame control moment gyroscope group (SGCMG) under nonsingular conditions. In Ref. [59], a controller was designed to avoid the deterioration of control effect and singularity of solution in case of failure of a CMG by adjusting parameters when taking pyramid CMGs (control moment gyroscope group) as the actuator. Wie studied the use of SGCMGs as actuator to provide strong moment output in agile satellite attitude control system. A nonlinear feedback controller was designed, and the singularity analysis and avoidance method of SGCMGs were given as well. The simulation results showed that the control law had good robustness [45,60,61].

1.4.2 Large angle maneuver attitude control for flexible spacecraft

After decades of research and development, the control methods and theories of flexible satellite generally include the following: traditional control theory, robust control theory, adaptive control theory, variable structure control theory, intelligent control theory, and so on.

Traditional control

The traditional control theory generally refers to the developed PID controller. Because of its simple structure and high reliability, it has been widely used in the field of aerospace engineering. In Ref. [62], a PID controller was designed for a flexible satellite with helical antenna. The effectiveness of the controller is proved by semi-physical simulations. The very long baseline interferometric satellite launched by Japan in 1997 also used PID controller and added disturbance torque compensator on the basis of PID. The dynamic effect brought by flexible appendages was regarded as disturbance, and then RMM (robust modal matching) and observation estimation method were used to design the disturbance torque compensator, while a filter was introduced to remove the neglected high-order vibration of flexible appendages. It was verified by ground experimental system in the design stage, and then was used on the satellite launched in 1997 [63]. In Ref. [64], the problem of attitude control of flexible satellite was solved by combining PD controller and Kalman filter, and the pointing of flexible antenna was successfully controlled as well. However, more and more complex space missions not only require attitude control to have certain robustness but also have higher requirements for its control accuracy. At this time, traditional controllers that obtain stability by sacrificing control accuracy cannot meet the needs of on orbit missions. Moreover, the low-frequency vibration characteristics of flexible appendages will further deteriorate the accuracy of traditional controller. Therefore, in view of the obvious characteristics of parameter uncertainty, disturbance sensitivity, and rigid–flexible coupling nonlinearity of flexible satellites, researchers in the aerospace field have developed a variety of new control methods, such as robust control, adaptive control, variable structure control, and intelligent control.

Robust control

The main characteristic of robust control algorithm is that it is insensitive to system uncertainties and disturbances and can ensure the stability of the system. Compared with rigid satellites, flexible satellites with large flexible appendages are subject to stronger structural uncertainty, unmodeled dynamics, and modal truncation errors caused by flexible appendages, so the robust control algorithm is more suitable and a large number of literatures have proposed attitude controller for flexible satellites based on robust control algorithm. Considering that the uncertainty of moment of inertia of the space station was obvious due to its many flexible appendages, Wie [65] designed a full state robust controller. Further considering the persistent disturbances, the robust control method was improved and better simulation results were obtained. The National Space Development Agency of Japan tested the controller on the sixth satellite engineering test satellite (Engineering Test Satellite VI) launched in 1994. The on orbit test showed that under the control of the two degree of freedom H_∞ controller, the flexible satellite with two wing solar panels can meet the control target requirements of the on orbit mission [66]. Chen [67] designed a hybrid controller based on fuzzy algorithm. However, because the designed control law increased the order of the controller and the dimension of the system, it is still debatable whether it is applicable to the model outside the literature itself. In Ref. [68], a robust attitude controller was designed by combining μ analysis and H_∞ coprime decomposition. The simulation showed that the controller was robust to the structural uncertainty caused by the turnover of solar panels. In Ref. [69], Boulet successfully designed experiments to verify the robustness of the μcontroller on the spatial flexible structure under the collocated arrangement of actuators and sensors. In Ref. [70], a robust controller was designed for flexible spacecraft with considering limitations of both flywheel speed and control torque. A compensator was used to solve the problem of control torque saturation, and the unwinding problem caused by quaternion was addressed as well. The inverse optimal algorithm was further introduced to avoid the difficulty of calculating Hamilton partial differential equation.

Adaptive control

Adaptive control methods are mainly divided into self-tuning controller (STC) and model reference adaptive controller (MRAC) according to the different ways of parameter identification and updating. The core feature of adaptive control method is to deal with the change of control object with the change of control parameters. Through the identification of model and parameters, a certain adaptive law is designed to update the control function, and finally achieve the control of the model with parameter and structural uncertainties. Therefore, the adaptive control method is robust to model uncertainties. Sobel [71] first proposed the model reference adaptive control method and successfully applied it to the control of simple models. In 1999, without considering model uncertainties, an adaptive attitude controller was designed for an aircraft with rotating flexible appendages [72]. Singh [73] proposed an adaptive attitude controller for flexible spacecraft, when all the model parameters of the system were completely unknown. Since the controller contains flexible modes, this cannot always be met for on orbit missions. Then without the need of mode truncation order and model structure, an adaptive attitude controller was designed for flexible spacecraft by Maganti and Singh. Because the nonlinear model was required to meet certain linearization conditions, its applicability was still need to be tested in future experiment [74]. In the condition of large-angle maneuver, an adaptive control method requiring only attitude angle and attitude angular velocity was proposed in [75]. Hu and Zhang [76] studied the active control of large flexible spacecraft using CMG. In this paper, two adaptive controllers were designed. The first controller was responsible for large angle maneuver. When the attitude error was very small, the system was linearized first, and then a simple adaptive controller was designed. At this time, the problem of gyro singularity was involved, and the use of the second controller can avoid gyro singularity. Thus, the first controller was used to approach the target attitude, and when the attitude error becomes smaller, the second adaptive controller was in charge to achieve good vibration suppression effect.

Sliding mode variable structure control

Variable structure control is mainly divided into sliding mode variable structure control with sliding mode characteristics and variable structure control without sliding mode characteristics (such as PID control with variable structure characteristics). Since the former is more widely used, the research status of sliding mode variable structure controller is mainly described below. Sliding mode variable structure control was first developed in the 1950s by former Soviet scientists Emelyanov [77], Utkin and Taran [78]. It was proposed. In the 1970s, it was introduced into Europe and America by Utki and Itkis, and the preliminary results were summarized, and the development direction was prospected, which laid a foundation for the development of sliding mode variable structure control method. The sliding mode variable structure control process is mainly divided into two parts [79]. The first part is to design the control law to make the system trajectory reach the sliding mode surface in a finite time, and the second part is that the system trajectory converges to the equilibrium point along the sliding mode surface. Because the sliding mode is independent of the model itself, it is robust to model uncertainties and disturbances. Finally, sliding mode variable structure control has gradually developed into an independent control method. In addition, the rigid-flexible coupling characteristics of flexible satellites cannot be ignored, and they are sensitive to uncertainties and disturbances including modal truncation and modal overflow, inaccurate parameters such as frequency and damping. The existing control methods cannot deal with this problem well, while sliding mode variable structure control has certain advantages over this, thus many literatures began to study sliding mode attitude control for flexible spacecraft. Singh [80] proposed a sliding mode attitude controller for flexible manipulator and the input shaping technology was incorporated to further reduce the flexible vibration. Hu et al. [81,82] designed a sliding mode controller with considering model uncertainties. The controller only used the output for feedback control, and the uncertainty does not meet the matching conditions. Finally, the input molding technology is introduced to suppress the vibration. Sinha [83] designed a sliding mode controller for the flexible satellite considering the random disturbance, and the Kalman filter was introduced to reconstruct the system state. Hu [84] carried out a two-stage control scheme for the flexible satellite considering external disturbances. First, an adaptive sliding mode controller was designed to control the attitude of the central rigid body. When the attitude error became small, the dynamic equation was decoupled into a quadratic linear decoupled dynamic equation. At this time, with the ceramic strain gauge installed, the strain rate feedback control was used to actively suppress the appendage vibration. Guan et al. [85] designed a fuzzy function-based sliding mode controller for flexible satellite and suppressed the vibration of the flexible appendages to a certain extent simultaneously. Xiao et al. [86] designed a sliding mode fault tolerant attitude controller in the condition of the partial failure of actuators. The output torque of the failed flywheel was regarded as a disturbance and estimated online. At the same time, the neural network was introduced to approximate the total perturbation.

For an ideal sliding mode controller, if its "variable structure" process is ideal, that is, there is no space and time lag, then the sliding mode variable structure controller has stronger robustness than other types of controllers, and the design is simple. However, the actual sliding mode control system will inevitably produce chattering due to its switching characteristics [87]. For satellites with large flexible appendages, the system chattering generated by sliding mode variable structure control will stimulate the vibration of the flexible appendages, deteriorate the attitude control performance, and even make the system unstable in serious cases. Therefore, a large number of scholars have put forward methods to solve the chattering phenomenon from different perspectives. Slotine et al. [88] divided the sliding mode variable structure control into two sections with the "boundary layer" close to both sides of the sliding mode surface. The normal sliding mode control was still used outside the boundary layer, and the saturation function instead of the sign function was used in the boundary layer, which effectively weakened the chattering phenomenon. While using the boundary layer method, Araujo [89], etc., further considered the influence of the boundary layer on the steady-state characteristics, and then reduced the chattering by appropriately adjusting the thickness of the boundary layer. Gao [90] tried to solve the system chattering phenomenon from the sliding mode arrival stage and proposed reaching law methods, such as constant velocity reaching law, exponential reaching law, and power reaching law. The chattering can be suppressed by adjusting the parameters of the reaching law. By combining the boundary layer method and the exponential reaching law, Zhou et al. [91] proposed a sliding mode attitude control for flexible spacecraft, and then introduced an elastic mode stabilizer in the boundary layer to weaken the chattering and suppress the vibration of flexible appendages at the same time. In recent years, some scholars have designed the high-order sliding mode controller, and by adding the discontinuous term in sliding mode control to the first-order differential term or high-order differential term of the controller, the controller itself became continuous and the chattering was suppressed [92]. In Ref. [93], a second-order sliding mode attitude controller was designed, which effectively eliminated the chattering phenomenon with realizing large-angle rapid maneuver. Ma [94] designed a second-order sliding mode controller for large spacecraft in large-angle attitude maneuver. With considering uncertainties in moment of inertia and external disturbance, the second-order sliding mode controller not only showed stronger robustness than traditional sliding mode controller but also effectively reduced the chattering phenomenon. Some scholars [95] also introduce intelligent control algorithms into the sliding mode controller to approximate the discontinuous term or disturbance term, so as to reduce the system chattering. The details will be described in the research status of intelligent control.

Intelligent control

Intelligent control mainly refers to the use of various intelligent algorithms to solve problems that cannot be solved by traditional control methods. It does not strictly rely on the accurate mathematical model of the controlled object, but can approximate the nonlinear and unknown terms in the controlled model adjust some parameters online based on its universal approximation property and learning ability. The common way is the combination of intelligent algorithm and other control methods. The two most widely used intelligent control algorithms are fuzzy control algorithm and neural network control algorithm. Neural network and fuzzy function can approximate any smooth function with arbitrary accuracy in a given region, so these two control algorithms are very effective in the field of system control, especially when the system contains strong nonlinearity and serious uncertainty [96]. In Ref. [97], a fuzzy strategy-based low-pass filter was introduced into the sliding mode controller to make the control signal continuous, so as to reduce the chattering. Huang et al. [98] designed an adaptive second-order sliding mode controller for flexible spacecraft and introduced a fuzzy function to approximate the rigid flexible coupling term in the system, which not only achieved attitude stability but also reduced the chattering and vibration of flexible appendages. In Ref. [99], the learning ability of Fourier neural network was used to compensate the system with flexible appendages, and the simulation verified the convergence of neural network and the effectiveness of the whole controller. In Ref. [100], an adaptive output feedback attitude controller was proposed, while the Chebyshev neural network was incorporated to approximate the nonlinear term and disturbance of the system, and a switching mechanism to switch between output feedback based on neural network and robust control was also designed to avoid over estimation of unknown terms by the Chebyshev neural network.

1.4.3 Finite time control for spacecraft attitude

Through the research and summary of the current situation at home and abroad, it is not difficult to find that most control strategies can only meet the asymptotic convergence. Since more and more space missions require flexible satellites to have the ability of large-angle rapid maneuver, finite time attitude control begins to enter people's vision and gradually becomes a research hotspot.

Convergence is an important criterion to evaluate a control system. In general controller design, if a system can be asymptotically stable, it can be considered that the control system has good convergence [101]. The fastest form of asymptotic stability is exponential stability, but under the Lipschitz condition, both control and stability are discussed in the sense of infinite convergence time. The trajectory of the system can increase its convergence speed, but can never reach the equilibrium point in finite time [102]. Under this background, some scholars proposed the concept of finite time stability and carried out relevant research. It is found that the finite time control can achieve finite time stability with better disturbance-rejection performance due to the existence of fractional power [103].

Finite time control first appeared in open-loop systems, but it was sensitive to initial values and had poor robustness. Then, the finite time control turned to the finite time control based on discontinuous state feedback. Bhat [104] studied the finite time control theory based on homogeneous method in 1997 and proposed the more widely used finite time control theory based on the Lyapunov stability theorem. The research of finite time control then began to develop fast on the basis of these theories.

Jin and Sun [105] designed a finite time controller for rigid spacecraft, which is robust to model uncertainties and external disturbances. In Ref. [106], a finite time attitude tracking controller is designed for a single rigid spacecraft considering external disturbance. On this basis, a finite time attitude cooperative controller is designed for multiple spacecraft. Zak [107] proposed the concept of terminal attractor in the process of studying neural networks. In Ref. [108], the concept of terminal attractor was used in the controller design by introducing the nonlinear term into the design of sliding mode surface. When the system trajectory reached the sliding mode surface, it can reach the origin along the sliding mode surface in a finite time. Since the system trajectory can reach the sliding mode surface in a limited time by designing the controller, a finite time attitude controller based on the terminal sliding mode surface was formed. Because the terminal sliding mode surface cannot guarantee a fast convergence speed when the system trajectory is far from the equilibrium point, Yu and Zhihong [109] designed a fast terminal sliding mode surface by introducing a linear term into the original terminal sliding mode surface. The linear term provides a fast convergence speed when the system trajectory is far from the equilibrium point. When the system trajectory is close to the equilibrium point, the nonlinear term can still maintain a fast convergence speed. Both traditional terminal sliding mode and fast terminal sliding mode are sensitive to the selection of initial value. When the initial value is not selected properly, singularity may occur. In order to solve this problem, Feng et al. [110] proposed the nonsingular terminal sliding mode surface, so that the controller can avoid singularity while ensuring finite time control. Yang and Yang [111] proposed the nonsingular fast terminal sliding mode by combining the traditional sliding mode and nonsingular sliding mode, so that system trajectory on the sliding mode surface can not only avoid singular problems but also maintain fast convergence speed. With the development of the terminal sliding mode theory, many scholars applied it to spacecraft control. Wu et al. [112] designed a terminal sliding mode controller for rigid spacecraft with considering the external disturbance, and the external disturbance was estimated by the adaptive method. In Ref. [113], a continuous finite time controller for rigid spacecraft was designed based on nonsingular fast terminal sliding mode and an adaptive estimation strategy was introduced when the upper bound of total perturbation including model uncertainty and external disturbance was unknown. In Ref. [114], a nonsingular terminal sliding mode controller was designed with a finite time observer to observe the disturbance. On the basis of the previous paper, considering the limitation of control input, a nonsingular fast terminal sliding mode controller was designed for rigid spacecraft, and an adaptive law is introduced to estimate external disturbances and model uncertainties.

1.5 Research status of hybrid control

The concept of hybrid control first appeared in Ref. [115]. In Refs. [116, 117], the author adopted the hybrid magnetic torquer and reaction wheel control to realize the maneuver of rotating around the instantaneous Euler axis and significantly reduced the saturation phenomenon of reaction wheel during maneuvering. In Ref. [118], the author used hybrid CMGs and magnetic torque control to realize the satellite attitude large angle maneuver. The magnetic torquer was mainly used to suppress external disturbances and kept the frame angle within the preset range to avoid singularity of CMG.

Based on Wie's attitude maneuver control algorithm of rotating around the Euler axis [45], Chen and Steyn [115] proposed hybrid thruster and flywheel control and used PD control to correct the open-loop control deviation caused by moment of inertia error and external disturbances. However, this simple hybrid control algorithm had strong limitations: the premise of rotating around the Euler axis was that the initial angular velocity of the satellite was collinear with the quaternion vector (the special case was that the initial angular velocity is 0) [45]. Later, Hall proposed hybrid control stratagem using the thruster to provide the large torque required during maneuver and the momentum wheel for high-precision adjustment to achieve high precision [119], but the problem of inaccurate output torque of the thruster was still remain unsolved.

When the satellite performs attitude maneuver, its maneuver speed and accuracy cannot be always guaranteed at the same time: if only the control moment gyro is used as the actuator, its maneuver speed can be guaranteed, but its tracking accuracy will decline to a certain extent; when using flywheel as actuator, the precision is often high, but the maneuvering speed is slow. In view of this situation, a series of hybrid control algorithms [120,121] have been proposed to integrate the advantages of the two control mechanisms and achieved good control performance. In Ref. [122], Verbin and Lappas proposed a hybrid CMG and flywheel attitude control algorithm. The algorithm integrated the advantages of CMG with large control torque and flywheel with high control accuracy. The CMG was used to improve the maneuver speed during maneuvering, while the flywheel was used to improve the control accuracy at the end of the maneuvering. In this paper, according to the ability of the satellite to decelerate along the Euler axis, the author designed the "brake line" related to angular velocity, and then derived the angular momentum control from the "brake line", so as to produce the control law for frame angular of the CMG [122]. Liden believed that the hybrid control algorithm was not paid attention mainly because of its complexity. Based on this, the author proposed a relatively concise hybrid control algorithm for rigid agile satellite, that is, the CMG was used to shorten the maneuver time during maneuver, and the flywheel was used to correct the attitude maneuver error after the maneuver [123]. Based on Kane method, the dynamic equation of satellite with SGCMG and flywheel hybrid actuator structure was derived in Ref. [124], and the required control torque was distributed to the two actuators through a simple distribution method. Subsequently, Skelton and Hall [125] proposed a scheme, in which SGCMGs were used for coarse attitude control, and reaction wheel was used for accurate attitude adjustment during attitude large-angle maneuver control. The corresponding controller was designed based on Lyapunov method, which had achieved good control performance, but problems such as singularity of CMG and saturation of flywheel torque and speed still remained unsolved. Leeghim et al. [126] proposed a hybrid control scheme that can distribute the control torque of each actuator through the optimization function and adjust the output torque of each actuator by adjusting the corresponding weight, and the flywheel saturation phenomenon can be effectively avoided [126]. Leve and Fitz [127] proposed a scheme to solve the singularity problem of CMG by using variable speed control moment gyroscope, that is, using CMG and flywheel to cogenerate three-axis control torque. Jin derived the control torque required for tracking control based on Lyapunov principle. When SGCMGs were singular, the torque command along the singular direction was distributed to the flywheel by singular value decomposition [128]. It can be seen that for different space missions, SGCMGs and flywheel can be reasonably allocated to achieve high-performance attitude control.

At the same time, as a new control actuator used in satellite, fluid ring has attracted certain attention in recent years. The control effect of fluid ring is similar to that of flywheel, but compared with flywheel, fluid ring has the advantages of light weight and large output torque. Maynard [129] first introduced the concept of liquid ring as control mechanism into satellite control system and proposed a stability control algorithm based on liquid ring to suppress disturbance torque. On the basis of the three-axis dynamic model of satellite with fluid ring, Nobari [130] presented a redundant conical configuration scheme of four fluid rings, and then designed a sliding mode controller based on fluid ring to eliminate the adverse effects of model uncertainty and system chattering. Then, considering the limitation that the magnetic torquer can only provide the control torque perpendicular to the earth's magnetic field, the author proposed a hybrid control scheme using a fluid ring to provide the control torque along the magnetic field direction and discussed the possible failure of this hybrid control scheme. The simulation results showed that the proposed control system had good control performance [131].

References

[1] Z.H. Li, The Study on the Rapid Maneuver and Stabilization of Small Satellite Attitude Control Approaches, Harbin Institute of Technology, Harbin, 2011, pp. 1–8.

[2] M. Steckling, U. Renner, H.P. Roser, DLR-TUBSAT, Qualification of High Precision Attitude Control in Orbit, Acta Astronaut. 39 (9–12) (1996) 951–960.

[3] B. Girouart, I. Sebbag, J.M. Lachiver, Pleiades-HR CMGs-based attitude control system design, development status and performances, in: Proceedings of 17th IFAC Symposium on Automatic Control in Aerospace, Toulouse, France, 2007, pp. 834–839.
[4] N. Longbotham, C. Bleiler, C. Chaapel, et al., Spectral classification of Worldview-2 multi-angle sequence, in: IEEE GRSS and ISPRS Joint Urban Remote Sensing Event, IEEE Computer Society, Munich, Germany, 2011, pp. 109–112.
[5] B. Wie, D. Bailey, C. Heiberg, Rapid multitarget acquisition and pointing control of agile spacecraft, J. Guid. Control Dynam. 25 (1) (2002) 96–104.
[6] K.W. Young, R. Muehlhaeusser, R.S.H. Piggin, et al., Agile control systems, Proc. Inst. Mech. Eng. D J. Automob. Eng. 215 (2) (2001) 189–195.
[7] M. Lemaitre, G. Verfaillie, F. Jouhaud, et al., Selecting and scheduling observations of agile satellites, Aerosp. Sci. Technol. 6 (5) (2002) 367–381.
[8] Y. Li, M. Xu, R. Wang, Scheduling observations of agile satellites with combined genetic algorithm, in: 3rd International Conference on Natural Computation, ICNC 2007, Inst. of Elec. and Elec. Eng. Computer Society, Haikou, Hainan, China, 2007, pp. 29–33.
[9] H.C. Hao, J. Wei, Y.J. Li, Research on agile satellite dynamic mission planning based on multi-agent, J. Natl. Univ. Def. Technol. 35 (01) (2013) 53–59.
[10] V.J. Lappas, A Control Moment Gyro (CMG) Based Attitude Control System (ACS) for Agile Small Satellites, University of Surrey, 2002, pp. 1–9.
[11] A. Bradford, L.M. Gomes, S.M. Sweeting, et al., Bilsat-1: a low-cost, agile, earth observation microsatellite for Turkey, Acta Astronaut. 53 (4–10) (2003) 761–769.
[12] J.R. Zhang, Estimation of MCMG parameters and torques for rapid slewing maneuver of agile spacecraft, J. Tsinghua Univ. Sci. Technol. (11) (2007) 2040–2043.
[13] L. Xing, C. Sun, T. Liang, VSCMGs-based steering law design for a high attitude stability agile satellite, Aerosp. Control and Appl. (06) (2008) 24–28.
[14] E. Arthur, J. Bryson, Y.C. Ho, Applied Optimal Control: Optimization, Estimation, and Control, Hemisphere Publishing Company, 1975.
[15] G.A. Boyarko, M. Romano, O.A. Yakimenko, Time-optimal reorientation of a spacecraft using an inverse dynamics optimization method, J. Guid. Control Dynam. 34 (4) (2011) 1197–1208.
[16] B.S. Attili, M.I. Syam, Efficient shooting method for solving two point boundary value problems, Chaos Solitons Fractals 35 (2008) 895–903.
[17] F. Li, P.M. Bainum, Use of the Quasilinearization Algorithm for the Simulation of LSS (Large Space Structures) Slewing, United States, 1989, pp. 1–25.
[18] F. Li, P.M. Bainum, Numerical approach for solving rigid spacecraft minimum time attitude maneuvers, J. Guid. Control Dynam. 13 (1) (1990) 38–45.
[19] F. Li, P.M. Bainum, N.G. Creamer, et al., 3-axis near-minimum-time Maneuvers of reshape-numerical and experimental results, J. Astronaut. Sci. 43 (2) (1995) 161–178.
[20] K.D. Bilimoria, B. Wie, Time-optimal three-axis reorientation of a rigid spacecraft, J. Guid. Control Dynam. 16 (3) (1993) 446–452.
[21] X. Bai, J.L. Junkins, New results for time-optimal three-Axis reorientation of a rigid spacecraft, J. Guid. Control Dynam. 32 (4) (2009) 1071–1076.
[22] S.L. Scrivener, R.C. Thompson, Survey of time-optimal attitude maneuvers, J. Guid. Control Dynam. 17 (2) (1994) 225–233.
[23] P.E. Gill, L.O. Jay, M.W. Leonard, et al., An SQP method for the optimal control of large-scale dynamical systems, J. Comput. Appl. Math. 120 (1–2) (2000) 197–213.
[24] P.J. Enright, B.A. Conway, Optimal finite-thrust spacecraft trajectories using collocation and nonlinear-programming, J. Guid. Control Dynam. 14 (5) (1991) 981–985.

[25] A.L. Herman, B.A. Conway, Direct optimization using collocation based on high-order gauss-lobatto quadrature rules, J. Guid. Control Dynam. 19 (3) (1996) 592–599.
[26] C.R. Hargraves, S.W. Paris, Direct trajectory optimization using nonlinear-programming and collocation, J. Guid. Control Dynam. 10 (4) (1987) 338–342.
[27] S.L. Scrivener, Time-Optimal Multi-Axis Attitude Maneuvers of Rigid Spacecraft Using Collocation and Nonlinear Programming, Pennsylvania State University, Pennsylvania, United States, 1993, pp. 1–175.
[28] D. Benson, A Gauss Pseudospectral Transcription for Optimal Control, Massachusetts Institute of Technology, Massachusetts, United States, 2005, pp. 1–25.
[29] G.T. Huntington, Advancement and Analysis of a Gauss Pseudospectral Transcription for Optimal Control Problems, Massachusetts Institute of Technology, Massachusetts, United States, 2007, pp. 13–16.
[30] G. Basset, Y. Xu, O.A. Yakimenko, Computing short-time aircraft maneuvers using direct methods, J. Comput. Syst. Sci. Int. 49 (3) (2010) 481–513.
[31] G. Boyarko, O. Yakimenko, M. Romano, Formulation and analysis of matching points of interest in two-spacecraft for optimal rendezvous, in: AIAA Guidance, Navigation, and Control Conference and Exhibit, American Institute of Aeronautics and Astronautics Inc., Chicago, IL, United states, 2009, pp. 1–22.
[32] I.D. Cowling, O.A. Yakimenko, J.F. Whidborne, et al., A prototype of an autonomous controller for a quadrotor UAV, in: European Control Conference, 2007, pp. 1–8.
[33] O.A. Yakimenko, Direct method for rapid prototyping of near-optimal aircraft trajectories, J. Guid. Control Dynam. 23 (5) (2000) 865–875.
[34] M.J. Kim, M.S. Kim, S.Y. Shin, A general construction scheme for unit quaternion curves with simple high order derivatives, in: Proceedings of the 22nd Annual Conference on Computer Graphics and Interactive Techniques, ACM, 1995, pp. 369–376.
[35] E.B. Meier, An Efficient Algorithm for Bang-Bang Control Systems Applied to a Two-Link Manipulator, Stanford University, California, United States, 1987, pp. 1–13.
[36] E.B. Meier, A.E. Bryson, Efficient algorithm for time-optimal control of a 2-link manipulator, J. Guid. Control Dynam. 13 (5) (1990) 859–866.
[37] S.W. Liu, T. Singh, Fuel/time optimal control of spacecraft maneuvers, J. Guid. Control Dynam. 20 (2) (1997) 394–397.
[38] R.M. Byers, Minimum Time Reorientation of an Arbitrary Rigid Body [C], in: Proceedings of the AAS/AIAA Astrodynamics Conference, Univelt Inc, Victoria, BC, Can, 1993, pp. 1889–1903.
[39] R.M. Byers, S.R. Vadali, Quasi-closed-form solution to the time-optimal rigid spacecraft reorientation problem, J. Guid. Control Dynam. 16 (3) (1993) 453–461.
[40] B.A. Joseph, J.A. Burns, E.M. Cliff, Time-optimal slewing of flexible spacecraft, J. Guid. Control Dynam. 15 (2) (1992) 360–367.
[41] G. Singh, P.T. Kabamba, N.H. McClamroch, Time optimal slewing of a rigid body with flexible appendages, in: Proceedings of the 26th IEEE Conference on Decision and Control, IEEE, Los Angeles, CA, USA, 1987, pp. 1441–1442.
[42] C.F. Cutforth, Analysis and Design of Fast Input Shapers for the Control of Flexible Structures, University of Colorado at Boulder, 2001, pp. 1–16.
[43] A. Ebrahimi, S.A. Moosavian, M. Mirshams, Control of space platforms with flexible links using command shaping method, Iran. J. Sci. Technol. Trans. B Eng. 32 (B1) (2008) 13–24.
[44] Q. Liu, B. Wie, Robust time-optimal control of uncertain flexible spacecraft, J. Guid. Control Dynam. 15 (3) (1992) 597–604.

[45] B. Wie, J.B. Lu, Feedback control logic for spacecraft Eigenaxis rotations under slew rate and control constraints, J. Guid. Control Dynam. 18 (6) (1995) 1372–1379.
[46] W. Feng, C. Xibin, Z. Shijie, Attitude tracking control of on-orbit-servicing spacecraft based on Eigenaxis maneuver, J. Astronaut. 39 (2) (2008) 570–575.
[47] C.Y. Dong, H. Ying, C. Yu, A cascade-saturation fuzzy variable structure control for spacecraft large angle attitude Maneuvers, J Astronaut. (05) (2006) 974–978.
[48] X.B. Cao, F. Wang, S.J. Zhang, Model independent near eigenaxis rotation for spacecraft maneuver under slew rate and control constraints, J. Harbin Inst. Technol. (09) (2006) 1413–1418.
[49] W.S. Newman, Robust near time-optimal control, IEEE Trans. Autom. Control 35 (7) (1990) 841–844.
[50] S.J. Dodds, J. Vittek, Spacecraft attitude control using an induction motor actuated reaction wheel with sensorless forced dynamic drive, in: Proceedings of 1998 IEE Colloquium on All Electric Aircraft, UK, London, 1998, pp. 1–14.
[51] J. Crassidis, S. Vadali, F. Markley, Optimal variable-structure control tracking of spacecraft maneuvers, J. Guid. Control Dynam. 23 (3) (2000) 564–565.
[52] J.L. Crassidis, F.L. Markley, Sliding mode control using modified Rodrigues parameters, J. Guid. Control Dynam. 19 (6) (1996) 1381–1383.
[53] K. Jongrae, C. John, A comparative study of sliding mode control and time-optimal control, in: AIAA/AAS Astrodynamics Specialist Conference and Exhibit, American Institute of Aeronautics and Astronautics, 1998, pp. 316–323.
[54] J. Lawton, R.W. Beard, Attitude regulation about a fixed rotation axis, J. Guid. Control Dynam. 26 (2) (2003) 253–258.
[55] A.Y. Han, H.L. Xu, A double-loop controller for the attitude of aerocraft based on the time-scale separation principle, Meas. Control Technol. (02) (2010) 52–54.
[56] D. Verbin, V.J. Lappas, J.Z. Ben-Asher, Time-efficient angular steering laws for rigid satellites, J. Guid. Control Dynam. 34 (3) (2011) 878–892.
[57] W.C. Luo, Y.C. Chu, K.V. Ling, Inverse optimal adaptive control for attitude tracking of spacecraft, IEEE Trans. Autom. Control 50 (11) (2005) 1639–1654.
[58] H. Liu, J.F. Li, SGCMG steering law design for agility spacecraft reorientation, J. Astronaut. (01) (2006) 113–116. +125.
[59] J.R. Zhang, A. Rachid, Y. Zhang, Attitude control for part actuator failure of agile small satellite, Acta Mech. Sinica 24 (4) (2008) 463–468.
[60] B. Wie, H. Weiss, A. Arapostathis, quaternion feedback regulator for spacecraft Eigenaxis rotations, J. Guid. Control Dynam. 12 (3) (1989) 375–380.
[61] B. Wie, D. Bailey, C. Heiberg, Singularity robust steering logic for redundant single-gimbal control moment Gyros, J. Guid. Control Dynam. 24 (5) (2001) 865–872.
[62] P.H. Mak, M.M. Tong, A.B. Jenkin, Dynamics and control analysis of a satellite with a large flexible spinning antenna, AAS 87 (482) (1987) 935–951.
[63] T. Yamashita, N. Ogura, T. Kurii, et al., Improved satellite attitude control using a disturbance compensator, Acta Astronaut. 55 (1) (2004) 15–25.
[64] N.E. Goodzeit, M.Q. Phan, System and Periodic Disturbance Identification for Feedforward-Feedback Control of Flexible Spacecraft, AIAA, 1997, pp. 1–15.
[65] B. Wie, K.W. Warren, New approach to attitude momentum control for the space station, J. Guid. Control Dynam. 12 (5) (1989) 714–722.
[66] Y. Chida, H. Soga, Y. Yamaguchi, On orbit attitude control experiments using ETS-VI by $H\infty$ control and two-degree-of-freedom $H\infty$ control, J. Guid. Control Dynam. 21 (6) (1998) 1109–1116.

[67] B.S. Chen, C.S. Wu, Y.W. Jan, Adaptive fuzzy mixed H2/H ∞ attitude control of spacecraft, IEEE Trans. Aerosp. Electron. Syst. 36 (4) (2000) 1343–1359.
[68] L.B. Sandrine, D. Gilles, Control of an earth observation satellite, J. Guid. Control Dynam. 19 (3) (1996) 628–635.
[69] B. Boulet, B.A. Francis, P.C. Hughes, et al., μ synthesis for a large flexible space structure experimental testbed, J. Guid. Control Dynam. 24 (4) (2001) 967–977.
[70] H. Long, F. Liu, J. Zhao, et al., Anti-unwinding constrained attitude control for flexible spacecraft with actuator saturation, Proc. Inst. Mech. Eng. G J. Aerosp. Eng. (2015).
[71] L. Behera, M. Gopal, S. Chaudhury, On adaptive trajectory tracking of a robot manippulator using inversion of its neural Rmulator, IEEE Trans. Neural Netw. 7 (6) (1996) 1401–1404.
[72] Y. Zeng, A. Araujo, S.N. Singh, Output feedback variable structure adaptive control of a flexible spacecraft, Acta Astronaut. 44 (1) (1999) 11–22.
[73] S.N. Singh, R. Zhang, Adaptive output feedback control of spacecraft with flexible appendages by modeling error compensation, Acta Astronaut. 54 (4) (2004) 229–243.
[74] G.B. Maganti, S.N. Singh, Simplified adaptive control of an orbiting flexible spacecraft, Acta Astronaut. 61 (7–8) (2007) 575–589.
[75] K.W. Lee, S.N. Singh, Adaptive control of spacecraft with flexible appendages using Nussbaum gain, in: AIAA Guidance, Navigation and Control Conference and Exhibit, AIAA Press, South Carolina, 2007, pp. 1288–1299.
[76] Q. Hu, J. Zhang, Attitude control and vibration suppression for flexible spacecraft using control moment gyroscopes, J. Aerosp. Eng. 29 (1) (2015).
[77] S.V. Emel, Variable structure control systems: synthesis of scalar and vector systems by state and output feedback, Comput. Math. Model. 21 (3) (2010) 253–274.
[78] V. Utkin, Variable structure systems with sliding modes, IEEE Trans. Autom. Control 22 (2) (1977) 212–222.
[79] J.D. Boskovic, S.M. Li, R.K. Mehra, Robust tracking control design for spacecraft under control input saturation, J. Guid. Control Dynam. 27 (4) (2004) 627–633.
[80] T. Singh, M.F. Golnaraghi, R.N. Dubly, Sliding mode shaped input control of flexible rigid link robots, J. Sound Vib. 171 (2) (1994) 185–200.
[81] Q. Hu, P. Shi, H. Gao, Adaptive variable structure and commanding shaped vibration control of flexible spacecraft, J. Guid. Control Dynam. 30 (3) (2007) 804–815.
[82] Q.L. Hu, G.F. Ma, Active vibration suppression in flexible spacecraft with mismatched uncertainty via sliding mode shaped input control, J. Vib. Shock 26 (6) (2007) 133–138.
[83] A. Sinha, D.W. Miller, Optimal sliding mode control of a flexible spacecraft under stochastic disturbances, J. Guid. Control Dynam. 18 (18) (1993) 188–194.
[84] Q. Hu, Sliding mode attitude control with L 2 -gain performance and vibration reduction of flexible spacecraft with actuator dynamics, Acta Astronaut. 67 (5–6) (2010) 572–583.
[85] P. Guan, X.J. Liu, J.Z. Liu, Adaptive fuzzy sliding mode control for flexible satellite, Eng. Appl. Artif. Intel. 18 (4) (2005) 451–459.
[86] B. Xiao, Q. Hu, Y. Zhang, Adaptive sliding mode fault tolerant attitude tracking control for flexible spacecraft under actuator saturation, IEEE Trans. Control Syst. Technol. 20 (6) (2011) 1605–1612.
[87] K.D. Young, V.I. Utkin, U. Ozguner, A control engineer's guide to sliding mode control, IEEE Trans. Control Syst. Technol. 7 (3) (1999) 328–342.
[88] J.E. Slotine, Sliding controller design for nonlinear systems, Int. J. Control. 40 (2) (1984) 421–434.
[89] A.D. Araujo, S.N. Singh, Variable structure adaptive control of wing-rock motion of slender delta wings, J. Guid. Control Dynam. 21 (2) (1998) 251–256.

[90] W. Gao, J.C. Hung, Variable structure control of nonlinear systems: a new approach, IEEE Trans. Ind. Electron. 40 (1) (1993) 45–55.
[91] L.W. Zhou, J. Zhou, W.H. Li, Sliding mode variable structure control for large angle maneuvering of flexible spacecraft, Flight Dyn. 22 (1) (2004) 71–73.
[92] A. Levant, Quasi-continuous high-order sliding-mode controllers, IEEE Trans. Autom. Control 50 (11) (2005) 1812–1816.
[93] G. Bartolini, A. Ferrara, E. Usai, et al., On multi-input chattering-free second-order sliding mode control, IEEE Trans. Autom. Control 45 (9) (2000) 1711–1717.
[94] K.M. Ma, Design of higher order sliding mode attitude control laws for large-scale spacecraft, Control Decis. 28 (2) (2013) 201–204.
[95] C.H. Tsai, H.Y. Chung, F.M. Yu, Neural-sliding mode control with its applications in seesaw systems, IEEE Trans. Neural Netw. 15 (1) (2004) 124–134.
[96] K. Hornik, M. Stinchcombe, H. White, Multilayer feedforward networks are universal approximators, Neural Netw. 2 (5) (1989) 359–366.
[97] Y.R. Hwang, M. Tomizuka, Fuzzy smoothing algorithms for variable structure systems, IEEE Trans. Fuzzy Syst. 2 (4) (1994) 277–284.
[98] X.Y. Huang, Q. Wang, C.Y. Dong, Adaptive second-order sliding mode attitude control of flexible spacecraft, in: Proceedings of SPIE-The International Society for Optical Engineering, Beijing, China, 2008, pp. 1185–1196.
[99] W. Zuo, Y. Zhu, L. Cai, Fourier-neural-network-based learning control for a class of nonlinear systems with flexible components, IEEE Trans. Neural Netw. 20 (1) (2009) 139–151.
[100] A.M. Zou, K.D. Kumar, Z.G. Hou, Quaternion-based adaptive output feedback attitude control of spacecraft using Chebyshev neural networks, IEEE Trans. Neural Netw. 21 (9) (2010) 1457–1471.
[101] S. Bhat, D.S. Bernstein, Continuous bounded finite-time stabilization of the translational and rotational double integrators, Clin. Exp. Dermatol. 30 (3) (2005) 272–276.
[102] S.H. Ding, S.H. Li, Q. Li, Stability analysis for a second-order continuous finite-time control system subject to a disturbance, J. Control Theory Appl. 7 (3) (2009) 271–276.
[103] Y. Hong, J. Huang, On an output feedback finite-time stabilisation problem, in: IEEE Conference on Decision & Control, IEEE, Phoenix, AZ, USA, 1999, pp. 1302–1307.
[104] S. Bhat, D.S. Bernstein, Finite-time stability of homogeneous systems, in: Proceedings of the 1997 American Control Conference. Part 3 (of 6). Albuquerque, NM, USA, 1997, pp. 2513–2514.
[105] E. Jin, Z. Sun, Robust controllers design with finite time convergence for rigid spacecraft attitude tracking control, Aerosp. Sci. Technol. 12 (4) (2008) 324–330.
[106] H. Du, S. Li, C. Qian, Finite-time attitude tracking control of spacecraft with application to attitude synchronization, IEEE Trans. Autom. Control 56 (11) (2011) 2711–2717.
[107] M. Zak, Terminal attractors for addressable memory in neural networks, Phys. Lett. A 133 (1–2) (1988) 18–22.
[108] Z.H. Man, X.H. Yu, Terminal sliding mode control of MIMO linear systems, IEEE Trans. Circuits Syst. I Fundam. Theory Appl. 44 (11) (1997) 1065–1070.
[109] X. Yu, Z. Man, Fast terminal sliding-mode control design for nonlinear dynamical systems, IEEE Trans. Circuits Syst. I Fundam. Theory Appl. 49 (2) (2002) 261–264.
[110] Y. Feng, X. Yu, Z. Man, Non-singular terminal sliding mode control of rigid manipulators, Automatica 38 (12) (2002) 2159–2167.
[111] L. Yang, J. Yang, Nonsingular fast terminal sliding-mode control for nonlinear dynamical systems, Int. J. Robust Nonlinear Control 21 (16) (2011) 1865–1879.

[112] S. Wu, G. Wu, S. Tan, et al., Quaternion-based adaptive terminal sliding mode control for spacecraft attitude tracking, IEEE Int. Conf. Control Autom. 45 (5) (2013) 913–917.
[113] K. Lu, Y. Xia, Adaptive attitude tracking control for rigid spacecraft with finite time convergence, Automatica 49 (12) (2013) 3591–3599.
[114] C. Pukdeboon, P. Siricharuanun, Nonsingular terminal sliding mode based finite-time control for spacecraft attitude tracking, Int. J. Control. Autom. Syst. 12 (3) (2014) 530–540.
[115] X.J. Chen, W.H. Steyn, Robust combined Eigenaxis slew manoeuvre, in: AIAA Guidance Navigation and Control Conference, 1999, pp. 521–529.
[116] Z.W. Sun, Y.H. Geng, G.D. Xu, et al., The combined control algorithm for large-angle maneuver of HITSAT-1 small satellite, Acta Astronaut. 54 (7) (2004) 463–469.
[117] Z.W. Sun, X.H. Lin, X.B. Cao, Combined control algorithm for large angle maneuver of small satellite, J. Harbin Inst. Technol. (06) (2003) 663–667. +670.
[118] V.J. Lappas, W.H. Steyn, C.I. Underwood, Control moment gyro (CMG) gimbal angle compensation using magnetic control during external disturbances, Electron. Lett. 37 (9) (2001) 603–604.
[119] C.D. Hall, P. Tsiotras, H.J. Shen, Tracking rigid body motion using thrusters and momentum wheels, J. Astronaut. Sci. 50 (3) (2002) 311–323.
[120] L.L. Hartter, S.C. Rybak, R.A. Mayo, et al., A hard-mounted ultrahigh-accuracy pointing control system, in: Guidance and Control Conference, 1972, pp. 1–17.
[121] T. Liang, Modeling and simulation of attitude control for space station with SGCMG systems, J. Astronaut. (02) (2003) 126–131.
[122] D. Verbin, V.J. Lappas, Rapid rotational maneuvering of rigid satellites with hybrid actuators configuration, J. Guid. Control Dynam. 36 (2) (2013) 532–547.
[123] S. Liden, A new fail operational control moment gyro configuration, in: Guidance, Control and Flight Mechanics Conference, American Institute of Aeronautics and Astronautics, 1971, pp. 1–9.
[124] C.M. Roithmayr, C.D. Karlgaar, R.R. Kumar, et al., Dynamics and Control of Attitude Power and Momentum for a Spacecraft Using Flywheels and Control Moment Gyroscopes, United States, 2003, pp. 1–91.
[125] C.E. Skelton, C.D. Hall, Mixed control moment gyro and momentum wheel attitude control strategies, Astrodynamics 116 (1) (2003) 887–899.
[126] H. Leeghim, D. Lee, H. Bang, et al., Spacecraft attitude control by combination of various Torquers, Int. J. Syst. Sci. 40 (10) (2009) 995–1008.
[127] F.A. Leve, N.G. Fitzcoy, Hybrid steering logic for single-gimbal control moment gyroscopes, J. Guid. Control Dynam. 33 (4) (2010) 1202–1212.
[128] J. Lei, S.J. Xu, et al., J. Astronaut. (03) (2008) 916–921. +951.
[129] R.S. Maynard, Fluidic momentum controller, US Patent 4,776,541, 1988.
[130] N.A. Nobari, A.K. Misra, Attitude dynamics and control of satellites with fluid ring actuators, J. Guid. Control Dynam. 35 (6) (2012) 1855–1864.
[131] N.A. Nobari, A.K. Misra, A hybrid attitude controller consisting of electromagnetic torque rods and an active fluid ring, Acta Astronaut. 94 (1) (2014) 470–479.

Part I

Preliminaries

Chapter 2

Mathematical model and basic control theory

The attitude of satellite refers to the relative position or pointing of the satellite with respect to the reference coordinate system. Satellites need to maintain high-precision pointing and positioning of their attitudes during missions, hence it is needed to apply attitude control to meet the attitude requirements of space missions, and the establishment of mathematical model for the rotation of the satellite around its mass center to describe is the basis for designing attitude control laws. In this chapter, the dynamic model for rigid and flexible satellites, kinetic model, and several attitude parameters are introduced, and some related control theories are also introduced for later chapters.

2.1 Spacecraft kinetic model and satellite parameters

2.1.1 Cosine matrix

The essence of directional cosine matrix is the coordinate transformation matrix between two reference systems, and in this book, it is used to describe the transformation matrix between inertia frame and body frame. Define inertia frame as $\boldsymbol{S}_a$ and its vector matrix as $\boldsymbol{f}_a$, define body frame as $\boldsymbol{S}_b$ and its vector matrix as $\boldsymbol{f}_b$, and the cosine matrix describing the transformation from inertia frame to body frame is

$$\boldsymbol{C} = \boldsymbol{f}_b \boldsymbol{f}_a^T = \begin{bmatrix} \boldsymbol{i}_b \cdot \boldsymbol{i}_a & \boldsymbol{i}_b \cdot \boldsymbol{j}_a & \boldsymbol{i}_b \cdot \boldsymbol{k}_a \\ \boldsymbol{j}_b \cdot \boldsymbol{i}_a & \boldsymbol{j}_b \cdot \boldsymbol{j}_a & \boldsymbol{j}_b \cdot \boldsymbol{k}_a \\ \boldsymbol{k}_b \cdot \boldsymbol{i}_a & \boldsymbol{k}_b \cdot \boldsymbol{j}_a & \boldsymbol{k}_b \cdot \boldsymbol{k}_a \end{bmatrix} \tag{2.1}$$

Matrix $\boldsymbol{C}$ is positive definite and symmetry matrix, and its eigenvalue satisfies

$$\lambda(\boldsymbol{C}) = 1 \tag{2.2}$$

The kinetic model based on cosine matrix is

$$\dot{\boldsymbol{C}} = -\boldsymbol{\omega}^{\times} \boldsymbol{C} \tag{2.3}$$

Fast Satellite Attitude Maneuver and Control. https://doi.org/10.1016/B978-0-323-95455-6.00014-7

Cosine matrix has the advantage of clear physical meaning, and its kinetic model is quite simple, but its disadvantage is too much variables, i.e., there is 9 variables to be controlled.

2.1.2 Euler angle/axis

Euler angle/axis parameter is derived via Euler's Rotation Theorem of rigid body, i.e., any rotation around a fixed point could be transformed to a rotation around a fixed axis with an angle, and the axis is called Euler axis $\boldsymbol{e}$, and the angle is called Euler angle φ.

The kinetic model based on Euler axis/angle could be written as

$$\begin{cases} \dot{\boldsymbol{e}} = \dfrac{1}{2}\boldsymbol{e}^{\times}\left(\boldsymbol{I}_3 - \cot\dfrac{\varphi}{2}\boldsymbol{e}^{\times}\right)\boldsymbol{\omega} \\ \dot{\varphi} = \boldsymbol{e}^T\boldsymbol{\omega} \end{cases} \tag{2.4}$$

This model also has the advantage of clear physical meaning, but when Euler angle φ tends to zero, the derivative of Euler Axis $\boldsymbol{e}$ tends to infinite, and this means model (2.10) has singularity issue near system equilibrium point.

2.1.3 Attitude quaternion

Attitude quaternion is an attitude parameter based on Euler angle/axis, its definition is as follows:

$$\boldsymbol{q} = \begin{bmatrix} \cos\dfrac{\varphi}{2} & \boldsymbol{e}^T\sin\dfrac{\varphi}{2} \end{bmatrix}^T = \begin{bmatrix} q_0 & \boldsymbol{q}_v^T \end{bmatrix}^T = \begin{bmatrix} q_0 & q_1 & q_2 & q_3 \end{bmatrix}^T \tag{2.5}$$

It is easily to see that attitude quaternion is a four-dimensional vector with such constraint

$$\boldsymbol{q}^T\boldsymbol{q} = q_0^2 + \boldsymbol{q}_v^T\boldsymbol{q}_v = q_0^2 + q_1^2 + q_2^2 + q_3^2 = 1 \tag{2.6}$$

Define quaternion $\boldsymbol{q}$ describes the rotation form frame $\boldsymbol{S}_a$ to frame $\boldsymbol{S}_b$, and the quaternion describing the rotation from $\boldsymbol{S}_b$ to$\boldsymbol{S}_a$ is the conjugation quaternion of $\boldsymbol{q}$, which is defined as follows:

$$\boldsymbol{q}^* = \begin{bmatrix} q_0 & -\boldsymbol{q}_v^T \end{bmatrix}^T \tag{2.7}$$

Also, define quaternion $\boldsymbol{q}$ describing the rotation from $\boldsymbol{S}_a$ to $\boldsymbol{S}_b$, quaternion $\boldsymbol{p}$ describing the rotation from $\boldsymbol{S}_b$ to $\boldsymbol{S}_c$, and the quaternion describing frame $\boldsymbol{S}_a$ to $\boldsymbol{S}_c$ could be written as

$$\boldsymbol{r} = \boldsymbol{q} \otimes \boldsymbol{p} = \begin{bmatrix} q_0p_0 - \boldsymbol{q}_v^T\boldsymbol{p}_v \\ q_0\boldsymbol{p}_v + p_0\boldsymbol{q}_v + \boldsymbol{q}_v \times \boldsymbol{p}_v \end{bmatrix} \tag{2.8}$$

with $\otimes$ is the rotation manipulator for attitude quaternion, and its definition is given in equation (2.14).

Noting that a rotation around axis $\boldsymbol{e}$ with angle φ and the rotation around axis $\boldsymbol{e}$ with angle $2\pi-\varphi$ are the same rotation, and if former rotation corresponds to quaternion $\boldsymbol{q}$, based on the definition of attitude quaternion it could be found the later rotation corresponds to quaternion $-\boldsymbol{q}$, this means $\boldsymbol{q}$ and $-\boldsymbol{q}$ describe the same rotation. Hence in this book, it could be assumed that $q_0 \geq 0$ if it is necessary.

System kinetic model based on attitude quaternion could be written as follows:

$$\begin{cases} \dot{q}_0 = -\dfrac{1}{2}\boldsymbol{q}_v^T\boldsymbol{\omega} \\ \boldsymbol{q}_v = \dfrac{1}{2}\left(q_0\boldsymbol{I}_3 + \boldsymbol{q}_v^{\times}\right)\boldsymbol{\omega} = \dfrac{1}{2}\boldsymbol{F}\boldsymbol{\omega} \end{cases} \tag{2.9}$$

with matrix $\boldsymbol{F}$ satisfies

$$\begin{cases} \lambda(\boldsymbol{F}) = |q_0|, 1 \\ \lambda_M(\boldsymbol{F}) = 1 \end{cases} \tag{2.10}$$

2.1.4 Modified Rodrigues parameter

Noting that attitude quanernion has a redundant component, Modified Rodrigues Parameter (MRP) is proposed. MRP is also defined based on Euler axis/angle as follows:

$$\boldsymbol{\sigma} = \boldsymbol{e}\tan\frac{\varphi}{4} = \frac{\sin(\varphi/2)\boldsymbol{e}}{1+\cos(\varphi/2)} = \frac{\boldsymbol{q}_v}{1+q_0} \tag{2.11}$$

It could be found that when $\varphi = 2\pi$ singularity issue occurs, but in this situation system rotates a complete cycle, and this could be ignored in engineering practice. Also, noting that rotation around axis $\boldsymbol{e}$ with angle φ and the rotation around axis $\boldsymbol{e}$ with angle $2\pi-\varphi$ are the same rotation, it could be found that $\boldsymbol{\sigma}$ and $-\boldsymbol{\sigma}/\|\boldsymbol{\sigma}\|^2$ describe same rotation, hence in this book it could be defined that $\|\boldsymbol{\sigma}\| \leq 1$ if it is necessary.

The kinetic model based on MRP is

$$\dot{\boldsymbol{\sigma}} = \frac{1}{2}\left(\frac{1-\boldsymbol{\sigma}^T\boldsymbol{\sigma}}{2}\boldsymbol{I}_3 + \boldsymbol{\sigma}^{\times} + \boldsymbol{\sigma}\boldsymbol{\sigma}^T\right)\boldsymbol{\omega} = \boldsymbol{G}\boldsymbol{\omega} \tag{2.12}$$

where matrix $\boldsymbol{G}$ satisfies

$$\boldsymbol{G}^T\boldsymbol{G} = \left(\frac{1+\boldsymbol{\sigma}^T\boldsymbol{\sigma}}{4}\right)^2\boldsymbol{I}_3, \boldsymbol{G}^{-1} = \frac{16}{\left(1+\boldsymbol{\sigma}^T\boldsymbol{\sigma}\right)^2}\boldsymbol{G}^T \tag{2.13}$$

And its eigenvalue satisfies

$$\lambda_M(\boldsymbol{G}) = \|\boldsymbol{G}\|_2 = \sqrt{\lambda_M\left(\boldsymbol{G}^T\boldsymbol{G}\right)} = \frac{1+\boldsymbol{\sigma}^T\boldsymbol{\sigma}}{4} \leq \frac{1}{2} \tag{2.14}$$

2.2 Spacecraft dynamic model

2.2.1 Rigid dynamic model

The dynamic model of rigid satellite could be written as follows:

$$J\dot{\omega} + \omega^{\times} J\omega = u + d \tag{2.15}$$

where ω is the angular velocity, J is the inertia matrix of satellite that is a symmetric matrix, and d is the unknown disturbance torque with norm upper bound $\|d\| < \overline{d}$. Product matrix $r^{\times}$ of vector r is defined as

$$r^{\times} = \begin{bmatrix} 0 & -r_3 & r_2 \\ r_3 & 0 & -r_1 \\ -r_2 & r_1 & 0 \end{bmatrix} \tag{2.16}$$

Generally inertia matrix J is not accurate known, and it is assumed that

$$J = \widehat{J} + \widetilde{J} \tag{2.17}$$

where $\widehat{J}$ is the inertia matrix best estimate and $\widetilde{J}$ is the error matrix. Product matrix has an important property that will be used in the later part that the eigenvalues of $r^{\times}$ satisfies

$$\begin{gathered} \lambda(r^{\times}) = 0, \pm \|r\| \\ \lambda_{\max}(r^{\times}) = \|r\| \end{gathered} \tag{2.18}$$

In order to simplify the text, the maximum and minimum eigenvalue of matrix **A** is described as $\lambda_M(A)$ and $\lambda_m(A)$.

Also, cross-product manipulator has following properties

$$(r^{\times})^T = -r^{\times}, r^{\times} r = r^T r^{\times} = 0, r^{\times} m = -m^{\times} r, m^T r^{\times} m = 0 \tag{2.19}$$

$$(a^{\times} b)^{\times} = ba^T - ab^T \tag{2.20}$$

with a, b, m are all three-dimensional vectors.

In satellite attitude tracking issue, error quaternion and error angular velocity are defined as follows:

$$\begin{gathered} q_e = \overset{*}{q}_d \otimes q = \begin{bmatrix} q_{d0} q_0 + q_{dv}^T q_v \\ q_{d0} q_v - q_0 q_{dv} - q_{dv} \times q_v \end{bmatrix} \\ R(q_e) = \left(q_{e0}^2 - q_{ev}^T q_{ev}\right) I_3 + 2 q_{ev} q_{ev}^T - 2 q_{e0} q_{ev}^{\times} \\ \omega_e = \omega - R(q_e)\omega_d \end{gathered} \tag{2.21}$$

where the eigenvalue of R satisfies $\lambda(R) = 1$, q_d is the desired attitude quaternion, ω_d is the desired angular velocity, and $\overset{*}{q}_d$ is the conjugated quaternion defined as follows:

$$\overset{*}{q}_d = \left[q_{d0} \ -q_{dv}^T\right]^T \tag{2.22}$$

Substitute (2.21) in dynamic model (2.1), the dynamic model of error angular velocity could be written as (2.23).

$$J\dot{\omega}_e + JR\dot{\omega}_d - J\omega_e^{\times}R\omega_d + (\omega_e + R\omega_d)^{\times}J(\omega_e + R\omega_d) = u + d \tag{2.23}$$

The kinetic model of error quaternion could be written as follows:

$$\begin{cases} \dot{q}_{e0} = -\dfrac{1}{2}q_{ev}^T\omega_e \\ \dot{q}_{ev} = \dfrac{1}{2}\left(q_{e0}I_3 + q_{ev}^{\times}\right)\omega_e = \dfrac{1}{2}F_e\omega_e \end{cases} \tag{2.24}$$

Define the unit direction vector e_e as follows:

$$e_e = q_{ev}/\|q_{ev}\| \tag{2.25}$$

Calculate the derivative of $\|q_{ev}\|$ and e_e, it can be got that

$$\frac{d\|q_{ev}\|}{dt} = \frac{d\sqrt{1-q_{e0}^2}}{dt} = \frac{1}{2}\frac{-2q_{e0}\dot{q}_{e0}}{\sqrt{1-q_{e0}^2}} = -\frac{q_{e0}}{\|q_{ev}\|}\left(-\frac{1}{2}q_{ev}^T\omega_e\right) = \frac{1}{2}q_{e0}e_e^T\omega_e \tag{2.26}$$

$$\dot{e}_e = \frac{\|q_{ev}\|\dot{q}_{ev}}{\|q_{ev}\|^2} - \frac{q_{ev}}{\|q_{ev}\|^2}\frac{d\|q_{ev}\|}{dt} = \frac{1}{2}e_e^{\times}\omega_e + \frac{1}{2}\frac{q_{e0}}{\|q_{ev}\|}\left(I_3 - e_e e_e^T\right)\omega_e \tag{2.27}$$

Assume that φ_e is the Euler angle of error quaternion q_{ev}, and φ_e satisfies

$$\cot\frac{\varphi_e}{2} = \frac{q_{e0}}{\|q_{ev}\|} \tag{2.28}$$

Considering that e_e is a unit vector and $I_3 - e_e e_e^T = -e_e^{\times}e_e^{\times}$, the kinetic model of e_e could be transformed to

$$\dot{e}_e = \frac{1}{2}e_e^{\times}\left(I_3 - \cot\frac{\varphi_e}{2}e_e^{\times}\right)\omega_e \tag{2.29}$$

Also, the error model based on MRP could be written as follows:

$$\dot{\sigma}_e = \frac{1}{2}\left(\frac{1-\sigma_e^T\sigma_e}{2}I_3 + \sigma_e^{\times} + \sigma_e\sigma_e^T\right)\omega_e = G_e\omega_e \tag{2.30}$$

Based on system kinetic model and its error model, it can be found that they have same structure and this property would bring benefits for the design of controllers in later text.

In cooperative object tracking issue, it is assumed $q_d, \omega_d, \dot{\omega}_d$ are all accurate known.

2.2.2 Flexible dynamic model

In the endeavor toward satisfying the demand for the modeling accuracy of rigid-flexible coupling satellites in rapid maneuver, the dynamic model of

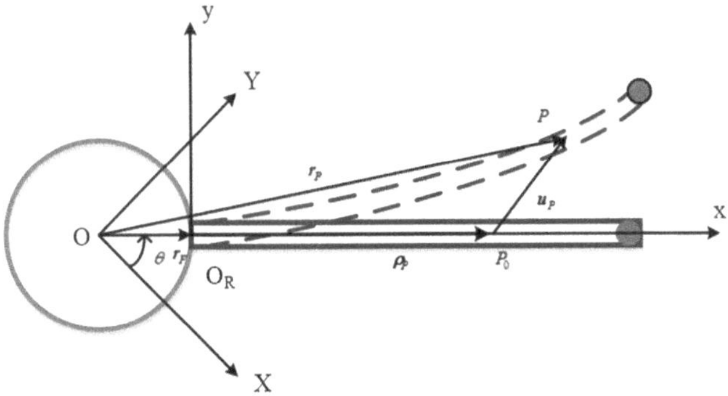

FIG. 2.1 Rigid-flexible coupling satellite model.

the considered satellite is set up based on the first-order approximate methodology in which the coupling effect between the transverse deformation and axial extension is incorporated in the strain displacement field.

The model of the rigid-flexible coupling satellite, as shown in Fig. 2.1, is constituted of a central rigid body, of which the radius is a, rotating around its center O; a uniform cantilever beam, whose length is L, as the flexible appendage attached radially to the central rigid body; and a tip mass weighting m_t, fixed to the end of the beam. OXY is the inertial coordinate system with its origin at the center of the central rigid body, and O_Rxy is the reference coordinate system, whose origin locates in the junction of the central rigid body and the uniform cantilever beam, and x axis is directed along the neutral axis of the undeformed configuration of the beam.

The flexible beam is divided into n finite elements whose length is $l=L/n$. Each finite element has its own coordinate system $i\widetilde{xy}$attached to its first node in its undeformed configuration, and the distance of the first node to the origin $O_R$of the reference coordinate system is $L_i=L(i-1)/n$. The node displacement coordinate vector of the ith element is defined by $p_i=[v_i,\beta_i]^T$, where v_i is the transverse deformation of the first node in the reference frame, and $\beta_i=\partial v_i/\partial\widetilde{x}$. Because the first node is at the junction, so we have $p_1=[v_1,\beta_1]^T=[0,0]^{\mathrm{T}}$. Starting from the second node, the generalized nodal coordinate matrix can be defined as follows:

$$\boldsymbol{p}(t)=\left[p_2^{\mathrm{T}}(t),\ldots,p_i^{\mathrm{T}}(t),\ldots p_{n+1}^{\mathrm{T}}(t)\right]^{\mathrm{T}}$$

It should be noticed that shear and rotary inertia effects of the beam are ignored for its slender shape, and the gravity effect is neglected since the motion of the beam is confined to the horizontal plane. Then according to the finite elements method and the Hamilton principle , the first-order approximation rigid-flexible coupling dynamics equation is set up as follows:

$$J_{\mathrm{mb}}\ddot{\theta}+F\dot{\theta}-\boldsymbol{U}\boldsymbol{M}_{\mathrm{s}}^{-1}\boldsymbol{C}_{\mathrm{s}}\chi-\boldsymbol{U}\boldsymbol{M}_{\mathrm{s}}^{-1}\left(\boldsymbol{K}_{\mathrm{s}}-\dot{\theta}^2\boldsymbol{G}\right)\boldsymbol{p}=T_{\mathrm{c}}+T_D \tag{2.31}$$

$$\dot{\chi} + M_s^{-1} C_s \dot{p} + M_s^{-1}\left(K_s + \dot{\theta}^2 G\right) p = 0 \tag{2.32}$$

$$\chi = \dot{p} + M_s^{-1} U^{\mathrm{T}} \dot{\theta} \tag{2.33}$$

whcrc

$$F = U M_s^{-1} C_s M_s^{-1} U^T \tag{2.34}$$

$$J_{\mathrm{mb}} = J_{\mathrm{H}} + J_{\mathrm{A}} + J_{\mathrm{T}} - U M_s^{-1} U^{\mathrm{T}} > 0 \tag{2.35}$$

$$G^{\mathrm{T}} = G = D_r - M_s > 0 \tag{2.36}$$

$$T_D = T_d + p^{\mathrm{T}} G p \ddot{\theta} + 2\dot{\theta} p^{\mathrm{T}} G \chi \tag{2.37}$$

and where χ is the reflection of the vibration speed of the flexible appendage. $J_{\mathrm{H}}, J_{\mathrm{A}}, J_{\mathrm{T}}$ are the moment of inertia of the central rigid body, flexible cantilever beam, and tip mass relative to the spin axis, respectively. The rotation angle of the central body is θ, and the control torque and disturbance torque are T_c and T_d, respectively. Symmetric positive definite matrixes M_s and K_s are the mass matrix and the stiffness matrix of the flexible cantilever beam, respectively. Symmetric positive definite matrix D_r is the rotational stiffness matrix of the flexible cantilever beam, while $D_r - M_s$ is also symmetric and positive definite. $C_s = b_1 M_s + b_2 K_s$ is the structural damping matrix designed by Rayleigh damping modal, where b_1 and b_2 are positive coefficients. U is the rigid-flexible coupling coefficient matrix.

Remark 2.1 It is worth mentioning that, during the deformation analysis, the second-order coupling term of deformation in the longitudinal and transverse deformation field is considered, which brings in the dynamic stiffness term $\dot{\theta}^2 D_r$. And this term is directly proportional to the square of the angular velocity, which means the bigger the rotational velocity of the satellite is, the larger proportion the dynamic stiffness term will occupy in the system stiffness matrix. As a result, the dynamic stiffening effect cannot be ignored, especially when the rigid-flexible satellite is in large angle rapid maneuver. In addition, the dynamic model described in this subsection is called the first-order approximation rigid-flexible coupling dynamic model for the consideration of the dynamic stiffening term, and it becomes the zero-order approximation rigid-flexible coupling model if the dynamic stiffening term is subtracted. Comparing to the zero-order approximation model and linear model, the dynamic model considered in this subsection is more accurate in the large angle maneuver mode.

Let the desired attitude angle beθ_d, the error attitude angle be $\theta_e = \theta - \theta_d$, and the error attitude angular velocity be $\dot{\theta}_e = \dot{\theta} - \dot{\theta}_d = \dot{\theta}$, then based on the dynamic model above, the relative attitude error dynamic equation can be obtained as

$$J_{\mathrm{mb}} \ddot{\theta}_e + F \dot{\theta}_e - U M_s^{-1} C_s \chi - U M_s^{-1}\left(K_s - \dot{\theta}_e^2 G\right) p = T_c + T_D \tag{2.38}$$

$$\dot{\chi} + M_s^{-1} C_s \dot{p} + M_s^{-1} \left(K_s + \dot{\theta}_e{}^2 G \right) p = 0 \tag{2.39}$$

With considering the uncertainty of the satellite moment of inertia, we have $J_{mo} = J_{mb} - \Delta J_{mo}$, where ΔJ_{mo} and J_{mo} are the uncertain part and known nominal part of the system moment of inertia, respectively. Let $(J_{mo} + \Delta J_{mo})^{-1} = J_{mo}^{-1} + \Delta \widetilde{J}_{mo}$ and $\Delta \widetilde{J}_{mo}$ represent the uncertain part of the system moment of inertia, then the relative attitude error dynamic model can be expressed as

$$\ddot{\theta}_e = -J_{mo}^{-1}\left(H + F\dot{\theta}_e\right) + J_{mo}^{-1} T_c + g + \Delta M \tag{2.40}$$

where

$$H = -UM_s^{-1} C_s \chi - UM_s^{-1} \left(K_s - \dot{\theta}_e^2 G \right) p \tag{2.41}$$

$$\Delta M = \Delta \widetilde{J}_{mo} \left(T_c - H - F\dot{\theta}_e \right) \tag{2.42}$$

$$g = J_{mo}^{-1} T_D + \Delta \widetilde{J}_{mo} T_D \tag{2.43}$$

where H is the coupling term, ΔM is the model uncertainty caused by the uncertain part of the system moment of inertia, and g is the external disturbance.

2.3 Relevant control theory

2.3.1 Stability theories

For autonomous system

$$\dot{x} = f(x) \tag{2.44}$$

with f is n-dimensional vector function, x is n-dimensional vector, if $x = 0$ is system's equilibrium point, we have:

(1) If for any $\varepsilon > 0$, there exists $\delta = \delta(\varepsilon) > 0$ to satisfy $\|x(0)\| < \delta \Rightarrow \|x(t)\| < \varepsilon$ with $\forall t \geq 0$, system is stable at its equilibrium point;
(2) If property (1) is not satisfied, system is unstable at its equilibrium point;
(3) If system is stable at an equilibrium point, and there exists $\delta > 0$ to satisfy $\|x(0)\| < \delta \Rightarrow \lim_{t \to \infty} \|x(t)\| = 0$, then system is asymptotic stability at the equilibrium point.

Theorem 2.1 (Lyapunov stability theorem): If $x = 0$ is an equilibrium point of system (2.31) and define positive scalar function $V(x)$ near the equilibrium point $U_0 \subset U$ with continuous derivative $\dot{V}(x)$, then we have (1) for any $\forall x \in U_0$ if $\dot{V}(x) \leq 0$ is satisfied, system is stable at $x = 0$; (2) for any $\forall x \in U_0$ if $\dot{V}(x) < 0$ is satisfied, system is asymptotic stable at $x = 0$;

(3) for global state space $\boldsymbol{U}$ and any $\forall \boldsymbol{x} \in \boldsymbol{U}$, if $V(\boldsymbol{x}) \geq 0$ is satisfied, system is global asymptotic stable at $\boldsymbol{x} = \boldsymbol{0}$.

Theorem 2.2 (Lasalle's invariant principle): If $f(t)$ is a continuous vector function defined in $\boldsymbol{U}$, scalar function $V(\boldsymbol{x})$ is positive definite and radial unbounded with continuous 1-order derivative, and for any $\forall \boldsymbol{x} \in \boldsymbol{U}$, $\dot{V}(\boldsymbol{x}) \leq 0$ is satisfied, and if $\boldsymbol{G}$ is the set that satisfies $\dot{V}(\boldsymbol{x}) = 0$ and $\overline{\boldsymbol{G}}$ is the maximum invariant set, then system converge to $\overline{\boldsymbol{G}}$.

Theorem 2.3 (Barbalat Lemma): If $f(t)$ is a continuous vector function defined in $t \in [0, \infty)$ with continuous 1-order derivative, and the limitation $\lim_{t\to\infty} f(t)$ exists, and $\dot{f}(t)$ is continuous for any $\forall t \in [0, \infty)$, then we have $\lim_{t\to\infty} \dot{f}(t) = 0$.

2.3.2 Sliding mode control

For general nonlinear system

$$\dot{\boldsymbol{x}} = f(\boldsymbol{x}, \boldsymbol{u}, t) \tag{2.45}$$

The basic idea to design sliding mode controller is that design sliding mode surface or switching function $s(\boldsymbol{x})$ and design controller

$$\boldsymbol{u}(\boldsymbol{x}) = \begin{cases} \boldsymbol{u}^+(\boldsymbol{x}), & s(\boldsymbol{x}) > 0 \\ \boldsymbol{u}^-(\boldsymbol{x}), & s(\boldsymbol{x}) < 0 \end{cases} \tag{2.46}$$

to satisfy following properties:

(1) With any initial state away from the sliding mode surface, system state could reach the sliding mode surface or the neighbor area around the surface with short enough time.
(2) System state is stable along the sliding mode surface and satisfy the demand control performance.
(3) The desired sliding mode surface $s(\boldsymbol{x})$ exists.

The above properties are the basic requirements for sliding mode controller design, and the process to design sliding mode controllers could be concluded as: (1) design the demanded sliding mode surface to meet system performance and (2) design controller to make system state reach the desired surface and converge along this surface.

Standard linear sliding mode surface could be written as follows:

$$s = \omega + k\boldsymbol{q}_v \tag{2.47}$$

with positive scalar k.

When reaching this sliding mode surface, we have

$$\omega = -k\boldsymbol{q}_v \tag{2.48}$$

Substitute equation (2.35) to kinetic model (2.15)

$$\dot{\boldsymbol{q}}_v = \frac{1}{2}\left(q_0 \boldsymbol{I}_3 + \boldsymbol{q}_v^{\times}\right)\boldsymbol{\omega} = -\frac{1}{2} k q_0 \boldsymbol{q}_v \tag{2.49}$$

Noting that $q_0 \geq 0$, it could be easily found attitude quaternion converges with exponential rate, and this means system is stable along sliding mode (2.34).

The next step is to design controller to achieve system reaching this sliding mode with any initial conditions. Select Lyapunov function as follows:

$$V_s = \frac{1}{2}\boldsymbol{s}^T \boldsymbol{J}\boldsymbol{s} \tag{2.50}$$

and calculate its derivative

$$\begin{aligned} \dot{V}_s &= \boldsymbol{s}^T \boldsymbol{J}\dot{\boldsymbol{s}} = \boldsymbol{s}^T \boldsymbol{J}\dot{\boldsymbol{\omega}} + k\boldsymbol{s}^T \boldsymbol{J}\dot{\boldsymbol{q}}_v \\ &= \boldsymbol{s}^T \boldsymbol{J}\boldsymbol{u} + \boldsymbol{s}^T \boldsymbol{J}\boldsymbol{d} - \boldsymbol{s}^T \boldsymbol{J}\boldsymbol{\omega}^{\times}\boldsymbol{\omega} + \frac{1}{2} k\boldsymbol{s}^T \boldsymbol{J}\boldsymbol{F}\boldsymbol{\omega} \end{aligned} \tag{2.51}$$

Hence, if following controller is designed, $\dot{V}_s \leq 0$ could be ensured.

$$\boldsymbol{u} = -k_p \boldsymbol{s} + \boldsymbol{\omega}^{\times}\boldsymbol{J}\boldsymbol{\omega} - \frac{1}{2} k\boldsymbol{J}\boldsymbol{F}\boldsymbol{\omega} - \overline{d}\,\boldsymbol{sgn}(\boldsymbol{s}) \tag{2.52}$$

with positive scalar k_p and sign function $\boldsymbol{sgn}(\boldsymbol{x})$ for vector $\boldsymbol{x}$.

It is obviously sliding mode control (2.34) and (2.39) has clear physical meaning and simple structure, and this is the reason sliding mode control is widely studied and applied.

2.3.3 Finite time control

In control problems, linear controllers always bring exponential convergence rate that means system state would reach the equilibrium point with infinite time. In order to improve system performance near the equilibrium point, finite time control algorithm is proposed. In order to introduce this control method, it is necessary to introduce the concept of finite time stability.

For nonlinear system (2.31), if system is stable at $\boldsymbol{x} = \boldsymbol{0}$ and could converge to this equilibrium point (in this book, this property means $\|\boldsymbol{x}(t)\| \leq \varepsilon$ for small enough positive scalar ε) within finite time, then system is finite time stable at $\boldsymbol{x} = \boldsymbol{0}$. For any initial state $\boldsymbol{x}_0 \in \boldsymbol{U}_0$ where $\boldsymbol{U}_0$ is a neighbor area around equilibrium point $\boldsymbol{x} = \boldsymbol{0}$, define the solution of system as $\boldsymbol{x}(t, \boldsymbol{x}_0)$. Finite time stability is that there exists a time function $\boldsymbol{T}(\boldsymbol{x})$ that when $t \in [0, T(\boldsymbol{x}_0)]$ we have $\lim_{t \to T(\mathbf{x}_0)} \boldsymbol{x}(t, \boldsymbol{x}_0) = \boldsymbol{0}$, and when $\lim_{t \to T(\mathbf{x}_0)} \boldsymbol{x}(t, \boldsymbol{x}_0) = \boldsymbol{0}$, we have $\boldsymbol{x}(t, \boldsymbol{x}_0) = \boldsymbol{0}$. And if this property still holds for global state space $\boldsymbol{U}$, then system is global finite time stable at $\boldsymbol{x} = \boldsymbol{0}$.

Theorem 2.4 (Finite time stability theorem): If $V(\mathbf{x})$ is positive definite scalar function and there exists positive scalars γ and $\alpha \in (0,1)$, and for any $\forall \boldsymbol{x} \in \boldsymbol{U}_0$ with $\boldsymbol{U}_0$ around the equilibrium point, inequality $\dot{V}(\boldsymbol{x}) \leq -\gamma V^{\alpha}(\boldsymbol{x})$ holds, then system is finite time stable at equilibrium point $\boldsymbol{x}=\boldsymbol{0}$. And if this property still holds for global state space $\boldsymbol{U}$, then system is global finite time stable at $\boldsymbol{x}=\boldsymbol{0}$. And system convergence time satisfies

$$T \leq \frac{V_0^{1-\alpha}}{\gamma(1-\alpha)} \tag{2.53}$$

Part II

Research on attitude control methods for fast maneuver satellites

Chapter 3

Near time-optimal control based on analytical trajectory planning

3.1 Introduction

Satellite attitude maneuver issue has been discussed by researchers for decades, and lot of work has been done on this field. Xiao [1–3] designed robust sliding mode controllers for satellite attitude maneuver considering perturbation such as system inertia matrix uncertainty, unknown disturbance torque, and actuator failure. The focus of his work was the system robustness and steady accuracy. Several standard sliding modes were given in his work, and the system performance on proposed sliding modes was analyzed. Wu [4–6] also designed sliding mode controllers for satellite maneuver. The chattering issue was discussed in his work, and the controller is designed under system uncertainty and disturbance torque. In sliding mode controllers, system convergence trajectory was designed by constructing the relationship between angular velocity and attitude quaternion. Generally, system convergence rate of standard sliding mode controller is exponential. In order to improve system convergence rate, researchers improved standard attitude controllers by several methods.

When the trajectory of convergence process is determined, the next step is to design attitude tracking controllers to ensure system maneuver along the desired trajectory. Zhang [7] proposed a satellite attitude tracking controller to track the optimized trajectory considering the failure of actuator. Cao [8] modeled the dynamic of actuator and designed attitude tracking controller considering system uncertainties. Guo [9] designed back-stepping controller for satellite attitude tracking control considering control torque constraint. The main focus of satellite attitude tracking issue is system stability analysis, actuator failure, system uncertainties, and so on.

Considering that current space missions need the capability of fast attitude maneuver, it is necessary to discuss how to improve system convergence rate of standard attitude maneuver controllers. Trajectory planning algorithm is an effective method to improve system convergence rate. The essence of trajectory

Fast Satellite Attitude Maneuver and Control. https://doi.org/10.1016/B978-0-323-95455-6.00001-9

planning algorithm is similar as guidance law to some extent: the manipulator generates the desired position and velocity information of next control cycle under some specific assumptions, and the system is chasing the desired state during the whole process. In fact, the relationship between trajectory planning algorithm and tracking controller could be treated as: the trajectory planning algorithm determines where to go for next control cycle, and tracking controller determines the norm and direction of control torque so that system could reach the desired position in every control cycle. Li [10,11] designed finite time controller and dynamic sliding mode controller, and standard sliding mode is modified in these papers. The maneuver stage was constructed hence the system could rotate with constant angular velocity and the deceleration process was optimized in these papers. In Li's work, he pointed out that the key to improve system convergence rate is to design angular velocity curve. Bai [12] optimized the trajectory of satellite attitude maneuver by Lagrange manipulator. He pointed out that the time-optimal trajectory is derived under control torque saturation condition and the rotation along Euler axis is time-optimal or near time-optimal. Li and Bainum [13] linearized the kinetic model and calculated the time-optimal trajectory of Euler-axis maneuver. Geogre [14] solved the nonlinear model by Gaussian pseudospectral method, and he pointed out that the trajectory optimization issue could be treated as the issue to determine the switching time of actuator since the control torque is locked to system upper bound. Besides the attitude maneuver issue, trajectory planning algorithm was also used in other control issues. Zhang [15] proposed a real-time attitude stabilization trajectory optimization method for an underactuated rigid spacecraft. Wang [16] optimized the trajectory of three-dimensional maneuver for UAV using an improved bat algorithm. Zhuang [17] optimized the trajectory of underactuated spacecraft using a hybrid particle swarm optimization algorithm. Ridder [18] and Xie [19] proposed trajectory planning algorithms for launch vehicles and systems with high nonlinearity are discussed in these papers. Generally, the optimal trajectory is derived under two conditions: (*a*) the total rotate angle should be minimum ornear minimum and (*b*) the system should maneuver under control torque saturation or velocity saturation during the whole process. Inspired by these two properties, a near-time optimal trajectory planning algorithm will be discussed in later section. Moreover, most of the algorithms listed above are numerical that means the computation is large. Considering that the capability of onboard computer is limited till now, it is necessary to develop an analytical algorithm to reduce the computation and make it easier for engineering practice.

The goal of this chapter could be concluded as (1) develop a near-time optimal trajectory planning algorithm that satisfies conditions ***a*** and ***b*** above for satellite attitude maneuver issue hence system convergence rate could be largely improved; (2) give the analytical solution of the desired trajectory hence the computation could be reduced; and (3) design attitude tracking controller hence system could maneuver along the designed trajectory. Goals 1 and 2 are the major contributions of this chapter.

3.2 Problem formulation

3.2.1 Physical constraints on spacecraft

Generally, larger control torque and angular velocity bring better convergence rate, but spacecraft system has some physical restrictions on these vectors. Considering that the capability of control actuators is limited, the restriction on control torque is expressed as continuous inequality constraint given by

$$\boldsymbol{u}(t) \leq \overline{u} \tag{3.1}$$

Moreover, excessive angular velocity would bring unexpected system characters. First, the excessive angular velocity would bring large deformations of flexible appendages and satellite could no longer be treated as rigid body. Second, considering that extend Kaiman filter (EKF) is the most mature and widely used attitude determination algorithm that is derived from Taylor formula, the attitude determination accuracy would drop since the ignored higher order terms are no longer infinitely small. Third, some star trackers could not image if the angular velocity exceeds its threshold. (The star tracker on Proba 2 could not image if the angular velocity exceeds 7 deg./s i.e., 0.122 rad/s.) Hence, it is necessary to ensure the angular velocity do not exceed system upper bound. The restriction on angular velocity is expressed as continuous inequality constraint given by

$$\boldsymbol{\omega}(t) \leq \overline{\omega} \tag{3.2}$$

It is necessary to ensure inequalities (3.1) and (3.2) are all satisfied during the whole maneuver process.

3.2.2 Control objective

The objective of satellite attitude control could be concluded as two issues: attitude stabilization and attitude tracking. Attitude stabilization issue could be treated as a special case of attitude tracking (desired angular velocity and attitude quaternion equals to zero), hence the attitude tracking issue is discussed in this chapter. In order to deal with all kinds of attitude maneuver, the initial and terminal angular velocity is assumed nonzero in this chapter.

Select proper coordinates and the initial angular velocity, attitude quaternion could be written as follows:

$$\boldsymbol{q}_i = \left[q_{i0}\ \boldsymbol{q}_{iv}^T\right]^T = \left[q_{i0}\ q_{i1}\ q_{i2}\ q_{i3}\right]^T, \boldsymbol{\omega}_i = [\omega\ 0\ 0]^T \tag{3.3}$$

Let the desired angular velocity and attitude quaternion be

$$\boldsymbol{q}_d = \left[q_{d0}\ \boldsymbol{q}_{dv}^T\right]^T = \left[q_{d0}\ q_{d1}\ q_{d2}\ q_{d3}\right]^T, \boldsymbol{\omega}_d = \left[\omega_{d1}\ \omega_{d2}\ \omega_{d3}\right]^T \tag{3.4}$$

Current trajectory planning algorithms could fully utilize system capability, but the cost is the numerical solution and large computation. It is necessary to balance system performance and computation for engineering practice.

Hence, the control objective of this chapter could be concluded as design analytical near time-optimal attitude maneuver trajectory and attitude tracking controller so that $\boldsymbol{q} \to \boldsymbol{q}_d$ and $\boldsymbol{\omega} \to \boldsymbol{\omega}_d$ as fast as possible with any initial condition $\mathbf{q}_i$ and $\boldsymbol{\omega}_i$; meanwhile, system constraints (3.1) and (3.2) are all satisfied.

3.3 Controller design

3.3.1 Trajectory planning algorithm

The kinetic model (2.15) could be re-written as follows:

$$\dot{\boldsymbol{q}} = \frac{1}{2}\begin{bmatrix} -\boldsymbol{q}_v^T \\ q_0\boldsymbol{I}_3 + \boldsymbol{q}_v^\times \end{bmatrix}\boldsymbol{\omega} = \frac{1}{2}\begin{bmatrix} 0 & -\boldsymbol{\omega}^T \\ \boldsymbol{\omega} & -\boldsymbol{\omega}^\times \end{bmatrix}\begin{bmatrix} q_0 \\ \boldsymbol{q}_v \end{bmatrix} = \frac{1}{2}\boldsymbol{\Omega}(\boldsymbol{\omega})\boldsymbol{q} \tag{3.5}$$

Based on (15), the set of attitude quaternion of fixed axis rotation from any initial condition $\boldsymbol{\omega} = [\omega_1\ \ \omega_2\ \ \omega_3]^T$ and $\boldsymbol{q} = [q_0\ \ q_1\ \ q_2\ \ q_3]^T$ could be written as follows:

$$\boldsymbol{T}(\boldsymbol{\omega}, \boldsymbol{q}) = \boldsymbol{T} = e^{\frac{1}{2}\boldsymbol{\Omega}(\boldsymbol{\omega})t}\boldsymbol{q} = \boldsymbol{L}(\boldsymbol{\omega})\boldsymbol{q} \tag{3.6}$$

$$\boldsymbol{L}(\boldsymbol{\omega}) = \begin{bmatrix} \cos\theta & -\frac{\omega_1}{2\omega_0}\sin\theta & -\frac{\omega_2}{2\omega_0}\sin\theta & -\frac{\omega_3}{2\omega_0}\sin\theta \\ \frac{\omega_1}{2\omega_0}\sin\theta & \cos\theta & \frac{\omega_3}{2\omega_0}\sin\theta & -\frac{\omega_2}{2\omega_0}\sin\theta \\ \frac{\omega_2}{2\omega_0}\sin\theta & -\frac{\omega_3}{2\omega_0}\sin\theta & \cos\theta & \frac{\omega_1}{2\omega_0}\sin\theta \\ \frac{\omega_3}{2\omega_0}\sin\theta & \frac{\omega_2}{2\omega_0}\sin\theta & -\frac{\omega_1}{2\omega_0}\sin\theta & \cos\theta \end{bmatrix} \tag{3.7}$$

$$\omega_0 = \|\boldsymbol{\omega}\|^2/2, \theta = \omega_0 t$$

where t is the maneuver time.

Substitute the initial condition (3.3) into (3.6), the set of initial attitude quaternion $\boldsymbol{T}_i$ of fixed axis rotation around $\boldsymbol{\omega}_i$ could be written as follows:

$$\boldsymbol{T}_i = \boldsymbol{L}_i(\boldsymbol{\omega}_i)\boldsymbol{q}_i = [\cos\theta_1\ \ \sin\theta_1\ \ 0\ \ 0]^T \tag{3.8}$$

$$\boldsymbol{L}_i(\boldsymbol{\omega}_i) = \begin{bmatrix} \cos\theta_1 & -\sin\theta_1 & 0 & 0 \\ \sin\theta_1 & \cos\theta_1 & 0 & 0 \\ 0 & 0 & \cos\theta_1 & \sin\theta_1 \\ 0 & 0 & -\sin\theta_1 & \cos\theta_1 \end{bmatrix}, \theta_1 = \frac{\omega t_1}{2} \tag{3.9}$$

Similar as discussed above, the set of desired attitude quaternion $\boldsymbol{T}_d$ of fixed axis rotation around $\boldsymbol{\omega}_d$ could be written as follows:

$$\boldsymbol{T}_d = \boldsymbol{L}_d(\boldsymbol{\omega}_d)\boldsymbol{q}_d \tag{3.10}$$

$$\boldsymbol{L}_d(\boldsymbol{\omega}_d) = \begin{bmatrix} \cos\theta_2 & -\frac{\omega_1}{2\omega_0}\sin\theta_2 & -\frac{\omega_2}{2\omega_0}\sin\theta_2 & -\frac{\omega_3}{2\omega_0}\sin\theta_2 \\ \frac{\omega_1}{2\omega_0}\sin\theta_2 & \cos\theta_2 & \frac{\omega_3}{2\omega_0}\sin\theta_2 & -\frac{\omega_2}{2\omega_0}\sin\theta_2 \\ \frac{\omega_2}{2\omega_0}\sin\theta_2 & -\frac{\omega_3}{2\omega_0}\sin\theta_2 & \cos\theta_2 & \frac{\omega_1}{2\omega_0}\sin\theta_2 \\ \frac{\omega_3}{2\omega_0}\sin\theta_2 & \frac{\omega_2}{2\omega_0}\sin\theta_2 & -\frac{\omega_1}{2\omega_0}\sin\theta_2 & \cos\theta_2 \end{bmatrix}$$
$$\omega_0 = \|\boldsymbol{\omega}_d\|/2, \theta_2 = \omega_0 t_2 \tag{3.11}$$

In (3.9) and (3.11), t_1 and t_2 are the rotation time around initial axis and desired axis.

In this chapter, the near time-optimal maneuver trajectory could be got by following steps: (1) satellite rotate around the initial axis $\boldsymbol{\omega}_i$ to a proper switching point P_1 in set $\boldsymbol{T}_i$, and when reaching P_1, the angular velocity should be zero; (2) a rest-to-rest maneuver form P_1 to a proper switching point P_2 in set $\boldsymbol{T}_d$, and when reaching P_2, the angular velocity should be zero; and (3) a rotation process around axis $\boldsymbol{\omega}_d$ and when reaching desired attitude quaternion $\boldsymbol{q}_d$, the angular velocity should be desired angular velocity $\boldsymbol{\omega}_d$. Based on the discussion above, the maneuver process is constructed of three successive maneuvers: a maneuver from initial condition to switching point P_1, a maneuver from P_1 to switching point P_2, and a maneuver from P_2 to desired state. The terminal angular velocity of the first maneuver is zero, the initial angular velocity of the third maneuver is zero, and the second maneuver is a rest-to-rest maneuver. The basic idea of this method is inspired by orbit transfer: the initial rotation and terminal rotation are treated as two "orbits," and the key is to find the transition orbit that connects those two orbits, and once the departure and rendezvous point of transition orbit is determined, the orbit transfer process could be determined. To find switching point P_i is actually to find the transition "orbit." The advantage of this method could be concluded as (1) the key to determine maneuver trajectory is to determine the switching points P_i, and P_i could be calculated directly by formula (3.8)–(3.11), hence the trajectory is analytical and (2) the first rotation parallels to initial angular velocity and the last rotation parallels to terminal angular velocity, hence the initial and terminal angular velocity could be fully utilized.

In order to achieve the objective of near time-optimal maneuver, the trajectory should satisfy following conditions: (1) the maneuver path should be the shortest, i.e., the total maneuver angle of three successive maneuvers should be minimum in such a path and (2) the maneuver time of every successive maneuver should be optimized.

First condition 1 is discussed. Define the attitude quaternion of P_1 and P_2 as $\boldsymbol{q}_{T_i}$ and $\boldsymbol{q}_{T_d}$, the quaternion described the rotation from P_1 to P_2 could be written as follows:

$$\boldsymbol{q}_e = \boldsymbol{q}_{T_i}^C \otimes \boldsymbol{q}_{T_d} \tag{3.12}$$

Hence the rotate angle form P_1 to P_2 could be got as follows:

$$\begin{aligned}
\cos(\theta_3/2) &= q_{e0} \\
&= (q_{i0}\cos\theta_1 - q_{i1}\sin\theta_1)\left(q_{d0}\cos\theta_2 - \frac{\omega_{d1}q_{d1} + \omega_{d2}q_{d2} + \omega_{d3}q_{d3}}{2\omega_0}\sin\theta_2\right) \\
&+ (q_{i0}\sin\theta_1 + q_{i1}\cos\theta_1)\left(q_{d1}\cos\theta_2 + \frac{\omega_{d1}q_{d0} - \omega_{d2}q_{d3} + \omega_{d3}q_{d3}}{2\omega_0}\sin\theta_2\right) \\
&+ (q_{i2}\cos\theta_1 + q_{i3}\sin\theta_1)\left(q_{d2}\cos\theta_2 + \frac{\omega_{d1}q_{d3} + \omega_{d2}q_{d0} - \omega_{d3}q_{d1}}{2\omega_0}\sin\theta_2\right) \\
&+ (-q_{i2}\sin\theta_1 + q_{i3}\cos\theta_1)\left(q_{d3}\cos\theta_2 + \frac{-\omega_{d1}q_{d2} + \omega_{d2}q_{d1} + \omega_{d3}q_{d0}}{2\omega_0}\sin\theta_2\right) \\
&= \begin{bmatrix}\cos\theta_1 \\ \sin\theta_1\end{bmatrix}^T \begin{bmatrix} a_1 & a_2 \\ a_3 & a_4 \end{bmatrix} \begin{bmatrix}\cos\theta_2 \\ \sin\theta_2\end{bmatrix} \\
&= f(\theta_1, \theta_2)
\end{aligned} \tag{3.13}$$

$$\begin{aligned}
a_1 &= q_{i0}q_{d0} + q_{i1}q_{d1} + q_{i2}q_{d2} + q_{i3}q_{d3} \\
a_2 &= \frac{\omega_1}{2\omega_0}(-q_{i0}q_{d1} + q_{i1}q_{d0} + q_{i2}q_{d3} - q_{i3}q_{d2}) \\
&- \frac{\omega_2}{2\omega_0}(q_{i0}q_{d2} + q_{i1}q_{d3} - q_{i2}q_{d0} - q_{i3}q_{d1}) \\
&+ \frac{\omega_3}{2\omega_0}(-q_{i0}q_{d3} + q_{i1}q_{d2} - q_{i2}q_{d1} + q_{i3}q_{d0}) \\
a_3 &= -q_{i1}q_{d0} + q_{i0}q_{d1} + q_{i3}q_{d2} - q_{i2}q_{d3} \\
a_4 &= \frac{\omega_1}{2\omega_0}(q_{i0}q_{d0} + q_{i1}q_{d1} + q_{i2}q_{d2} + q_{i3}q_{d3}) \\
&+ \frac{\omega_2}{2\omega_0}(-q_{i0}q_{d3} + q_{i1}q_{d2} - q_{i2}q_{d1} + q_{i3}q_{d0}) \\
&+ \frac{\omega_3}{2\omega_0}(q_{i0}q_{d2} + q_{i1}q_{d3} - q_{i2}q_{d0} - q_{i3}q_{d1})
\end{aligned} \tag{3.14}$$

Hence the total maneuver angle of three successive maneuvers could be got as follows:

$$\varphi = 2\theta_1 + 2\theta_2 + 2\theta_3 = 2\theta_1 + 2\theta_2 + 2\arccos(f(\theta_1, \theta_2)) \tag{3.15}$$

Based on (3.15) it could be found that the total rotate angle φ is the function of θ_1 and θ_2, hence if θ_1 and θ_2 are determined, the maneuver path could be uniquely determined. In order to calculate the minimal value of φ, calculate the partial derivative of φ

$$\frac{\partial \varphi}{\partial \theta_1} = 2\left(1 - \frac{\partial f / \partial \theta_1}{\sqrt{1 - f^2(\theta_1, \theta_2)}}\right), \frac{\partial \varphi}{\partial \theta_2} = 2\left(1 - \frac{\partial f / \partial \theta_2}{\sqrt{1 - f^2(\theta_1, \theta_2)}}\right) \tag{3.16}$$

Hence, if following equation is satisfied, φ reaches its minimum value.

$$\begin{aligned} \frac{\partial \varphi}{\partial \theta_1} &= 2\left(1 - \frac{\partial f / \partial \theta_1}{\sqrt{1 - f^2(\theta_1, \theta_2)}}\right) = 0 \Rightarrow f^2(\theta_1, \theta_2) + \left(\frac{\partial f}{\partial \theta_1}\right)^2 = 1 \\ \frac{\partial \varphi}{\partial \theta_2} &= 2\left(1 - \frac{\partial f / \partial \theta_2}{\sqrt{1 - f^2(\theta_1, \theta_2)}}\right) = 0 \Rightarrow f^2(\theta_1, \theta_2) + \left(\frac{\partial f}{\partial \theta_2}\right)^2 = 1 \end{aligned} \tag{3.17}$$

Substitute (3.13) into (3.17), it could be got that

$$\begin{cases} f^2 + \left(\frac{\partial f}{\partial \theta_1}\right)^2 = \left(a_1^2 + a_3^2\right)\cos^2\theta_2 + \left(a_2^2 + a_4^2\right)\sin^2\theta_2 + 2(a_1a_2 + a_3a_4)\cos\theta_2\sin\theta_2 = 1 \\ f^2 + \left(\frac{\partial f}{\partial \theta_2}\right)^2 = \left(a_1^2 + a_2^2\right)\cos^2\theta_1 + \left(a_3^2 + a_4^2\right)\sin^2\theta_1 + 2(a_1a_3 + a_2a_4)\cos\theta_1\sin\theta_1 = 1 \end{cases}$$
$$\Rightarrow \begin{cases} \frac{a_1^2 - a_2^2 + a_3^2 - a_4^2}{2}\cos 2\theta_2 + (a_1a_2 + a_3a_4)\sin 2\theta_2 + \frac{a_1^2 + a_2^2 + a_3^2 + a_4^2}{2} = 1 \\ \frac{a_1^2 + a_2^2 - a_3^2 - a_4^2}{2}\cos 2\theta_1 + (a_1a_3 + a_2a_4)\sin 2\theta_1 + \frac{a_1^2 + a_2^2 + a_3^2 + a_4^2}{2} = 1 \end{cases} \tag{3.18}$$

Hence, if following equation is satisfied, φ reaches its minimum value.

$$\begin{cases} \theta_1 = \frac{1}{2}\left(\arcsin\frac{2 - \left(a_1^2 + a_2^2 + a_3^2 + a_4^2\right)}{\sqrt{\left(a_1^2 + a_2^2 - a_3^2 - a_4^2\right)^2 + 4(a_1a_3 + a_2a_4)^2}} - \gamma_1\right) \\ \theta_2 = \frac{1}{2}\left(\arcsin\frac{2 - \left(a_1^2 + a_2^2 + a_3^2 + a_4^2\right)}{\sqrt{\left(a_1^2 - a_2^2 + a_3^2 - a_4^2\right)^2 + 4(a_1a_2 + a_3a_4)^2}} - \gamma_2\right) \end{cases} \tag{3.19}$$

where

$$\tan\gamma_1 = \frac{a_1^2 + a_2^2 - a_3^2 - a_4^2}{2(a_1a_3 + a_2a_4)}, \quad \tan\gamma_2 = \frac{a_1^2 - a_2^2 + a_3^2 - a_4^2}{2(a_1a_2 + a_3a_4)} \tag{3.20}$$

When the initial and desired state is determined, parameters a_i could be calculated by (3.14), and based on (3.19) and (3.20) the rotate angle θ_i under the shortest path could be calculated, substitute θ_i into (3.8) and (3.10), the switching point P_i could be determined, hence the shortest maneuver path could be uniquely determined and condition 1 in previous text could be satisfied.

The next step is to give the near time-optimal trajectory of every successive maneuver. Considering that bang-bang control is a time-optimal open-loop control algorithm, and system capability could be fully utilized by this method, the near time-optimal trajectory is given based on bang-bang control. For standard single-axis maneuver, let the Euler axis and rotate angle be $\boldsymbol{e}$ and $\overline{\varphi}$, assume that the initial and terminal angular velocity vector parallel to $\boldsymbol{e}$ and their norm are $\overline{\omega}_i$ and $\overline{\omega}_d$, let the maximum norm of angular velocity and angular acceleration be $\overline{\omega}$and α, and the near time-optimal trajectory of angular velocity based on Bang-Bang control could be got as follows:

$$\boldsymbol{\omega} = \begin{cases} \sqrt{2\alpha\varphi + \overline{\omega}_i^2}\,\boldsymbol{e}, & 0 \le \varphi \le \dfrac{\overline{\omega}^2 - \overline{\omega}_i^2}{2\alpha} \\ \overline{\omega}\boldsymbol{e}, & \dfrac{\overline{\omega}^2 - \overline{\omega}_i^2}{2\alpha} < \varphi < \overline{\varphi} - \dfrac{\overline{\omega}^2 - \overline{\omega}_d^2}{2\alpha} \\ \sqrt{2\alpha(\overline{\varphi} - \varphi) + \overline{\omega}_d^2}\,\boldsymbol{e}, & \overline{\varphi} - \dfrac{\overline{\omega}^2 - \overline{\omega}_d^2}{2\alpha} \le \varphi \le \overline{\varphi} \end{cases} \tag{3.21}$$

The essence of (3.21) is that: first the system accelerates with α till system reaches the maximum angular velocity, and it could be calculated that the rotate angle of acceleration process is $\left(\overline{\omega}^2 - \overline{\omega}_i^2\right)/2\alpha$, then system should rotate with its maximum angular velocity; when approaching the desired position, system should decelerate with α till system reaches the desired angular velocity, and it could be calculated that the rotate angle of deceleration process is $\left(\overline{\omega}^2 - \overline{\omega}_d^2\right)/2\alpha$.

Based on (3.21), the optimized attitude quaternion and angular velocity in every control cycle could be got as follows:

$$\boldsymbol{\omega}(t + \Delta t) = \begin{cases} (\|\boldsymbol{\omega}(t)\| + \alpha\Delta t)\boldsymbol{e}, & 0 \le \varphi \le \dfrac{\overline{\omega}^2 - \overline{\omega}_i^2}{2\alpha} \\ \overline{\omega}\boldsymbol{e}, & \dfrac{\overline{\omega}^2 - \overline{\omega}_i^2}{2\alpha} < \varphi < \overline{\varphi} - \dfrac{\overline{\omega}^2 - \overline{\omega}_d^2}{2\alpha} \\ (\|\boldsymbol{\omega}(t)\| - \alpha\Delta t)\boldsymbol{e}, & \overline{\varphi} - \dfrac{\overline{\omega}^2 - \overline{\omega}_d^2}{2\alpha} \le \varphi \le \overline{\varphi} \end{cases} \tag{3.22}$$

$$\dot{\boldsymbol{\omega}}(t + \Delta t) = \begin{cases} \alpha\boldsymbol{e}, & 0 \le \varphi \le \dfrac{\overline{\omega}^2 - \overline{\omega}_i^2}{2\alpha} \\ 0, & \dfrac{\overline{\omega}^2 - \overline{\omega}_i^2}{2\alpha} < \varphi < \overline{\varphi} - \dfrac{\overline{\omega}^2 - \overline{\omega}_d^2}{2\alpha} \\ -\alpha\boldsymbol{e}, & \overline{\varphi} - \dfrac{\overline{\omega}^2 - \overline{\omega}_d^2}{2\alpha} \le \varphi \le \overline{\varphi} \end{cases} \tag{3.23}$$

$$\boldsymbol{q}(t + \Delta t) = \left[\cos\frac{\varphi(t + \Delta t)}{2} \quad \boldsymbol{e}^T \sin\frac{\varphi(t + \Delta t)}{2}\right]^T \tag{3.24}$$

$$\varphi(t+\Delta t)=\begin{cases}\varphi(t)+\|\omega(t)\|\Delta t+\dfrac{1}{2}\alpha\Delta t^2, & 0\le\varphi\le\dfrac{\overline{\omega}^2-\overline{\omega}_i^2}{2\alpha}\\ \varphi(t)+\overline{\omega}\Delta t, & \dfrac{\overline{\omega}^2-\overline{\omega}_i^2}{2\alpha}<\varphi<\overline{\varphi}-\dfrac{\overline{\omega}^2-\overline{\omega}_d^2}{2\alpha}\\ \varphi(t)+\|\omega(t)\|\Delta t-\dfrac{1}{2}\alpha\Delta t^2, & \overline{\varphi}-\dfrac{\overline{\omega}^2-\overline{\omega}_d^2}{2\alpha}\le\varphi\le\overline{\varphi}\end{cases} \tag{3.25}$$

where $\boldsymbol{x}(t)$ is the current state, $\boldsymbol{x}(t+\Delta t)$ is the optimized state for next control cycle, and Δt is the time of control cycle.

Noted that the Euler axis and rotate angle of the three successive maneuver in previous text could be written as follows:

$$\begin{aligned}\boldsymbol{e}_1&=\boldsymbol{\omega}_i/\|\boldsymbol{\omega}_i\|,\boldsymbol{e}_2=\boldsymbol{q}_{ev}/\|\boldsymbol{q}_{ev}\|,\boldsymbol{e}_3=\boldsymbol{\omega}_d/\|\boldsymbol{\omega}_d\|,\overline{\varphi}_1=\theta_1,\overline{\varphi}_2\\&=2\arccos\left(q_{e0}\right),\overline{\varphi}_3=\theta_2\end{aligned} \tag{3.26}$$

$$\boldsymbol{q}_e=\boldsymbol{q}_{T_i}^c\otimes\boldsymbol{q}_{T_d} \tag{3.27}$$

Hence, if the rotate angles θ_1 and θ_2 are determined based on (3.19), the near time-optimal trajectory based on bang-bang control of three successive maneuvers could be got by (3.22)–(3.27).

The next step is to discuss the physical constraints on control parameters. Based on the discussion in previous text, it could be found that the angular velocity constraint is satisfied during the whole maneuver process, and it is necessary to discuss the relationship between control torque upper bound $\overline{u}$ and angular acceleration α to ensure the required control torque do not exceed system upper bound.

Based on the definition of α, it could be got that

$$\begin{aligned}\alpha&=\frac{d\|\boldsymbol{\omega}\|}{dt}=\frac{d\sqrt{\boldsymbol{\omega}^T\boldsymbol{\omega}}}{dt}=\frac{1}{2}\frac{1}{\|\boldsymbol{\omega}\|}2\boldsymbol{\omega}^T\dot{\boldsymbol{\omega}}=\frac{\boldsymbol{\omega}^T}{\|\boldsymbol{\omega}\|}\boldsymbol{J}^{-1}\left(\boldsymbol{u}-\boldsymbol{\omega}^\times\boldsymbol{J}\boldsymbol{\omega}+\boldsymbol{d}\right)\\&\approx\frac{\boldsymbol{\omega}^T}{\|\boldsymbol{\omega}\|}\boldsymbol{J}^{-1}\left(\boldsymbol{u}-\boldsymbol{\omega}^\times\boldsymbol{J}\boldsymbol{\omega}\right)\end{aligned} \tag{3.28}$$

Noted that $\boldsymbol{\omega}^T/\|\boldsymbol{\omega}\|$ is a unit vector

$$\begin{aligned}\alpha+\frac{\boldsymbol{\omega}^T}{\|\boldsymbol{\omega}\|}\boldsymbol{J}^{-1}\boldsymbol{\omega}^\times\boldsymbol{J}\boldsymbol{\omega}&=\frac{\boldsymbol{\omega}^T}{\|\boldsymbol{\omega}\|}\boldsymbol{J}^{-1}\boldsymbol{u}\\ \alpha+\frac{\lambda_M(\boldsymbol{J})\overline{\omega}^2}{\lambda_m(\boldsymbol{J})}&\le\frac{\overline{u}}{\lambda_m(\boldsymbol{J})}\end{aligned} \tag{3.29}$$

Hence, if following inequality is satisfied, the required control torque could be ensured less than system upper bound.

$$\alpha\lambda_m(\boldsymbol{J})+\lambda_M(\boldsymbol{J})\overline{\omega}^2\le\overline{u} \tag{3.30}$$

In fact, the accurate value of inertia matrix $\boldsymbol{J}$ is hard to get in engineering practice, hence (3.30) is a theoretical constraint and the practical constraint will be given in later section.

In this section, the attitude maneuver issue with nonzero initial and terminal angular velocity is discussed, and the maneuver process is constructed as three successive maneuvers. By calculating the minimum value of total rotate angle and designing trajectory with bang-bang structure for each maneuver, the near time-optimal trajectory is got. Another contribution of this section is that the trajectory is analytical, hence the trajectory could be easily got by on-board computer and the computation is largely reduced.

3.3.2 Hohmann transfer-like maneuver

The trajectory discussed in previous section has a drawback that system would accelerate three times and decelerate three times, both accelerate and decelerate processes would slower system convergence rate. Hence, if these processes could be reduced in some feasible conditions, system convergence rate could be improved.

In orbit transfer issues, the Hohmann transfer orbit is an elliptical orbit used to transfer between two circular orbits of different radii in the same plane. In general, a Hohmann transfer orbit uses the lowest possible amount of energy in traveling between two objects orbiting at these radii. The key of Hohmann transfer is to calculate the point of departure and rendezvous of the circular orbit and elliptical orbit. Although the energy consumption is not the main focus in attitude control issue, the method of Hohmann transfer could be used in trajectory planning.

As discussed in precious section, if the set of initial attitude quaternion $\boldsymbol{T}_i$ and the set of desired attitude quaternion $\boldsymbol{T}_d$ have an point of intersection, define this point as P_0, and the near time-optimal maneuver trajectory could be got by following steps: (1) satellite rotate around the initial axis $\boldsymbol{\omega}_i$ to the switching point P_0, and when reaching P_0, the angular velocity should be zero; (2) a rotation process around axis $\boldsymbol{\omega}_d$ from P_0 and when reaching desired attitude quaternion $\boldsymbol{q}_d$, the angular velocity should be desired angular velocity $\boldsymbol{\omega}_d$. Comparing this trajectory with that in (3.1), it could be found that the maneuver between two switching points vanishes hence an acceleration process and a deceleration process is reduced. Although the minimum rotate angle condition does not hold in this section, the acceleration and deceleration processes are optimized hence system could also have better convergence rate.

Based on the discussion previously, if following equation is satisfied, $\boldsymbol{T}_i$ and $\boldsymbol{T}_d$ have a point of intersection.

$$\begin{aligned} q_{e0} &= 1 \\ \boldsymbol{q}_e &= \boldsymbol{q}_{T_i}^c \otimes \boldsymbol{q}_{T_d} \end{aligned} \tag{3.31}$$

The next step is to discuss in what condition Eq. (3.31) could be satisfied. Based on the definition of $\boldsymbol{q}_e$, it could be got that

$$q_{e0} = \begin{bmatrix} \cos\theta_1 \\ \sin\theta_1 \end{bmatrix}^T \begin{bmatrix} a_1 & a_2 \\ a_3 & a_4 \end{bmatrix} \begin{bmatrix} \cos\theta_2 \\ \sin\theta_2 \end{bmatrix} = f(\theta_1, \theta_2) \tag{3.32}$$

Noted that the maximum value of q_{e0} is 1, hence if (3.31) is satisfied, q_{e0} reaches its maximum value. Calculate the partial derivative of q_{e0} it could be got that

$$\begin{aligned} \frac{\partial f}{\partial \theta_1} &= -a_1 \sin\theta_1 \cos\theta_2 - a_2 \sin\theta_1 \sin\theta_2 + a_3 \cos\theta_1 \cos\theta_2 + a_4 \cos\theta_1 \sin\theta_2 \\ \frac{\partial f}{\partial \theta_2} &= -a_1 \cos\theta_1 \sin\theta_2 + a_2 \cos\theta_1 \cos\theta_2 - a_3 \sin\theta_1 \sin\theta_2 + a_4 \sin\theta_1 \cos\theta_2 \end{aligned} \tag{3.33}$$

Noted that if (3.31) is satisfied, $\partial f/\partial\theta_1 = \partial f/\partial\theta_2 = 0$ holds. Hence,

$$\theta_1 = \frac{1}{2}\arctan\left(\frac{2a_1a_3 + 2a_2a_4}{a_1^2 + a_2^2 - a_3^2 - a_4^2}\right), \theta_2 = \frac{1}{2}\arctan\left(\frac{2a_1a_2 + 2a_3a_4}{a_1^2 - a_2^2 + a_3^2 - a_4^2}\right) \tag{3.34}$$

$$\begin{aligned} q_{e0} &= a_1 \cos\theta_1 \cos\theta_2 + a_2 \cos\theta_1 \sin\theta_2 + a_3 \sin\theta_1 \cos\theta_2 + a_4 \sin\theta_1 \sin\theta_2 \\ &= \cos\theta_2(a_1 \cos\theta_1 + a_3 \sin\theta_1) + \cos\theta_2(a_2 \cos\theta_1 + a_4 \sin\theta_1) \\ &= \sqrt{(a_1 \cos\theta_1 + a_3 \sin\theta_1)^2 + (a_2 \cos\theta_1 + a_4 \sin\theta_1)^2} \sin(\theta_2 + \theta') \\ &\le \sqrt{\frac{a_1^2 + a_2^2 + a_3^2 + a_4^2}{2} + \frac{a_1^2 + a_2^2 - a_3^2 - a_4^2}{2} \cos 2\theta_2 + (a_1a_3 + a_2a_4)\sin 2\theta_2} \\ &= \sqrt{\frac{a_1^2 + a_2^2 + a_3^2 + a_4^2}{2} + \sqrt{\left(\frac{a_1^2 + a_2^2 - a_3^2 - a_4^2}{2}\right)^2 + (a_1a_3 + a_2a_4)^2} \sin(2\theta_2 + \theta'')} \\ &\le \sqrt{\frac{a_1^2 + a_2^2 + a_3^2 + a_4^2}{2} + \sqrt{\left(\frac{a_1^2 + a_2^2 - a_3^2 - a_4^2}{2}\right)^2 + (a_1a_3 + a_2a_4)^2}} \end{aligned} \tag{3.35}$$

$$\begin{aligned} \max(q_{e0}) &= \sqrt{\frac{a_1^2 + a_2^2 + a_3^2 + a_4^2}{2} + \sqrt{\left(\frac{a_1^2 + a_2^2 - a_3^2 - a_4^2}{2}\right)^2 + (a_1a_3 + a_2a_4)^2}} \\ &= \sqrt{\frac{a_1^2 + a_2^2 + a_3^2 + a_4^2}{2} + \sqrt{\left(\frac{a_1^2 - a_2^2 + a_3^2 - a_4^2}{2}\right)^2 + (a_1a_2 + a_3a_4)^2}} \end{aligned} \tag{3.36}$$

Hence if following equation is satisfied, (3.31) could be satisfied.

$$\begin{cases} \dfrac{a_1^2+a_2^2+a_3^2+a_4^2}{2}+\sqrt{\left(\dfrac{a_1^2+a_2^2-a_3^2-a_4^2}{2}\right)^2+(a_1a_3+a_2a_4)^2}=1 \\ \dfrac{a_1^2+a_2^2+a_3^2+a_4^2}{2}+\sqrt{\left(\dfrac{a_1^2-a_2^2+a_3^2-a_4^2}{2}\right)^2+(a_1a_2+a_3a_4)^2}=1 \end{cases} \tag{3.37}$$

Once the initial state and desired state are determined, parameter a_i could be determined, hence whether Eq. (3.37) holds could be checked, and if (3.37) holds, the attitude maneuver issue could be treated as two successive maneuvers: a maneuver form initial state to switching point P_0 and a maneuver form P_0 to desired state. Hence, the number of acceleration and deceleration process is reduced twice by this method. Based on the bang-bang structure trajectory discussed in Section 3.3.1, the near time-optimal trajectory could also be determined.

In this section, a special case in satellite attitude tracking issue is discussed and inspired by the Hohmann transfer in orbit control, a novel path is presented, and the acceleration and deceleration process is optimized.

3.3.3 Attitude tracking controller

In this section, the attitude tracking controller is proposed to ensure the satellite would maneuver along the trajectory designed in previous section. Considering that the main focus of this chapter is the trajectory planning algorithm, the standard PD attitude tracking controller is designed in this section.

The attitude tracking controller could be written as follows:

$$\boldsymbol{u}=-\rho\left(k_d\boldsymbol{\omega}_e+k_p\boldsymbol{q}_{ev}\right)+\hat{\boldsymbol{J}}\boldsymbol{R}\dot{\boldsymbol{\omega}}_d^*-\hat{\boldsymbol{J}}\boldsymbol{\omega}_e^{\times}\boldsymbol{R}\boldsymbol{\omega}_d^*+\boldsymbol{\omega}^{\times}(t)\hat{\boldsymbol{J}}\boldsymbol{\omega}(t)-l\,\boldsymbol{sgn}\,(\boldsymbol{\omega}_e) \tag{3.38}$$

$$\begin{gathered} \boldsymbol{q}_e=\boldsymbol{q}_d^*\otimes\boldsymbol{q}(t) \\ \boldsymbol{R}(\boldsymbol{q}_e)=\left(q_{e0}^2-\boldsymbol{q}_{ev}^T\boldsymbol{q}_{ev}\right)\boldsymbol{I}_3+2\boldsymbol{q}_{ev}\boldsymbol{q}_{ev}^T-2q_{e0}\boldsymbol{q}_{ev}^{\times} \\ \boldsymbol{\omega}_e=\boldsymbol{\omega}(t)-\boldsymbol{R}(\boldsymbol{q}_e)\boldsymbol{\omega}_d^* \end{gathered} \tag{3.39}$$

where k_d and k_p are all positive scalars, and $\boldsymbol{q}_d^*$ and $\boldsymbol{\omega}_d^*$ are the optimized state calculated by the trajectory planning algorithm. It is worth noticing that the state feedback in control (3.38) is error state $\boldsymbol{q}_e$, $\boldsymbol{R}(\boldsymbol{q}_e)$, $\boldsymbol{\omega}_e$, and Eq. (3.39) defines the error state. The unclaimed variables in (3.48) and (3.39) are defined as follows:

$$\rho=\begin{cases} \dfrac{-\boldsymbol{r}_1^T\boldsymbol{r}_2+\sqrt{\left(\boldsymbol{r}_1^T\boldsymbol{r}_2\right)^2+\|\boldsymbol{r}_1\|^2\left(\bar{u}^2-\|\boldsymbol{r}_2\|^2\right)}}{\|\boldsymbol{r}_1\|^2}, & \|\boldsymbol{u}\|\geq\bar{u} \\ 1, & \|\boldsymbol{u}\|<\bar{u} \end{cases} \tag{3.40}$$

$$\begin{cases} l = \lambda\left(\left\|\dot{\boldsymbol{\omega}}_d^*\right\| + \left\|\boldsymbol{\omega}_e^\times \boldsymbol{R}\boldsymbol{\omega}_d^*\right\| + \left\|\boldsymbol{\omega}(t)\right\|^2\right) + \overline{d} \\ \boldsymbol{r}_1 = -k_d\boldsymbol{\omega}_e - k_p\boldsymbol{q}_{ev} \\ \boldsymbol{r}_2 = \widehat{\boldsymbol{J}}\boldsymbol{R}\dot{\boldsymbol{\omega}}_d^* - \widehat{\boldsymbol{J}}\boldsymbol{\omega}_e^\times \boldsymbol{R}\boldsymbol{\omega}_d^* + \boldsymbol{\omega}^\times(t)\widehat{\boldsymbol{J}}\boldsymbol{\omega}(t) - l\,\boldsymbol{sgn}\left(\boldsymbol{\omega}_e\right) \end{cases} \tag{3.41}$$

where λ is a positive scalar which satisfies $\lambda \geq \lambda_M\left(\widetilde{\boldsymbol{J}}\right)$, $\overline{u}$ and $\overline{d}$ are the norm upper bound of control torque and disturbance torque. Gain factor ρ is between 0 and 1, and its function is to ensure control torque do not exceed system upper bound. When the norm of desired control torque is larger than system upper bound, i.e., $\|\boldsymbol{u}\| \geq \overline{u}$, the P and D parts are lessened proportionally, and the control torque is locked to system upper bound, hence the key to solve control torque saturation issue is to ensure the later part of controller (3.38) less than system upper bound. Define $\boldsymbol{u}_0$ as follows:

$$\boldsymbol{u}_0 = \widehat{\boldsymbol{J}}\boldsymbol{R}\dot{\boldsymbol{\omega}}_d^* - \widehat{\boldsymbol{J}}\boldsymbol{\omega}_e^\times \boldsymbol{R}\boldsymbol{\omega}_d^* + \boldsymbol{\omega}^\times(t)\widehat{\boldsymbol{J}}\boldsymbol{\omega}(t) - l\,\boldsymbol{sgn}\left(\boldsymbol{\omega}_e\right) \tag{3.42}$$

It could be got that

$$\|\boldsymbol{u}_0\| \leq \left(\lambda_M\left(\widehat{\boldsymbol{J}}\right) + \lambda\right)\left(\left\|\dot{\boldsymbol{\omega}}_d^*\right\| + \left\|\boldsymbol{\omega}_e^\times \boldsymbol{R}\boldsymbol{\omega}_d^*\right\| + \|\boldsymbol{\omega}\|^2\right) + \overline{d} \tag{3.43}$$

And noticing that

$$\begin{gathered} \left\|\dot{\boldsymbol{\omega}}_d^*\right\| \leq \alpha, \|\boldsymbol{\omega}\| \leq \overline{\omega}, \left\|\boldsymbol{\omega}_d^*\right\| \leq \overline{\omega}, \|\boldsymbol{R}\| \leq 1 \\ \|\boldsymbol{\omega}_e\| = \left\|\boldsymbol{\omega} - \boldsymbol{R}\boldsymbol{\omega}_d^*\right\| \leq \|\boldsymbol{\omega}\| + \|\boldsymbol{R}\|\left\|\boldsymbol{\omega}_d^*\right\| \leq 2\overline{\omega} \end{gathered} \tag{3.44}$$

Hence, if follow inequality is satisfied, the control torque saturation issue could be solved.

$$\left(\lambda_M\left(\widehat{\boldsymbol{J}}\right) + \lambda\right)\left(\alpha + 3\overline{\omega}^2\right) + \overline{d} \leq \overline{u} \tag{3.45}$$

It is worth noticing that (3.45) is a strict constraint since the worst condition is considered in the discussion, and (3.45) does not need system state reaching the desired trajectory.

In fact, this method narrowed the span of control parameters, and noticing that

$$\|\boldsymbol{J}\| = \left\|\widehat{\boldsymbol{J}} + \widetilde{\boldsymbol{J}}\right\| \leq \lambda_M\left(\widehat{\boldsymbol{J}}\right) + \lambda \tag{3.46}$$

Hence substitute (3.46) into (3.30), the loose constraint for control parameters could be written as inequality (3.47).

$$\alpha\left(\lambda_m\left(\widehat{\boldsymbol{J}}\right) + \lambda\right) + \overline{\omega}^2\left(\lambda_M\left(\widehat{\boldsymbol{J}}\right) + \lambda\right) \leq \overline{u} \tag{3.47}$$

In fact, inequality (3.47) is derived based on (3.30) that discussed the maneuver on desired trajectory, and the maneuver from initial condition to desired trajectory is omitted hence inequality (3.47) could be treated as a loose constraint.

In order to prove system stability, select Lyapunov function as follows and calculate its derivative, it could be got that

$$\begin{aligned}
V &= \frac{1}{2}\boldsymbol{\omega}_e^T\boldsymbol{J}\boldsymbol{\omega}_e + 2\rho k_p(1-q_{e0})\\
\dot{V} &= \boldsymbol{\omega}_e^T\boldsymbol{J}\dot{\boldsymbol{\omega}}_e - 2\rho k_p\dot{q}_{e0}\\
&= \boldsymbol{\omega}_e^T\left(\boldsymbol{u}+\boldsymbol{d}-\boldsymbol{J}\boldsymbol{R}\dot{\boldsymbol{\omega}}_d^* + \boldsymbol{J}\boldsymbol{\omega}_e^\times\boldsymbol{R}\boldsymbol{\omega}_d^* - \boldsymbol{\omega}^\times\boldsymbol{J}\boldsymbol{\omega}\right) + \rho k_p\boldsymbol{\omega}_e^T\boldsymbol{q}_{ev}\\
&= -\rho\boldsymbol{\omega}_e^T\left(k_d\boldsymbol{\omega}_e + k_p\boldsymbol{q}_{ev}\right) + \boldsymbol{\omega}_e^T\widehat{\boldsymbol{J}}\boldsymbol{R}\dot{\boldsymbol{\omega}}_d^* - \boldsymbol{\omega}_e^T\widehat{\boldsymbol{J}}\boldsymbol{\omega}_e^\times\boldsymbol{R}\boldsymbol{\omega}_d^* + \boldsymbol{\omega}_e^T\boldsymbol{\omega}^\times\widehat{\boldsymbol{J}}\boldsymbol{\omega}\\
&\quad -l\boldsymbol{\omega}_e^T\,\mathbf{sgn}\left(\boldsymbol{\omega}_e\right) - \boldsymbol{\omega}_e^T\boldsymbol{J}\boldsymbol{R}\dot{\boldsymbol{\omega}}_d^* + \boldsymbol{\omega}_e^T\boldsymbol{J}\boldsymbol{\omega}_e^\times\boldsymbol{R}\boldsymbol{\omega}_d^* - \boldsymbol{\omega}_e^T\boldsymbol{\omega}^\times\boldsymbol{J}\boldsymbol{\omega} + \boldsymbol{\omega}_e^T\boldsymbol{d}\\
&= -\rho k_d\boldsymbol{\omega}_e^T\boldsymbol{\omega}_e - \boldsymbol{\omega}_e^T\widetilde{\boldsymbol{J}}\boldsymbol{R}\dot{\boldsymbol{\omega}}_d^* + \boldsymbol{\omega}_e^T\widetilde{\boldsymbol{J}}\boldsymbol{\omega}_e^\times\boldsymbol{R}\boldsymbol{\omega}_d^* - \boldsymbol{\omega}_e^T\boldsymbol{\omega}^\times\widetilde{\boldsymbol{J}}\boldsymbol{\omega} + \boldsymbol{\omega}_e^T\boldsymbol{d}\\
&\quad -\left(\lambda\left(\left\|\dot{\boldsymbol{\omega}}_d^*\right\| + \left\|\boldsymbol{\omega}_e^\times\boldsymbol{R}\boldsymbol{\omega}_d^*\right\| + \|\boldsymbol{\omega}\|^2\right) + \overline{d}\right)\boldsymbol{\omega}_e^T\,\mathbf{sgn}\left(\boldsymbol{\omega}_e\right)\\
&\le -\rho k_d\boldsymbol{\omega}_e^T\boldsymbol{\omega}_e + \lambda_M\left(\widetilde{\boldsymbol{J}}\right)\|\boldsymbol{\omega}_e\|\left(\left\|\dot{\boldsymbol{\omega}}_d^*\right\| + \left\|\boldsymbol{\omega}_e^\times\boldsymbol{R}\boldsymbol{\omega}_d^*\right\| + \|\boldsymbol{\omega}\|^2 + \overline{d}\right)\\
&\quad -\left(\lambda\left(\left\|\dot{\boldsymbol{\omega}}_d^*\right\| + \left\|\boldsymbol{\omega}_e^\times\boldsymbol{R}\boldsymbol{\omega}_d^*\right\| + \|\boldsymbol{\omega}\|^2\right) + \overline{d}\right)\boldsymbol{\omega}_e^T\,\mathbf{sgn}\left(\boldsymbol{\omega}_e\right)\\
&\le -\rho k_d\boldsymbol{\omega}_e^T\boldsymbol{\omega}_e \le 0
\end{aligned} \tag{3.48}$$

In inequality (3.48), following property is used.

$$\begin{gathered}
\left|\boldsymbol{a}^T\boldsymbol{K}\boldsymbol{b}\right| \le \lambda_M(\boldsymbol{K})\|\boldsymbol{a}\|\|\boldsymbol{b}\|\\
\lambda \ge \lambda_M\left(\widetilde{\boldsymbol{J}}\right), \lambda\|\boldsymbol{a}\|\|\boldsymbol{b}\| \ge \lambda_M(\boldsymbol{K})\|\boldsymbol{a}\|\|\boldsymbol{b}\|
\end{gathered} \tag{3.49}$$

where $\boldsymbol{a}$, $\boldsymbol{b}$ are vectors and $\boldsymbol{K}$ is matrix.

Hence, system (2.1) and (2.15) governed by controller (3.38) is globally asymptotically stable and satellite could maneuver along the designed trajectory.

3.4 Simulation results

In this section, simulation results will be given to demonstrate the performance of the controller proposed in this chapter. Three situations will be discussed in this section: (1) attitude maneuver with zero terminal angular velocity (regular attitude stabilization issue); (2) regular attitude maneuver issue (non-zero initial and terminal angular velocity); and (3) Hohmann transfer like maneuver issue (the special case discussed in Section 3.3.2).

3.4.1 Regular attitude stabilization issue

In order to demonstrate the superiority of the control algorithm presented in this chapter, finite-time controller (3.50) proposed by Li [10] is compared. The finite-time sliding mode of controller (3.50) has a maneuver stage with constant angular velocity, hence the convergence rate of attitude quaternion could be maintained at a high level; meanwhile, the desired control torque is small. When approaching the system equilibrium point, the angular velocity with the fraction order norm of attitude quaternion is constructed hence system could have finite convergence time. These two stages are the main improvement of standard sliding mode controller, and system convergence time is less than 60 s for a 120 degree maneuver; in contrast, the convergence time of standard sliding mode controller is more than 300 s under same simulation configuration.

$$
\boldsymbol{u}=\begin{cases}-k\|\boldsymbol{s}_e\|^p\boldsymbol{s}_e+\boldsymbol{r}_1 & \|\boldsymbol{q}_{ev}\|>\alpha\\ -k\|\boldsymbol{s}_e\|^p\boldsymbol{s}_e+\boldsymbol{r}_2 & \beta<\|\boldsymbol{q}_{ev}\|\le\alpha\\ -k\|\boldsymbol{s}_e\|^p\boldsymbol{s}_e+\boldsymbol{r}_3 & \|\boldsymbol{q}_{ev}\|\le\beta\end{cases},\quad \boldsymbol{s}_e=\begin{cases}\boldsymbol{\omega}_e+k_1\boldsymbol{e}_e, & \|\boldsymbol{q}_{ev}\|>\alpha\\ \boldsymbol{\omega}_e+k_2\boldsymbol{q}_{ev}, & \beta<\|\boldsymbol{q}_{ev}\|\le\alpha\\ \boldsymbol{\omega}_e+k_3\|\boldsymbol{q}_{ev}\|^r\boldsymbol{e}_e, & \|\boldsymbol{q}_{ev}\|\le\beta\end{cases}
$$

$$
\begin{cases}
\boldsymbol{r}_1=\hat{\boldsymbol{J}}\boldsymbol{R}\dot{\boldsymbol{\omega}}_d-\hat{\boldsymbol{J}}\boldsymbol{\omega}_e^{\times}\boldsymbol{R}\boldsymbol{\omega}_d-\frac{1}{2}k_1\hat{\boldsymbol{J}}\boldsymbol{e}_e^{\times}\left(\boldsymbol{I}_3-\cot\frac{\varphi_e}{2}\boldsymbol{e}_e^{\times}\right)\boldsymbol{\omega}_e\\
\qquad+(\boldsymbol{\omega}_e+\boldsymbol{R}\boldsymbol{\omega}_d)^{\times}\hat{\boldsymbol{J}}(\boldsymbol{\omega}_e+\boldsymbol{R}\boldsymbol{\omega}_d)-l_1\,\mathrm{sgn}(\boldsymbol{s}_e)\\
\boldsymbol{r}_2=\hat{\boldsymbol{J}}\boldsymbol{R}\dot{\boldsymbol{\omega}}_d-\hat{\boldsymbol{J}}\boldsymbol{\omega}_e^{\times}\boldsymbol{R}\boldsymbol{\omega}_d-\frac{1}{2}k_2\hat{\boldsymbol{J}}\boldsymbol{F}_e\boldsymbol{\omega}_e+(\boldsymbol{\omega}_e+\boldsymbol{R}\boldsymbol{\omega}_d)^{\times}\hat{\boldsymbol{J}}(\boldsymbol{\omega}_e+\boldsymbol{R}\boldsymbol{\omega}_d)-l_2\,\mathrm{sgn}(\boldsymbol{s}_e)\\
\boldsymbol{r}_3=\hat{\boldsymbol{J}}\boldsymbol{R}\dot{\boldsymbol{\omega}}_d-\hat{\boldsymbol{J}}\boldsymbol{\omega}_e^{\times}\boldsymbol{R}\boldsymbol{\omega}_d+\frac{k_3^2}{2}rq_{e0}\|\boldsymbol{q}_{ev}\|^{2r-1}\hat{\boldsymbol{J}}\boldsymbol{e}_e\\
\qquad+(\boldsymbol{\omega}_e+\boldsymbol{R}\boldsymbol{\omega}_d)^{\times}\hat{\boldsymbol{J}}(\boldsymbol{\omega}_e+\boldsymbol{R}\boldsymbol{\omega}_d)-l_3\,\mathrm{sgn}(\boldsymbol{s}_e)
\end{cases}
\tag{3.50}
$$

Generally, $r, p \in (0, 1)$ could guarantee system have finite-time convergence time, and smaller α and larger β could bring better convergence time. Considering the maximum angular velocity of current satellite is about 0.1 rad/s, and angular velocity should be continuous at the switch points, select control parameters as follows:

$$
r=2/3, p=1/2, k_1=0.1, k_2=0.5, k_3=0.136, k=10, \alpha=0.2, \beta=0.02 \tag{3.51}
$$

Set the simulation parameters as follows:

$$
\begin{gathered}
\boldsymbol{J}=diag([30\ \ 25\ \ 20])kg\cdot m^2, \hat{\boldsymbol{J}}=diag([28\ \ 24\ \ 21])kg\cdot m^2, \lambda=2\\
\boldsymbol{\omega}_i=0.06\begin{bmatrix}\frac{1}{2} & \frac{\sqrt{3}}{2} & 0\end{bmatrix}^T \mathrm{rad/s}, \boldsymbol{q}_i=[0.8104\ \ -0.5587\ \ -0.1547\ \ 0.0845]^T\\
\boldsymbol{\omega}_d=\boldsymbol{0}_{3\times1}, \boldsymbol{q}_d=[1000]^T\\
\overline{u}=0.5N\cdot m, \overline{\varpi}=0.1\mathrm{rad/s}, \overline{d}=10^{-4}N\cdot m
\end{gathered}
\tag{3.52}
$$

The simulation results of finite-time controller (3.50) are given as follows: Based on Figs. 3.1–3.3, it could be found that system converges within 30 s, which is largely improved comparing with standard sliding mode controller.

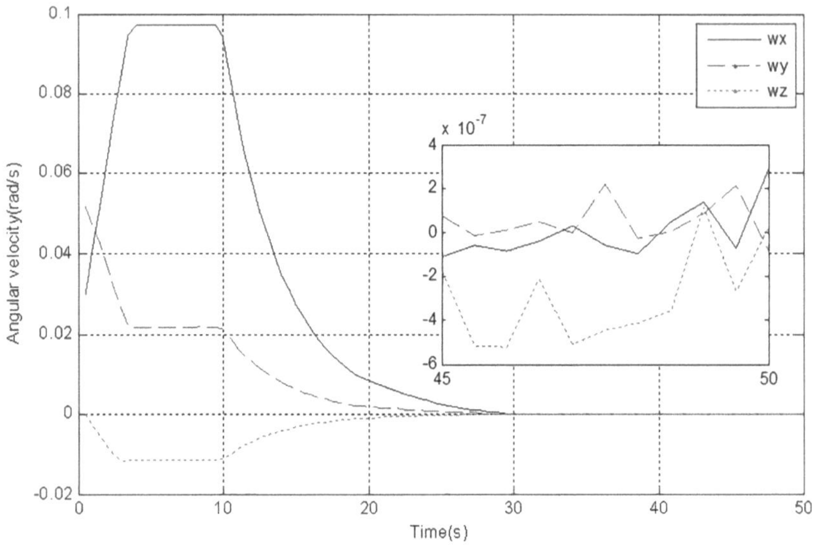

FIG. 3.1 Curve of angular velocity.

Based on Fig. 3.1, it could be found that during the maneuver stage with constant angular velocity, the convergence rate of attitude quaternion is at a high level and the control torque is small. Generally, controller (3.50) could improve system convergence rate for large-angle attitude maneuver (Fig. 3.4).

The next step is to demonstrate the simulation results of the controller (3.38) presented in this chapter. As discussed in previous section, the maximum

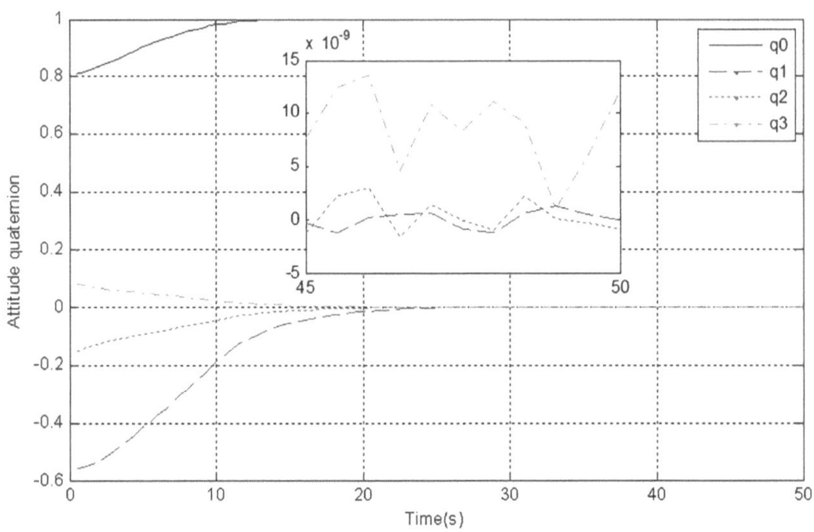

FIG. 3.2 Curve of attitude quaternion.

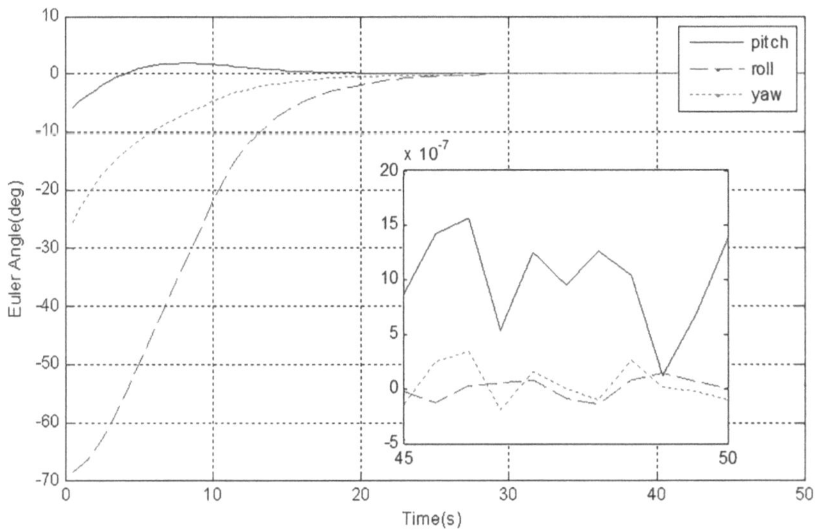

FIG. 3.3 Curve of Euler angle.

angular velocity is about 0.1 rad/s, hence select $\overline{\omega} = 0.1$. In order to satisfy the loose control torque constraint (3.47), and considering that select larger α could bring better convergence rate, hence select $\alpha = 0.02$. Noticing the estimation value of error matrix, select the control parameters as follows:

$$\overline{\omega} = 0.1\text{rad/s}, \alpha = 0.02, \lambda = 3, k_d = 10, k_p = 2 \tag{3.53}$$

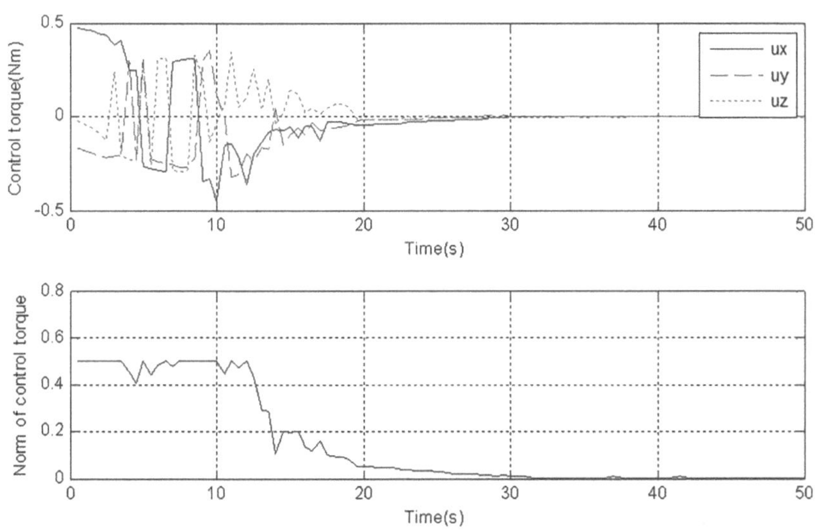

FIG. 3.4 Curve of control torque and its norm.

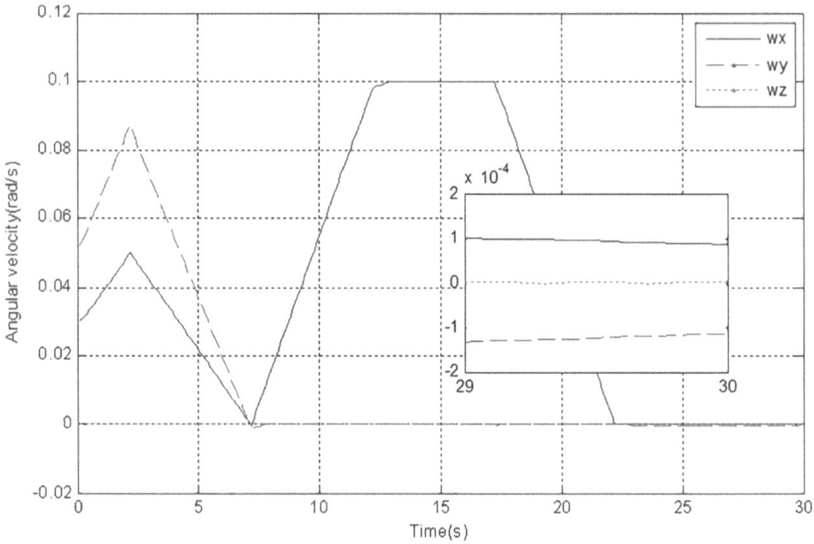

FIG. 3.5 Curve of angular velocity.

The simulation results of controller (3.38) are given as follows:

Based on Figs. 3.5–3.8, it could be found that system convergence time is about 22 s, which is less than 26% comparing with controller (3.50). It could be found that system has two acceleration processes and two deceleration processes, during these processes, system accelerates and decelerates with constant

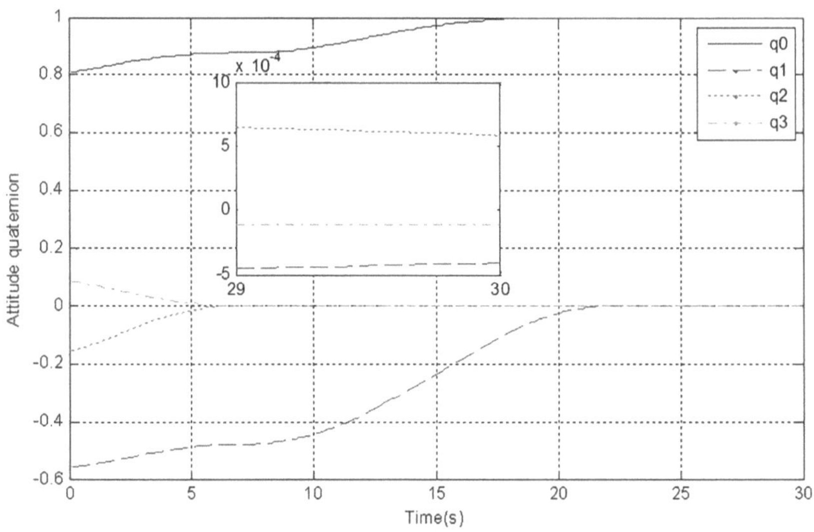

FIG. 3.6 Curve of attitude quaternion.

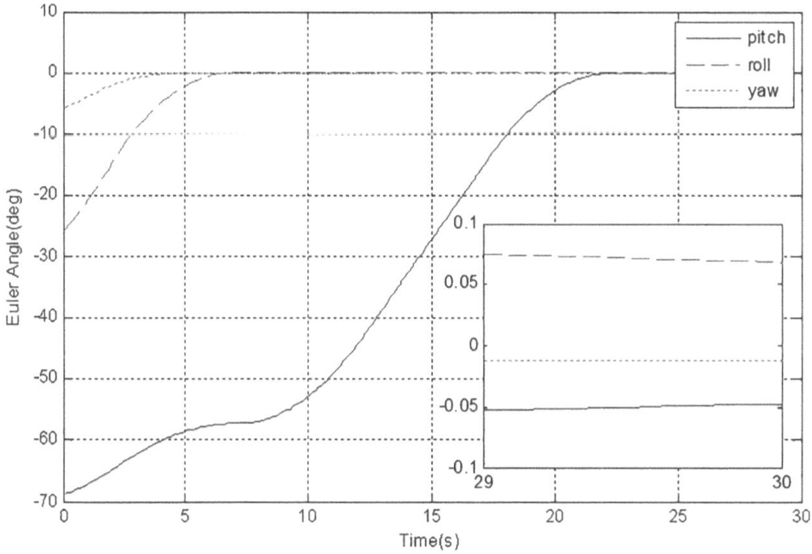

FIG. 3.7 Curve of Euler angle.

angular acceleration and the control torque is locked to system upper bound. Besides these processes, system has a maneuver stage with constant angular velocity and during this process, attitude quaternion convergence rate is maintained at a high level and desired control torque is small. Based on the simulation results, it could be found that the key to improve system

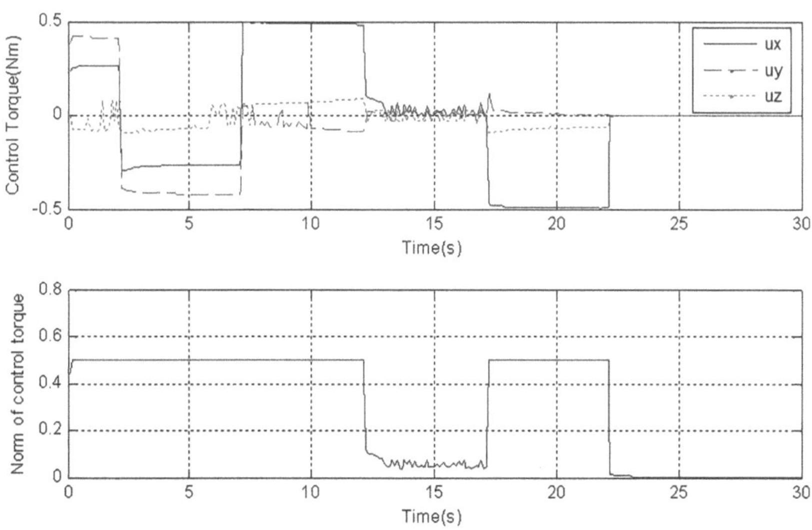

FIG. 3.8 Curve of control torque and its norm.

convergence rate is to design angular velocity trajectory properly. The bang-bang structure trajectory brings better convergence rate comparing with finite-time controller (3.50). Moreover, based on the discussion in previous section, the rotate angle is minimum under this path and this also improves system convergence time. Moreover, the trajectory presented in this chapter is analytical hence the computation is largely reduced. However, it is worth noticing that system accelerates and decelerates too many times hence system performance is sacrificed to some extent, and this is the cost to get analytical solution.

3.4.2 Regular attitude maneuver issue

In this section, simulation results of regular attitude maneuver issue, i.e., maneuver with nonzero initial and terminal angular velocity are given. Considering that the principle to select the control parameters is similar as discussed in Section 3.4.1, hence the system parameters and control parameters are the same as (3.51)–(3.53). The only difference is the initial and terminal states and set the system parameters as follows:

$$\begin{aligned}\boldsymbol{\omega}_i &= 0.04\begin{bmatrix}\frac{\sqrt{3}}{3} & 0 & \frac{\sqrt{6}}{3}\end{bmatrix}^T \text{rad/s}, \boldsymbol{q}_i = \begin{bmatrix}0.800 & -0.132 & -0.156 & -0.564\end{bmatrix}^T \\ \boldsymbol{\omega}_d &= 0.04\begin{bmatrix}-\frac{\sqrt{3}}{2} & \frac{1}{2} & 0\end{bmatrix}^T \text{rad/s}, \boldsymbol{q}_d = \begin{bmatrix}0.914 & 0.371 & -0.165 & 0.010\end{bmatrix}^T\end{aligned} \tag{3.54}$$

The simulation results of finite-time controller (3.50) are given as follows:

Based on Figs. 3.9–3.11, it can be found that system state converges to zero within 35 s. Based on (3.51) and Fig. 3.9, it could be found that $k_1 = 0.1$, i.e., the error angular velocity has reached system upper bound. The advantage of controller (3.50) has been described in Section 3.4.1, and the key to improve system convergence rate is also to design angular velocity trajectory (Fig. 3.12).

The next step is to demonstrate the simulation results of the controller (3.38) presented in this chapter. Select control parameters as (3.53), and the simulation results of controller (3.38) are given as follows:

Based on Figs. 3.13–3.16, system convergence time is about 25 s, which is improved more than 25% comparing with finite time controller (3.50). Moreover, based on Figs. 3.13 and 3.16, it could be found that during the acceleration and deceleration processes, the control torque is near system upper bound, and when system is rotating with constant angular velocity, the control torque is relatively small. The minimum rotate angle condition and bang-bang-like trajectory largely improves system convergence rate and system capability is utilized efficiently. Based on Fig. 3.15, it could be found that the velocity of pitch, roll, and yaw angle is about 3.4°/s, 2.6°/s and 1.2°/s, the three-axis attitude maneuver capability is also largely improved.

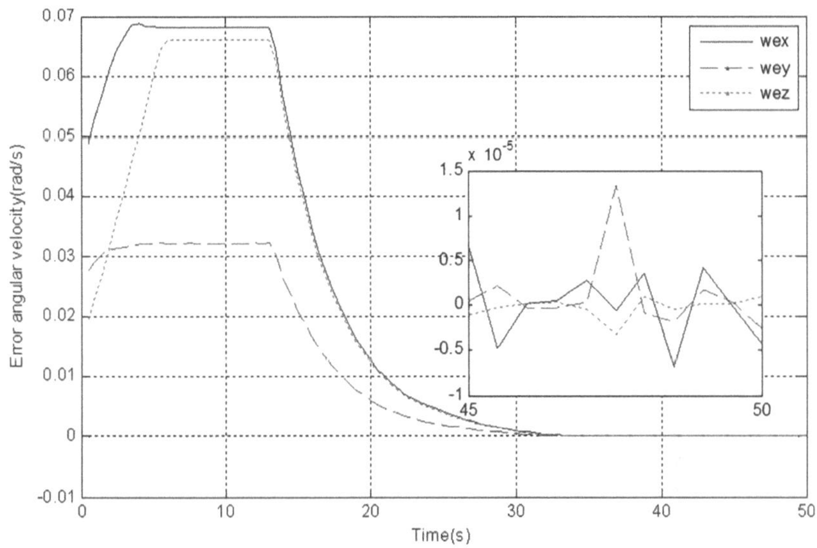

FIG. 3.9 Curve of error angular velocity.

It is worth noticing that the maneuver process has three acceleration processes and three deceleration processes, which is once more than the trajectory in Section 3.4.1. Similar as discussed in Section 3.4.1, too many acceleration and deceleration processes slow down the system.

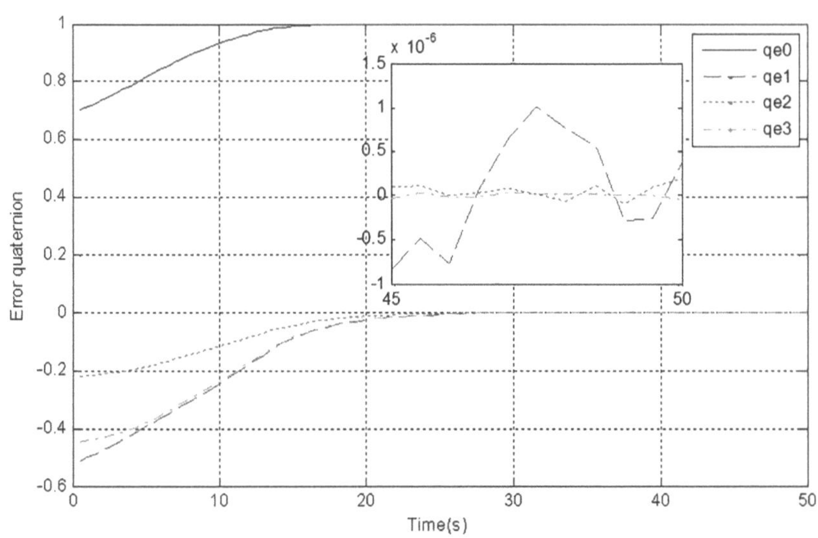

FIG. 3.10 Curve of error attitude quaternion.

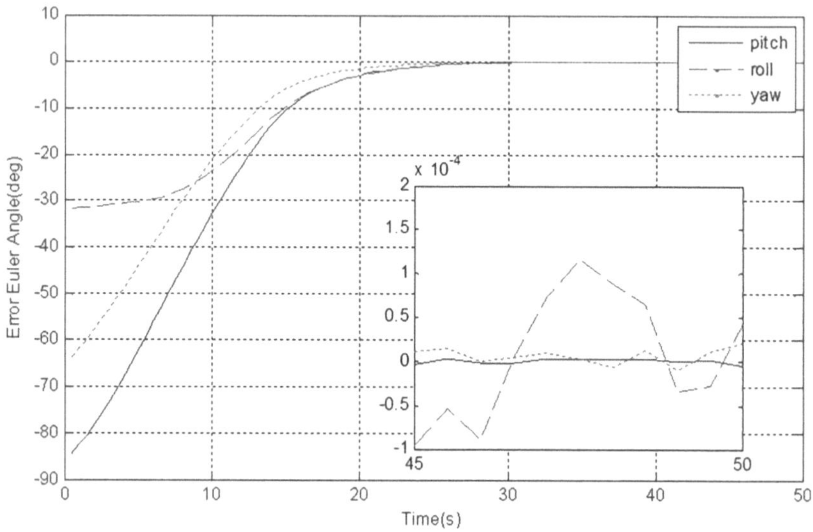

FIG. 3.11 Curve of error Euler angle.

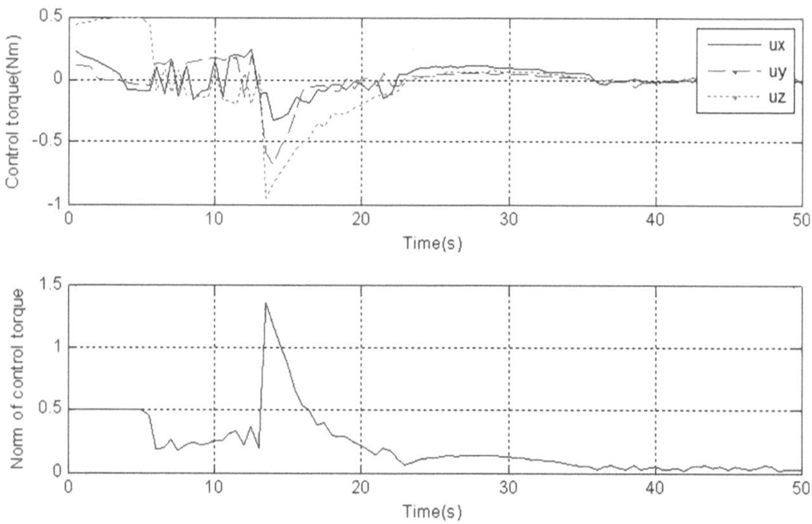

FIG. 3.12 Curve of control torque and its norm.

3.4.3 Hohmann transfer-like maneuver issue

In this section, simulation for a special case discussed in Section 3.3.2 is given.

Set system parameters as follows:

$$\boldsymbol{\omega} = 0.02\left[-\sqrt{3}/2\ \ 1/2\ \ -0.1\right]^T \text{rad/s}, \boldsymbol{q} = [0.735\ \ 0.587\ \ -0.339\ \ 0]^T$$
$$\boldsymbol{\omega}_d = 0.06\left[\sqrt{3}/3\ \ 0\ \ \sqrt{6}/3\right]^T \text{rad/s}, \boldsymbol{q}_d = [0.817\ \ -0.333\ \ 0\ \ -0.417]^T \tag{3.55}$$

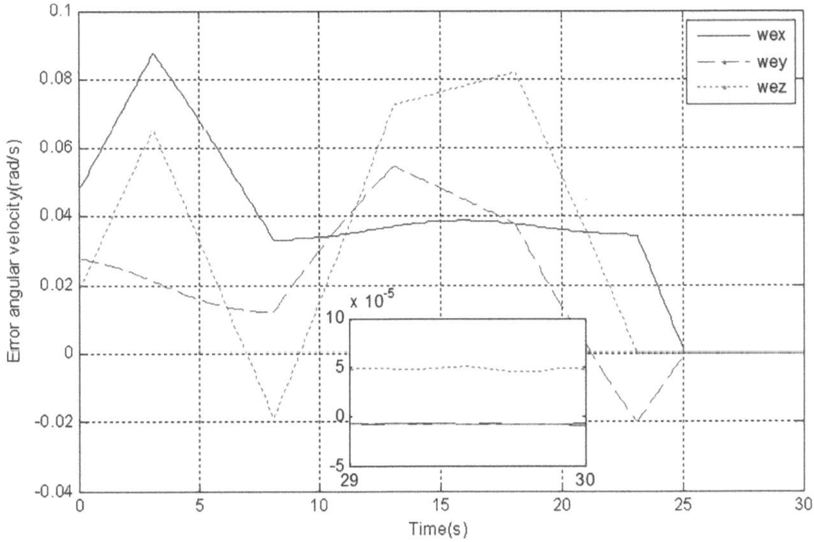

FIG. 3.13 Curve of error angular velocity.

Based on (3.37), it could be found that T_i and T_d have a point of intersection.

The simulation results of finite time controller (3.50) are given from Figs. 3.17–3.20.

Based on Figs. 3.17–3.20, it could be found that system convergence time is about 40 s, and the angular velocity has reached system upper bound.

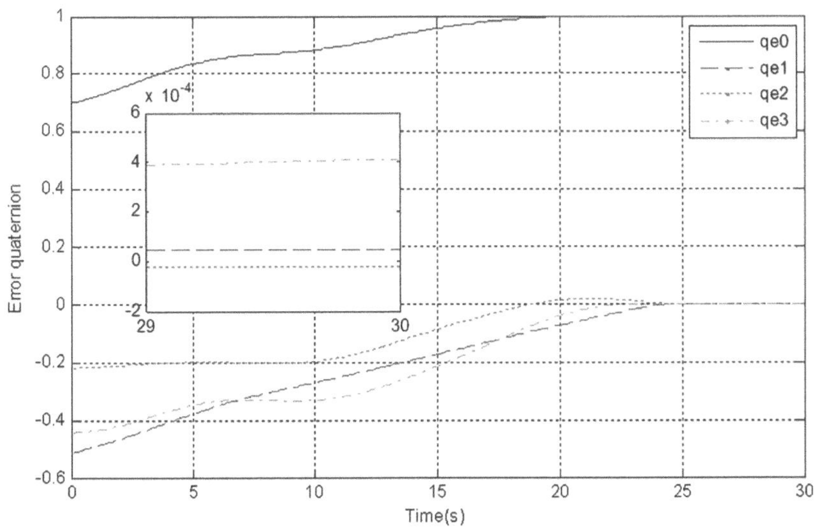

FIG. 3.14 Curve of error quaternion.

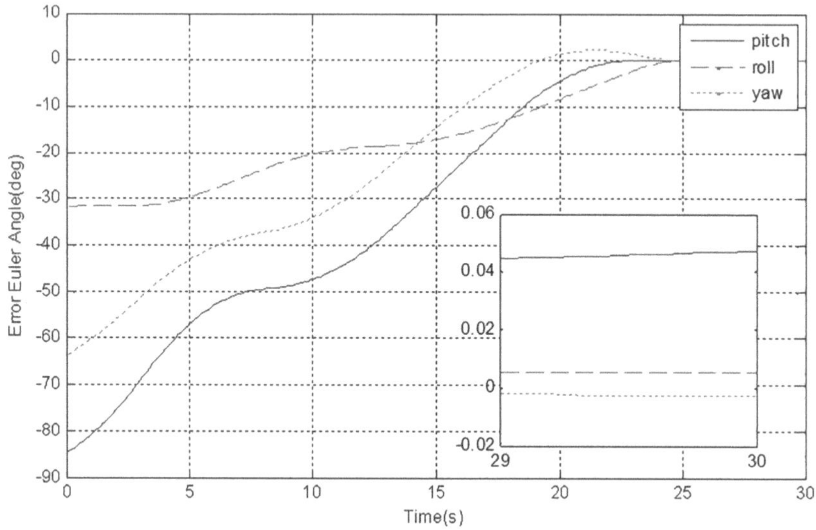

FIG. 3.15 Curve of Euler angle.

In order to demonstrate the performance of the algorithm presented in Section 3.3.2, the trajectory planning algorithm in (3.1) is also compared. Next, the simulation results of minimum rotate angle in (3.1) are given as follows:

Based on Figs. 3.21–3.24, system convergence time is about 28 s, and system has three acceleration and three deceleration processes.

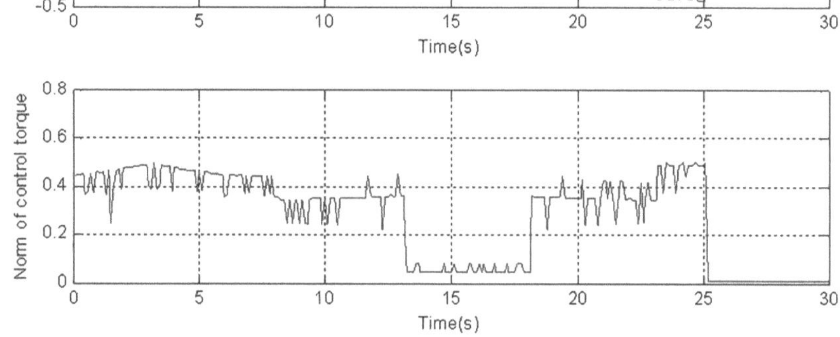

FIG. 3.16 Curve of control torque and its norm.

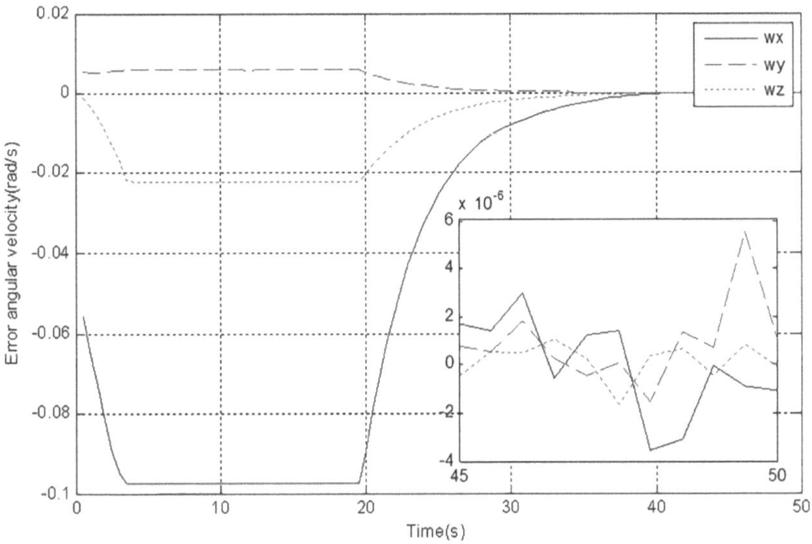

FIG. 3.17 Curve of error angular velocity.

Next the simulation results of Hohmann transfer like trajectory is given as Figs. 3.25–3.28.

Based on Figs. 3.25–3.28, it could be found that system convergence time is about 23 s, which is less than that of finite time controller and the controller in Section 3.3.1. Although the minimum rotate angle does not hold in this

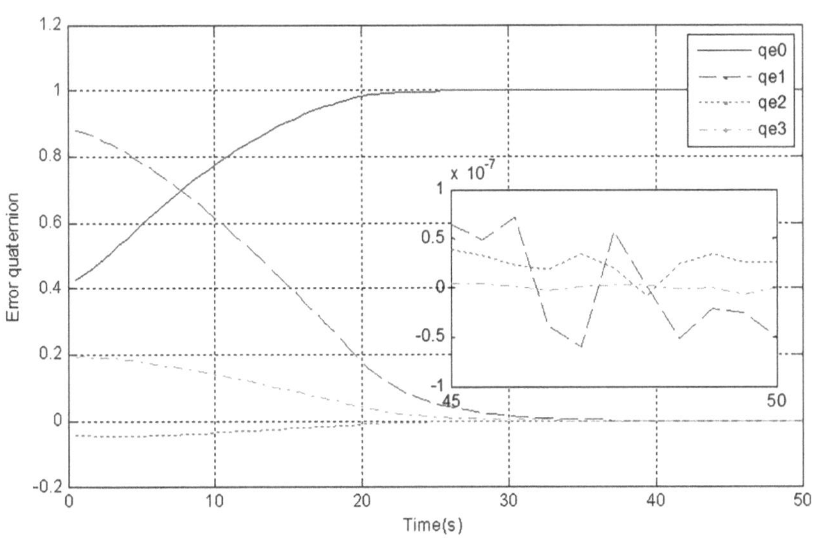

FIG. 3.18 Curve of error quaternion.

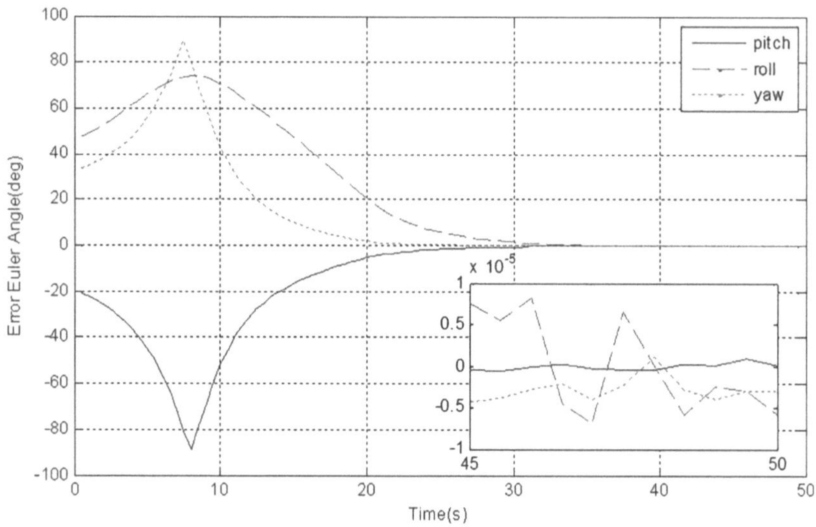

FIG. 3.19 Curve of error Euler angle.

situation, but the convergence time is still improved. The key to improve system convergence rate is to optimize the acceleration and deceleration process. The Hohmann transfer like maneuver has two acceleration and deceleration processes, which is less than the minimum rotate angle maneuver, and the acceleration time is lessened about 10s. However, it is worth noticing that the convergence time of Hohmann transfer like maneuver could be larger than that of minimum rotate angle, when the total rotate angle differs too much.

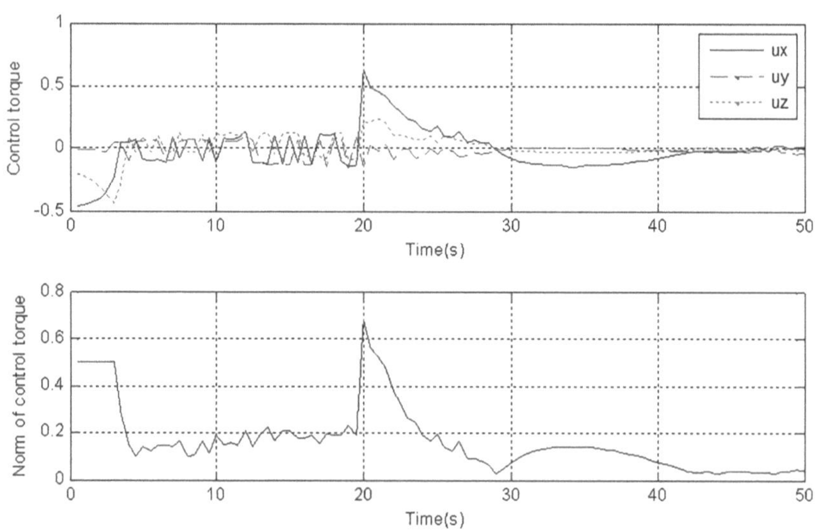

FIG. 3.20 Curve of control torque and its norm.

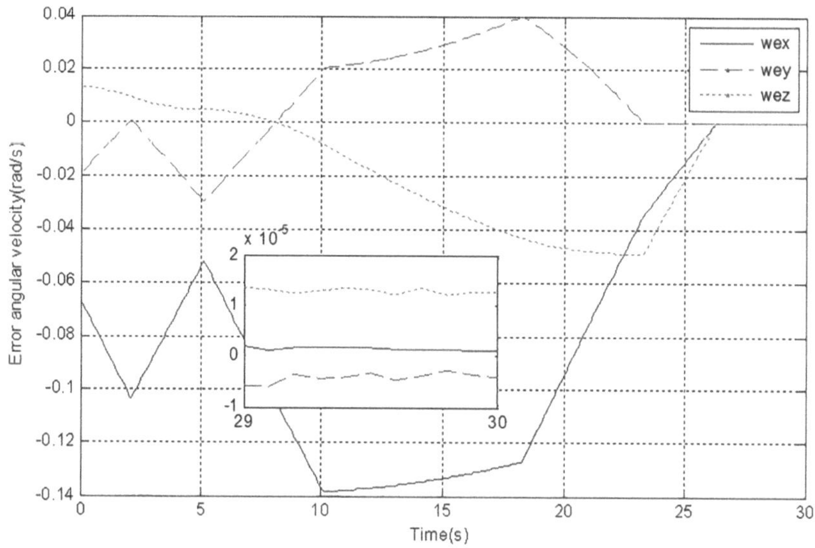

FIG. 3.21 Curve of error angular velocity.

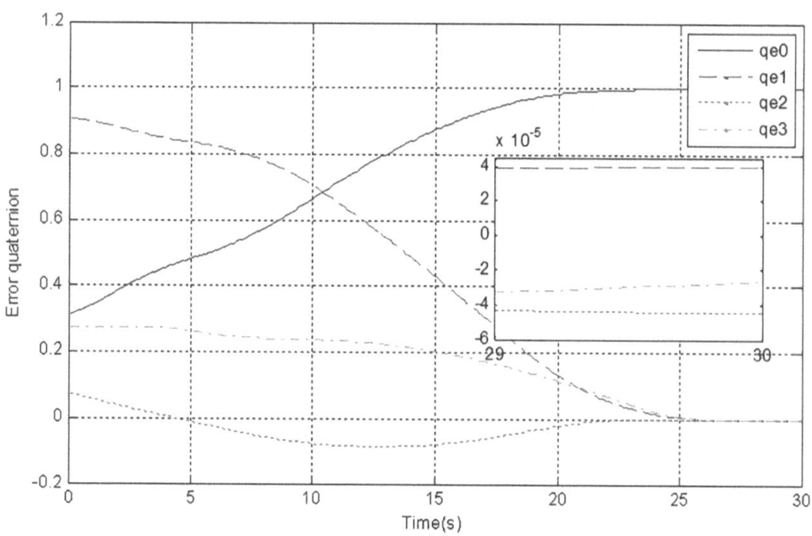

FIG. 3.22 Curve of error quaternion.

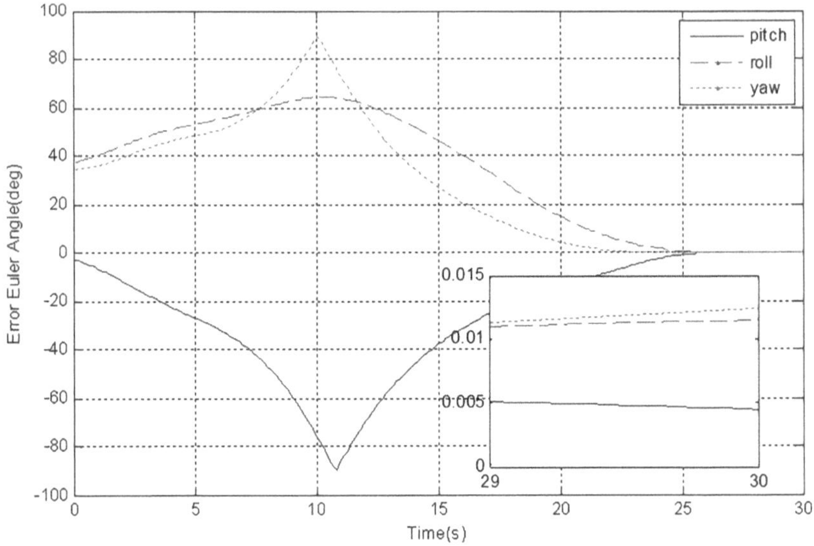

FIG. 3.23 Curve of error Euler angle.

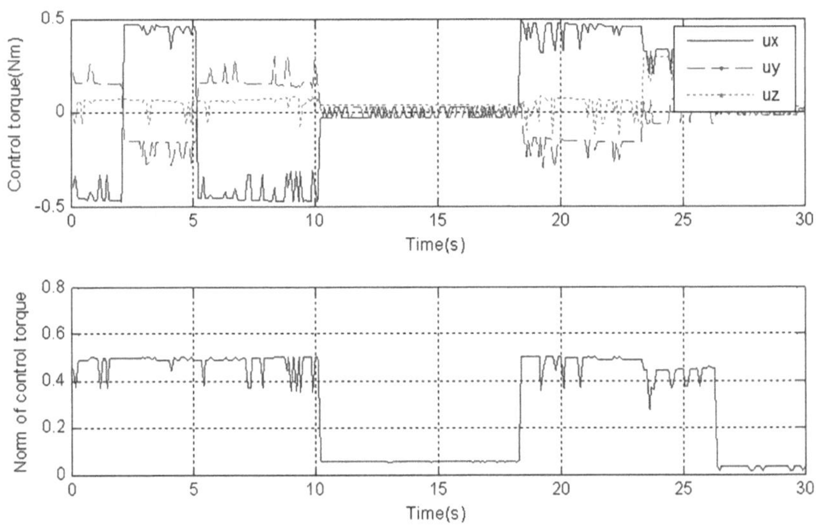

FIG. 3.24 Curve of control torque and its norm.

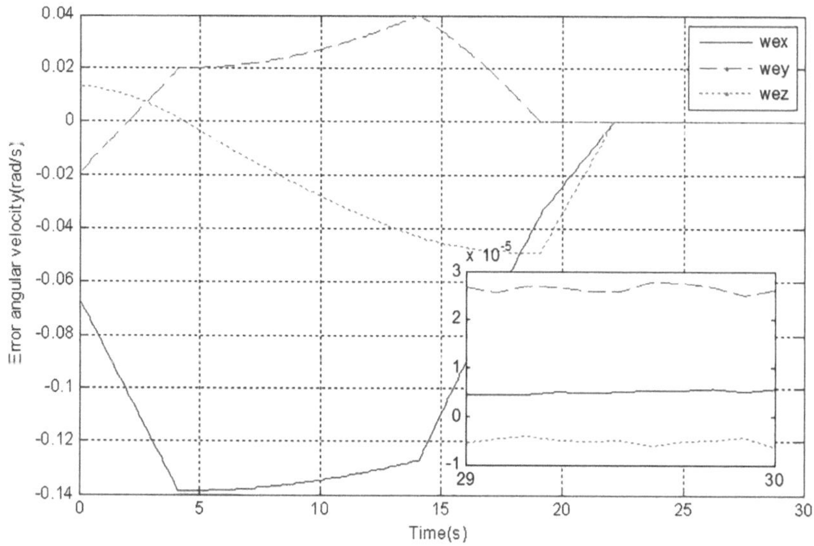

FIG. 3.25 Curve of error angular velocity.

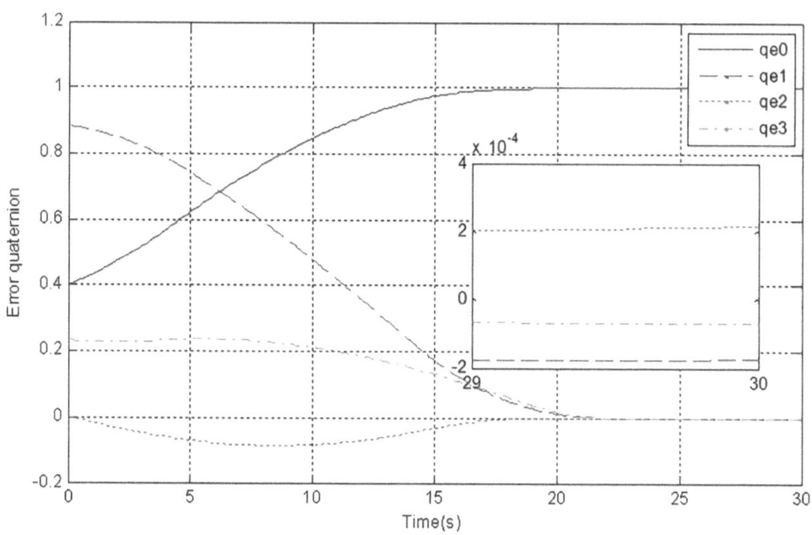

FIG. 3.26 Curve of error quaternion.

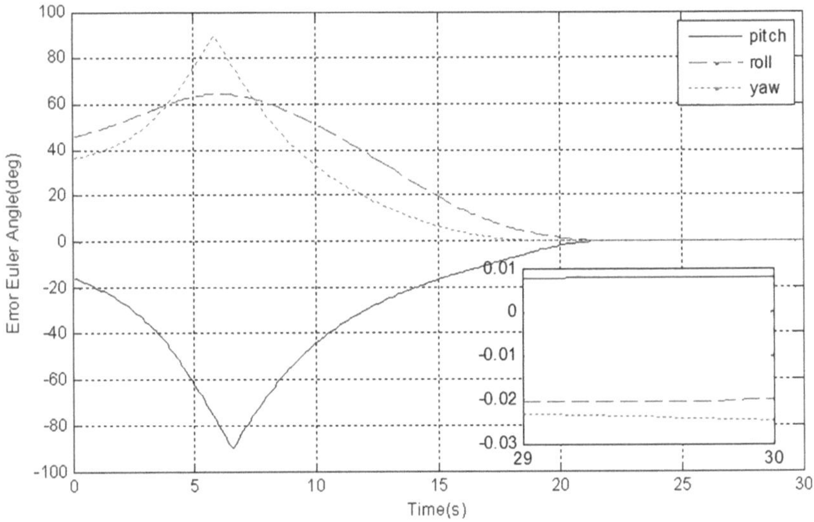

FIG. 3.27 Curve of error Euler angle.

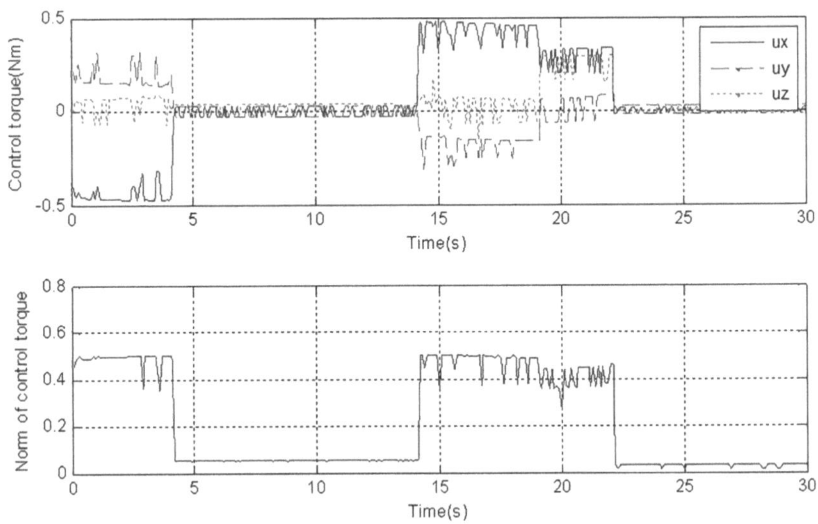

FIG. 3.28 Curve of control torque and its norm.

3.5 Conclusion

In this chapter, a near-time optimal analytical trajectory was proposed for satellite attitude maneuver. The attitude maneuver process was constructed as three successive maneuvers, and the minimum total rotate angle was derived. Each maneuver was optimized by open-loop bang-bang control algorithm. The system convergence rate was largely improved comparing with finite-time controller and standard sliding mode controller. Inspired by Hohmann transfer, a special case was discussed, and the number of acceleration and deceleration processes was both reduced once, hence the system convergence time could still be improved. The trajectory planning algorithm presented in this chapter was analytical hence the computation was largely reduced, and the algorithm is more suitable for engineering practice. In fact, the key to improve convergence time could be concluded as following three aspects: (1) minimun rotate angle; (2) bang-bang like trajectory; and (3) less accelerate and decelerate processes.

It is worth noticing that in order to get the analytical solution of maneuver trajectory, the algorithm presented in this chapter was not time optimal. System would accelerate and decelerate 4–6 times, and the system capability was not fully used. Later work could focus on the issue of analytical solution of time-optimal attitude maneuver.

References

[1] B. Xiao, S. Yin, L.G. Wu, A structure simple controller for satellite attitude tracking maneuver, IEEE Trans. Ind. Electron. 64 (2) (2017) 1436–1446.

[2] B. Xiao, S. Yin, H.J. Gao, Tracking control of robotic manipulators with uncertain kinematics and dynamics, IEEE Trans. Ind. Electron. 63 (10) (2016) 6439–6449.

[3] B. Xiao, S. Yin, Velocity-free fault tolerant and uncertainty attenuation control for a class of nonlinear systems, IEEE Trans. Ind. Electron. 63 (7) (2016) 4400–4411.

[4] S.N. Wu, Z.W. Sun, H. Deng, Robust sliding mode controller design for globally fast attitude tracking of target spacecraft, Int. J. Innov. Comput. Inf. Control 7 (5) (2011) 2087–2098.

[5] S.N. Wu, R. Wang, G. Radice, Z.G. Wu, Robust attitude maneuver control of spacecraft with reaction wheel low-speed friction compensation, Aerosp. Sci. Technol. 43 (2015) 213–218.

[6] S.N. Wu, Z.G. Wu, G. Radice, R. Wang, Adaptive control for spacecraft relative reanslation with parametric uncertainty, Aerosp. Sci. Technol. 31 (1) (2013) 53–58.

[7] J.R. Zhang, X.Y. Chu, Y. Zhang, Q. Hu, G. Zhai, Y.Y. Li, Safe-trajectory optimization and tracking control in ultra-close proximity to a failed satellite, Acta Astronaut. 144 (144) (2018) 339–352.

[8] X.B. Cao, B.L. Wu, Robust spacecraft attitude tracking control using hybrid actuators with uncertainties, Acta Astronaut. 136 (2017) 1–8.

[9] Y. Guo, J.H. Guo, S.M. Song, Backstepping control for attitude tracking of the spacecraft under input saturation, Acta Astronaut. 138 (2017) 318–325.

[10] Y. Li, D. Ye, Z.W. Sun, Robust finite time control algorithm for satellite attitude control, Aerosp. Sci. Technol. 68 (2017) 46–57.

[11] Y. Li, D. Ye, Time efficient sliding mode controller based on bang-bang logic for satellite attitude control, Aerosp. Sci. Technol. 75 (2018) 342–352.

[12] X.L. Bai, J.L. Junkins, New results for time-optimal three-axis reorientation of a rigid spacecraft, J. Guid. Control. Dyn. 32 (4) (2009) 1071–1076.
[13] F. Li, P.M. Bainum, Numerical approach for solving rigid spacecraft minimum time attitude maneuvers, J. Guid. Control. Dyn. 13 (1) (1990) 38–45.
[14] A.B. Geogre, M. Romano, O.A. Yakimenko, Time-optimal reorientation of a spacecraft using an inverse dynamic optimization method, J. Guid. Control. Dyn. 34 (4) (2011) 1197–1208.
[15] Y.F. Zhang, G.F. Ma, H.B. Huang, C.J. Li, Real-time trajectory optimization of an underactuated rigid spacecraft using differential flatness, Aerosp. Sci. Technol. 23 (1) (2012) 132–139.
[16] G.G. Wang, H.E. Chu, S. Mirjalili, Three-dimensional path planning for UCAV using an improved bat algorithm, Aerosp. Sci. Technol. 49 (2016) 231–238.
[17] Y.F. Zhuang, H.B. Huang, Time-optimal trajectory planning for underactuated spacecraft using a hybrid particle swarm optimization algorithm, Acta Astronaut. 94 (2) (2014) 690–698.
[18] De Ridder S. Mooji E., Terminal area trajectory planning using the energy-tube concept for reusable launch vehicles, Acta Astronaut. 68 (7–8) (2011) 915–930.
[19] Y. Xie, L.H. Liu, J. Liu, G.J. Tang, W. Zheng, Rapid generation of entry trajectories with waypoint and no-fly zone constraints, Acta Astronaut. 77 (2012) 167–181.

Chapter 4

Time-efficient robust PID plus control with constraints

4.1 Introduction

PID (proportional, integral, differential) control is one of the most mature and widely used control algorithms in satellite attitude control system. The advantages of PID control could be concluded as simple structure, robustness of system inertia matrix, specific physical meaning, and so on. A lot of work has been done on PID control algorithm. Jin [1,2] designed PID controllers for single satellite and satellite formation attitude control. Kamesh [3] and Clarke et al. [4] designed PID controller for single satellite attitude stabilization and attitude tracking control. Wie [5–9] has done a lot of fundamental and inspiring work in the design of PID controllers. Some widely used Lyapunov functions were given in his papers. All the stability analysis in these papers was done by Lyapunov method.

Although PID control has those advantages described above, low convergence rate is a serious problem of PID control. Moreover, it is difficult to analyze the convergence time of PID controller. Comparing with PID control algorithm, sliding mode control is a control algorithm convenient to improve the convergence rate. Lu et al. [10] designed a time-delay adaptive sliding mode for satellite formation control. The sliding mode designed in this paper has the exponential convergence rate. Yang et al. [11] designed a fuzzy sliding mode, and Yang and Shi [12] designed a high-order sliding mode. In these two papers, standard sliding mode is modified to improve the accuracy at the steady stage and robustness to system uncertainty. Wu et al. [13] designed an adaptive sliding mode for satellite attitude control and the system uncertainty is fixed. In this work, the convergence rate is also proved to be exponential. All these sliding modes by Lu, Yang, and Wu are proved to be asymptotic stable, which means the system state converges to zero with infinite time. In order to improve the terminal convergence rate, Wu et al. [14] and Liang et al. [15] proposed finite sliding modes for single satellite attitude and satellite formation attitude control. Verbin [16–19] has done a lot of important works in time-efficient attitude control. In these papers, the desired angular velocity is designed to improve the convergence rate of standard PID controllers.

Fast Satellite Attitude Maneuver and Control. https://doi.org/10.1016/B978-0-323-95455-6.00004-4

Although sliding mode control brings a better performance on convergence rate and its analysis, sliding mode control needs the system inertia matrix, which could not be accurate known in real engineering. Ma and Ma [20] and Wang et al. [21] designed controllers robust to system inertia matrix uncertainty and disturbance torque for satellite attitude control. In these papers, the system uncertainty was treated as norm upper bounded disturbance and was fixed by sign function. In this paper, this method is used and the controller is robust to system uncertainty and disturbance torque.

Considering that the control torque of satellite control system is finite, it is necessary to ensure the norm of control torque does not exceed the system upper bound. Moreover, extend Kalman filter (EKF) is the most widely used attitude determination algorithm on satellite and its fundamental is to linearize nonlinear system based on Taylor formula. However, the high-order terms in Taylor Formula become noninfinitesimal when the angular velocity is large, thus the accuracy of EKF descends severely in this situation. Furthermore, the star tracker cannot image when the angular velocity exceeds its threshold. So it is reasonable to deem that excessive angular velocity would harm the attitude determination system. Meanwhile excessive angular would cause the flexible deformation of flexible appendages. Thus is it necessary to consider the constraint of control torque and angular velocity when designing controllers. Hu [22–25] has designed controllers under the control torque and angular velocity constraint.

In this chapter, a robust PID plus controller is proposed to improve the convergence rate. A time-efficient sliding mode is constructed, and the controller is designed to ensure system state could reach the sliding mode. Moreover, a novel method for the estimation of norm of angular velocity is proposed. The constraint of control torque and angular velocity is added into the controller to ensure the system state do not exceed their upper bound. The globally asymptotic stability of all the conditions considered in this chapter is proved and the performance of the controller is demonstrated by simulation.

4.2 Time-efficient PID plus controller

Standard PID controller consists of proportional term, integral term, and derivative term. Generally, larger proportional term brings faster convergence rate, but the controller with only proportional term cannot stabilize the system. Thus, the derivative term is added into the controller to enhance the stability. However, the negative angular velocity feedback would slowdown the system and the angular velocity would decrease. The integral term is added into the controller to reduce the steady-state error. However, the system governed by the controller with integral term is highly nonlinear, and the stability of the system is hard to analyze. Moreover, convergence rate is slowed down, and state vibration is strengthened by the integral term. Considering that the convergence rate of attitude quaternion is determined by angular velocity, PID control has the issue that the angular velocity decreases quickly so that the convergence rate of attitude quaternion slows down. The ideal condition is that when the system state is away from the equilibrium point, the controller should have larger proportional gain factor so that the convergence rate could be maintained; and when the system state approaches the equilibrium point, the differential gain factor should be larger so that the system could be slowed down, and the integral term should be added into the controller to reduce the steady state error. Based on the discussion above, standard PID controller is modified in this section to improve the convergence rate of the system.

The PID plus controller developed in this chapter could be written as follows:

$$\boldsymbol{u} = \begin{cases} \varsigma_1 + z_1 & \|\boldsymbol{q}_v\| \geq \alpha \\ \varsigma_2 + z_2 & \alpha > \|\boldsymbol{q}_v\| \geq \beta \\ \varsigma_3 + z_3 & \beta > \|\boldsymbol{q}_v\| \end{cases} \tag{4.1}$$

$$\begin{cases} \varsigma_1 = -k_d\boldsymbol{\omega} - (k_1 k_d / \|\boldsymbol{q}_v\|)\boldsymbol{q}_v \\ \varsigma_2 = -k_d\boldsymbol{\omega} - k_2 k_d \boldsymbol{q}_v - \dfrac{k_2}{2} q_0 \widehat{\boldsymbol{J}}\boldsymbol{\omega} \\ \varsigma_3 = -k_d\boldsymbol{\omega} - k_2 k_d \boldsymbol{q}_v - k_I \boldsymbol{\upsilon} - \dfrac{k_2}{2} q_0 \widehat{\boldsymbol{J}}\boldsymbol{\omega} \end{cases} \tag{4.2}$$

$$\begin{cases} z_1 = -\overline{d}\,\boldsymbol{sgn}(\boldsymbol{\omega}) + \boldsymbol{\omega}^{\times}\boldsymbol{r}_1 \\ z_2 = -\overline{d}\,\boldsymbol{sgn}(\boldsymbol{\omega}) + \boldsymbol{\omega}^{\times}\boldsymbol{r}_2 - \dfrac{k_2}{2}\lambda |q_0| \|\omega\|\, \boldsymbol{sgn}(\boldsymbol{s}) \\ z_3 = -\overline{d}\,\boldsymbol{sgn}(c\boldsymbol{\omega} + l_2\boldsymbol{q}_v) + (c\boldsymbol{\omega} + l_2\boldsymbol{q}_v)^{\times}\boldsymbol{r}_3 - \dfrac{k_2}{2}\lambda |q_0| \|\omega\|\, \boldsymbol{sgn}(\boldsymbol{s}) \end{cases} \tag{4.3}$$

where $\boldsymbol{sgn}(\boldsymbol{x})$ is the sign function of vector $\boldsymbol{x}$, k_1, k_2, k_I, k_d, k_p, α, β, c, l_2 are positive scalars, and k_1, k_2, $\boldsymbol{s}$, $\boldsymbol{r}_i$, $\boldsymbol{\upsilon}$ are defined as follows:

$$\boldsymbol{s} = \begin{cases} \boldsymbol{\omega} + k_1\boldsymbol{e} & \|\boldsymbol{q}_v\| \geq \alpha \\ \boldsymbol{\omega} + k_2\boldsymbol{q}_v & \|\boldsymbol{q}_v\| < \alpha \end{cases} \tag{4.4}$$

$$k_2 = k_1/\alpha \tag{4.5}$$

$$\begin{cases} \boldsymbol{r}_1 = \left(\widehat{\boldsymbol{J}}\boldsymbol{\omega} - \dfrac{1}{2}\widehat{\boldsymbol{J}}\boldsymbol{s} - \dfrac{1}{2}\cot\dfrac{\varphi}{2}\boldsymbol{e}^{\times}\widehat{\boldsymbol{J}}\boldsymbol{s}\right) - \lambda\left(\|\omega\| + \dfrac{1}{2}\left(1 + \cot\dfrac{\varphi}{2}\right)\|s\|\right)\boldsymbol{sgn}\left(\left(\boldsymbol{e}^T\boldsymbol{\omega}^{\times}\right)^T\right) \\ \boldsymbol{r}_2 = \left(\widehat{\boldsymbol{J}}\boldsymbol{\omega} - \dfrac{1}{2}\widehat{\boldsymbol{J}}\boldsymbol{s}\right) - \lambda\left(\|\omega\| + \dfrac{1}{2}\|s\|\right)\boldsymbol{sgn}\left(\left(\boldsymbol{q}_v^T\boldsymbol{\omega}^{\times}\right)^T\right) \\ \boldsymbol{r}_3 = \dfrac{k_2}{ck_2 - l_2}\left(\widehat{\boldsymbol{J}}\boldsymbol{\omega} - \dfrac{1}{2}\widehat{\boldsymbol{J}}\boldsymbol{s}\right) - \dfrac{k_2}{ck_2 - l_2}\lambda\left(\|\omega\| + \dfrac{1}{2}\|s\|\right)\boldsymbol{sgn}\left(\left(\boldsymbol{q}_v^T\boldsymbol{\omega}^{\times}\right)^T\right) \end{cases} \tag{4.6}$$

$$\dot{\boldsymbol{\upsilon}} = \begin{cases} 0 & \beta \leq \|\boldsymbol{q}_v\| \\ c_1\boldsymbol{\omega} + c_2\boldsymbol{q}_v - \dfrac{1}{2}c_3\boldsymbol{F}\boldsymbol{\omega} & \beta > \|\boldsymbol{q}_v\| \end{cases} \tag{4.7}$$

where λ, c_1, c_2, c_3 are positive scalars and λ satisfy (4.8).

$$\lambda \geq \lambda\left(\tilde{J}\right) \tag{4.8}$$

The relationship between k_1 and k_2, i.e., Eq. (4.12) guarantees the angular velocity is continuous at the switch point of $\boldsymbol{s}$.

Based on Eqs. (4.1) and (4.4), it can be found that the converge process consists of three stages: the maneuver stage, i.e., the stage of $\|\boldsymbol{q}_v\| \geq \alpha$, the slow-down stage; i.e., the stage of $\alpha > \|\boldsymbol{q}_v\| \geq \beta$; and the converge stage, i.e., the stage of $\beta > \|\boldsymbol{q}_v\|$. According to (4.2), it can be found that ς_i is a standard PID controller. Except for the converge stage, the integral term is not brought into the controller to avoid the disadvantages of integral term. During the maneuver stage, the proportional gain factor is enlarged to maintain the convergence rate. In controller (4.1), z_i is the plus term designed in this chapter.

Controller (4.1) has the property that (1) controller (4.1) does not change the globally asymptotic stability of standard PID controller and (2) the sliding mode state is globally asymptotic stabile, i.e., $\boldsymbol{s} \to \boldsymbol{0}$, and the norm of angular velocity could be maintained constant at maneuver stage, the convergence rate at slow down and convergence stage could be maintained at exponential. These two properties will be discussed in later text.

4.2.1 Stability analysis

The goal of this section is to prove the globally asymptotic stability of the system governed by controller (4.1).

The globally asymptotic stability of maneuver stage and slow down stage is proved first. In order to analyze the stability, it is necessary to present two lemmas.

Lemma 4.1 The system governed by PD (proportional, differential) controller described as (4.9) with positive scalars k_p, k_d is globally asymptotic stable.

$$\boldsymbol{u} = -k_d\boldsymbol{\omega} - k_p\boldsymbol{q}_v - \overline{d}\,\boldsymbol{sgn}\left(\boldsymbol{\omega}\right) \tag{4.9}$$

Proof: Choose Lyapunov function as (4.17) and calculate its derivative it can be got that

$$V_1 = \frac{1}{2}\boldsymbol{\omega}^T\boldsymbol{J}\boldsymbol{\omega} + 2k_p(1 - q_0) \tag{4.10}$$

$$\dot{V}_1 = \boldsymbol{\omega}^T\boldsymbol{J}\dot{\boldsymbol{\omega}} - 2k_p\dot{q}_0 = -k_d\boldsymbol{\omega}^T\boldsymbol{\omega} + \boldsymbol{\omega}^T\boldsymbol{d} - \overline{d}\boldsymbol{\omega}^T\,\boldsymbol{sgn}\left(\boldsymbol{\omega}\right) \leq 0 \tag{4.11}$$

Based on Lasalle invariance principle, the system state converges to $\boldsymbol{\omega} \to \boldsymbol{0}$, thus $\dot{\boldsymbol{\omega}} \to \boldsymbol{0}$, $\boldsymbol{u} \to \boldsymbol{0}$, thus $\boldsymbol{q}_v \to \boldsymbol{0}$. Thus, the system is globally asymptotic stable.

Lemma 4.2 The system governed by the controller described as (4.19) is globally asymptotic stable.

$$\boldsymbol{u} = -k_d\boldsymbol{\omega} - k_p\boldsymbol{q}_v - \overline{d}\,\boldsymbol{sgn}(\boldsymbol{\omega}) + \boldsymbol{\omega}^{\times}\boldsymbol{r} \tag{4.12}$$

where $\boldsymbol{r}$ is a three-dimensional vector.

Proof: Select Lyapunov function as (4.10), calculate its derivative and substitute control law (4.12), it can be got that

$$\dot{V}_1 = \boldsymbol{\omega}^T\boldsymbol{J}\dot{\boldsymbol{\omega}} - 2k_p\dot{q}_0 = -k_d\boldsymbol{\omega}^T\boldsymbol{\omega} + \boldsymbol{\omega}^T\boldsymbol{d} - \overline{d}\boldsymbol{\omega}^T\,\boldsymbol{sgn}(\boldsymbol{\omega}) \le 0 \tag{4.13}$$

Similar to the proof of Lemma 4.1, it is easy to find that the system governed by controller (4.12) is globally asymptotic stable.

It is easy to find that the structure of the controller (4.18) at maneuver stage is the same as the structure of controller (4.12), thus the system is globally asymptotic stable based on Lemmas 4.1 and 4.2. Comparing the controller (4.11) at slow down stage with (4.12), the only difference is the q_0 term in (4.11). Considering that

$$\frac{k_2}{2}\left(\lambda + \lambda\left(\hat{\boldsymbol{J}}\right)\right)|q_0|\|\boldsymbol{\omega}\|\boldsymbol{\omega}^T\,\boldsymbol{sgn}(\boldsymbol{s}) \le \frac{k_2}{2}\left(\lambda + \lambda\left(\hat{\boldsymbol{J}}\right)\right)\boldsymbol{\omega}^T\boldsymbol{\omega}$$

$$-k_d\boldsymbol{\omega}^T\boldsymbol{\omega} + \frac{k_2}{2}\left(\lambda + \lambda\left(\hat{\boldsymbol{J}}\right)\right)|q_0|\|\boldsymbol{\omega}\|\boldsymbol{\omega}^T\,\boldsymbol{sgn}(\boldsymbol{s}) \le -\left(k_d - \frac{k_2}{2}\left(\lambda + \lambda\left(\hat{\boldsymbol{J}}\right)\right)\right)\boldsymbol{\omega}^T\boldsymbol{\omega} \tag{4.14}$$

Thus, if (4.15) is satisfied, the structure of controller (4.1) at slow down stage could be treated as the same structure of (4.12) with k_d' defined as follows:

$$\begin{aligned} k_d &> k_2\left(\lambda + \lambda\left(\hat{\boldsymbol{J}}\right)\right)/2 \\ k_d' &= k_d - \frac{k_2}{2}\left(\lambda + \lambda\left(\hat{\boldsymbol{J}}\right)\right) \end{aligned} \tag{4.15}$$

Therefore, based on Lemma 4.1, it is easy to find that the system is globally asymptotic stable at these two stages.

The following step is to prove the globally asymptotic stability of the system governed by the converge stage of controller (4.1). First, it is necessary to present Lemma 4.3.

Lemma 4.3 The system governed by controller described as (4.23) with control parameters claimed in previous section is globally asymptotic stable by selecting proper parameters.

$$\boldsymbol{u} = -k_d\boldsymbol{\omega} - k_p\boldsymbol{q}_v - k_I\boldsymbol{v} - \overline{d}\,\boldsymbol{sgn}(c\boldsymbol{\omega} + l_2\boldsymbol{q}_v) \tag{4.16}$$

Proof: Select Lyapunov function as follows:

$$V_2 = 2l_1(1-q_0) + \frac{1}{2}c\boldsymbol{\omega}^T\boldsymbol{J}\boldsymbol{\omega} + l_2\boldsymbol{q}_v^T\boldsymbol{J}\boldsymbol{\omega} + l_3(\boldsymbol{\upsilon}+c_3\boldsymbol{q}_v)^T(\boldsymbol{\upsilon}+c_3\boldsymbol{q}_v) \tag{4.17}$$

where l_1, l_2, l_3, c, c_3 are all positive scalars.

Considering that

$$1-q_0 \geq \frac{1}{2}\left(1-q_0^2\right) = \frac{1}{2}\boldsymbol{q}_v^T\boldsymbol{q}_v, \quad \forall q_0 \in [-1,1] \tag{4.18}$$

It can be got that

$$\begin{aligned} V_2 &\geq l_1\boldsymbol{q}_v^T\boldsymbol{q}_v + \frac{1}{2}c\boldsymbol{\omega}^T\boldsymbol{J}\boldsymbol{\omega} + l_2\boldsymbol{q}_v^T\boldsymbol{J}\boldsymbol{\omega} + l_3(\boldsymbol{\upsilon}+c_3\boldsymbol{q}_v)^T(\boldsymbol{\upsilon}+c_3\boldsymbol{q}_v) \\ &\geq l_1\boldsymbol{q}_v^T\boldsymbol{q}_v + \frac{1}{2}c\boldsymbol{\omega}^T\boldsymbol{J}\boldsymbol{\omega} + l_2\boldsymbol{q}_v^T\boldsymbol{J}\boldsymbol{\omega} = \begin{bmatrix}\boldsymbol{\omega}^T & \boldsymbol{q}_v^T\end{bmatrix}\begin{bmatrix}\frac{1}{2}c\lambda_{\min}(\boldsymbol{J})I_3 & \frac{1}{2}l_2\lambda(\boldsymbol{J})I_3 \\ \frac{1}{2}l_2\lambda(\boldsymbol{J})I_3 & l_1I_3\end{bmatrix}\begin{bmatrix}\boldsymbol{\omega} \\ \boldsymbol{q}_v\end{bmatrix} \end{aligned} \tag{4.19}$$

Thus, if the following equation is satisfied, V_2 is positively definite.

$$\frac{1}{2}cl_1\lambda_{\min}(\boldsymbol{J}) - \frac{1}{4}l_2^2\lambda^2(\boldsymbol{J}) > 0 \tag{4.20}$$

Calculate the derivative of V_2 and substitute the control law (4.16) it can be got that

$$\begin{aligned} \dot{V}_2 &= l_1\boldsymbol{\omega}^T\boldsymbol{q}_v + c\boldsymbol{\omega}^T(-\boldsymbol{\omega}^\times\boldsymbol{J}\boldsymbol{\omega}+\boldsymbol{u}+\boldsymbol{d}) + \frac{1}{2}l_2\boldsymbol{\omega}^T\boldsymbol{J}\boldsymbol{F}\boldsymbol{\omega} + l_2\boldsymbol{q}_v^T(-\boldsymbol{\omega}^\times\boldsymbol{J}\boldsymbol{\omega}+\boldsymbol{u}+\boldsymbol{d}) \\ &\quad + 2l_3(\boldsymbol{\upsilon}+c_3\boldsymbol{q}_v)^T\left(c_1\boldsymbol{\omega}+c_2\boldsymbol{q}_v - \frac{1}{2}c_3\boldsymbol{F}\boldsymbol{\omega} + \frac{1}{2}c_3\boldsymbol{F}\boldsymbol{\omega}\right) \\ &= \left(c\boldsymbol{\omega}^T + l_2\boldsymbol{q}_v^T\right)\left(-k_d\boldsymbol{\omega} - k_p\boldsymbol{q}_v - k_I\boldsymbol{\upsilon}\right) + \left(c\boldsymbol{\omega}^T + l_2\boldsymbol{q}_v^T\right)\left(-\bar{d}\,\mathbf{sgn}\left(c\boldsymbol{\omega}^T + l_2\boldsymbol{q}_v^T\right) + \boldsymbol{d}\right) \\ &\quad + \frac{1}{2}l_2\boldsymbol{\omega}^T\boldsymbol{J}\boldsymbol{F}\boldsymbol{\omega} - l_2\boldsymbol{q}_v^T\boldsymbol{\omega}^\times\boldsymbol{J}\boldsymbol{\omega} + 2l_3(\boldsymbol{\upsilon}+c_3\boldsymbol{q}_v)^T(c_1\boldsymbol{\omega}+c_2\boldsymbol{q}_v) + l_1\boldsymbol{\omega}^T\boldsymbol{q}_v \\ &\leq -\boldsymbol{\omega}^T\left(ck_d\boldsymbol{I}_3 - \frac{1}{2}l_2\boldsymbol{J}\boldsymbol{F}\right)\boldsymbol{\omega} - \boldsymbol{q}_v^T\left(l_2k_p\boldsymbol{I}_3 - 2c_2c_3l_3\boldsymbol{I}_3\right)q_v + (2c_1l_3 - ck_I)\boldsymbol{\upsilon}^T\boldsymbol{\omega} \\ &\quad + (2c_2l_3 - l_2k_I)\boldsymbol{\upsilon}^T\boldsymbol{q}_v + \left(l_1 - ck_p - l_2k_d + 2c_1c_3l_3\right)\boldsymbol{q}_v^T\boldsymbol{\omega} - l_2\boldsymbol{q}_v^T\boldsymbol{\omega}^\times\boldsymbol{J}\boldsymbol{\omega} \end{aligned} \tag{4.21}$$

Select proper parameters to satisfy the following equation

$$2l_3c_1 - ck_I = 2l_3c_2 - l_2k_I = l_1 - ck_p - l_2k_d + 2c_1c_3l_3 = 0 \tag{4.22}$$

Considering property (4.4), Eq. (4.28) could be transformed to

$$\dot{V}_2 \leq -\left[\boldsymbol{\omega}^T \boldsymbol{q}_v^T\right]\left[\begin{array}{cc} ck_d\boldsymbol{I}_3 - \frac{1}{2}l_2\lambda(\boldsymbol{J})\boldsymbol{I}_3 & \frac{1}{2}l_2\bar{\omega}\lambda(\boldsymbol{J})\boldsymbol{I}_3 \\ \frac{1}{2}l_2\bar{\omega}\lambda(\boldsymbol{J})\boldsymbol{I}_3 & l_2\left(k_p - c_3k_I\right)\boldsymbol{I}_3 \end{array}\right]\left[\begin{array}{c} \boldsymbol{\omega} \\ \boldsymbol{q}_v \end{array}\right] \tag{4.23}$$

where $\bar{\omega}$ is a positive scalar that satisfies $\bar{\omega} \geq \|\boldsymbol{\omega}\|$.

Thus, if the following equation is satisfied, $\dot{V}_2$ is negative definite.

$$l_2\left(ck_d - \frac{1}{2}l_2\lambda(\boldsymbol{J})\right)\left(k_2k_d - c_3k_I\right) - \frac{1}{4}l_2^2\bar{\omega}^2\lambda^2(\boldsymbol{J}) > 0 \tag{4.24}$$

Therefore, if (4.20) and (4.24) are satisfied, $V_2 \geq 0, \dot{V}_2 \leq 0$ could be guaranteed and the globally asymptotic stability is proved. Thus, Lemma 4.3 has been proved.

Comparing the structure of controller (4.1) at converge stage, the only difference is the q_0 term in (4.1). Similar as discussed in previous text, define k_d' as

$$k_d' = k_d - \left(k_2\left(\lambda + \lambda\left(\hat{\boldsymbol{J}}\right)\right)/2\right) \tag{4.25}$$

Thus, if following equation is satisfied, the system governed by controller (4.1) at the converge stage is globally asymptotic stable based on Lemma 4.3.

$$\begin{aligned} &\frac{1}{2}cl_1\lambda_{\min}(\boldsymbol{J}) - \frac{1}{4}l_2^2\lambda^2(\boldsymbol{J}) > 0 \\ &l_2\left(ck_d' - \frac{1}{2}l_2\lambda(\boldsymbol{J})\right)\left(k_2k_d' - c_3k_I\right) - \frac{1}{4}l_2^2\bar{\omega}^2\lambda^2(\boldsymbol{J}) > 0 \end{aligned} \tag{4.26}$$

Therefore, the stability of all the three stages of controller (4.1) has been proved, and the system governed by controller (4.1) is globally asymptotic stable.

4.2.2 Controller performance analysis

As describes in previous section, the goal of controller (4.1) is to achieve $\boldsymbol{s} \to \boldsymbol{0}$, i.e., the system reaches the sliding mode (4.4). In order to prove this property, select Lyapunov function as follow

$$V_3 = \frac{1}{2}\boldsymbol{s}^T\boldsymbol{J}\boldsymbol{s} \tag{4.27}$$

In this section, it is assumed that disturbance torque $\boldsymbol{d}$ is much less than the other terms in controller (4.1) thus $\boldsymbol{d}$ and $\overline{d}$ could be ignored in this section.

At the maneuver stage ($\|\boldsymbol{q}_v\| \geq \alpha$), calculate the derivative of V_3 it can be got that

$$\begin{aligned}
\dot{V}_3 &= \boldsymbol{s}^T\boldsymbol{J}\dot{\boldsymbol{s}} = \boldsymbol{s}^T\boldsymbol{J}\dot{\boldsymbol{\omega}} + k_1\boldsymbol{s}^T\boldsymbol{J}\dot{\boldsymbol{e}} = \boldsymbol{s}^T\left(\boldsymbol{u} + \boldsymbol{d} - \boldsymbol{\omega}^\times\boldsymbol{J}\boldsymbol{\omega}\right) + \frac{k_1}{2}\boldsymbol{s}^T\boldsymbol{J}\boldsymbol{e}^\times\left(\boldsymbol{I}_3 - \cot\frac{\varphi}{2}\boldsymbol{e}^\times\right)\boldsymbol{\omega} \\
&\approx \boldsymbol{s}^T\left(-k_d\boldsymbol{\omega} - (k_1k_d/\|\boldsymbol{q}_v\|)\boldsymbol{q}_v + \boldsymbol{\omega}^\times\boldsymbol{r}_1 - \boldsymbol{\omega}^\times\boldsymbol{J}\boldsymbol{\omega}\right) + \frac{k_1}{2}\boldsymbol{s}^T\boldsymbol{J}\boldsymbol{e}^\times\left(\boldsymbol{I}_3 - \cot\frac{\varphi}{2}\boldsymbol{e}^\times\right)\boldsymbol{\omega} \\
&= -k_d\boldsymbol{s}^T(\boldsymbol{\omega} + k_1\boldsymbol{e}) + \boldsymbol{s}^T\boldsymbol{\omega}^\times\boldsymbol{r}_1 - \boldsymbol{s}^T\boldsymbol{\omega}^\times\boldsymbol{J}\boldsymbol{\omega} + \frac{k_1}{2}\boldsymbol{s}^T\boldsymbol{J}\boldsymbol{e}^\times\boldsymbol{\omega} - \frac{k_1}{2}\cot\frac{\varphi}{2}\boldsymbol{s}^T\boldsymbol{J}\boldsymbol{e}^\times\boldsymbol{e}^\times\boldsymbol{\omega} \\
&= -k_d\boldsymbol{s}^T\boldsymbol{s} + k_1\boldsymbol{e}^T\boldsymbol{\omega}^\times\boldsymbol{r}_1 - k_1\boldsymbol{e}^T\boldsymbol{\omega}^\times\boldsymbol{J}\boldsymbol{\omega} + \frac{k_1}{2}\boldsymbol{e}^T\boldsymbol{\omega}^\times\boldsymbol{J}\boldsymbol{s} + \frac{k_1}{2}\cot\frac{\varphi}{2}\boldsymbol{e}^T\boldsymbol{\omega}^\times\boldsymbol{e}^\times\boldsymbol{J}\boldsymbol{s} \\
&= -k_d\boldsymbol{s}^T\boldsymbol{s} - k_1\lambda_2\left(\|\boldsymbol{\omega}\| + \frac{1}{2}\left(1 + \cot\frac{\varphi}{2}\right)\|\boldsymbol{s}\|\right)\boldsymbol{e}^T\boldsymbol{\omega}^\times\,\mathbf{sgn}\left(\left(\boldsymbol{e}^T\boldsymbol{\omega}^\times\right)^T\right) \\
&\quad + k_1\boldsymbol{e}^T\boldsymbol{\omega}^\times\left(-\widetilde{\boldsymbol{J}}\boldsymbol{\omega} + \frac{1}{2}\widetilde{\boldsymbol{J}}\boldsymbol{s} + \frac{1}{2}\cot\frac{\varphi}{2}\boldsymbol{e}^\times\widetilde{\boldsymbol{J}}\boldsymbol{s}\right) \\
&\leq -k_d\boldsymbol{s}^T\boldsymbol{s} \leq 0
\end{aligned} \tag{4.28}$$

where property (4.8) is used in (4.28).

At the slow down stage ($\alpha > \|\boldsymbol{q}_v\| \geq \beta$), calculate the derivative of V_3 it can be got that

$$\begin{aligned}
\dot{V}_3 &= \boldsymbol{s}^T\boldsymbol{J}\dot{\boldsymbol{s}} = \boldsymbol{s}^T\boldsymbol{J}\dot{\boldsymbol{\omega}} + k_2\boldsymbol{s}^T\boldsymbol{J}\dot{\boldsymbol{q}}_v = \boldsymbol{s}^T\left(\boldsymbol{u} + \boldsymbol{d} - \boldsymbol{\omega}^\times\boldsymbol{J}\boldsymbol{\omega}\right) + \frac{k_2}{2}\boldsymbol{s}^T\boldsymbol{J}\left(q_0\boldsymbol{I}_3 + \boldsymbol{q}_v^\times\right)\boldsymbol{\omega} \\
&\approx \boldsymbol{s}^T\left(-k_d\boldsymbol{\omega} - k_2k_d\boldsymbol{q}_v + \boldsymbol{\omega}^\times\boldsymbol{r}_2 - \frac{k_2}{2}\lambda_1|q_0|\|\boldsymbol{\omega}\|\,\mathbf{sgn}(\boldsymbol{s}) - \boldsymbol{\omega}^\times\boldsymbol{J}\boldsymbol{\omega}\right) \\
&\quad + \frac{k_2}{2}q_0\boldsymbol{s}^T\boldsymbol{J}\boldsymbol{\omega} + \frac{k_2}{2}\boldsymbol{s}^T\boldsymbol{J}\boldsymbol{q}_v^\times\boldsymbol{\omega} \\
&= -k_d\boldsymbol{s}^T(\boldsymbol{\omega} + k_2\boldsymbol{q}_v) + \boldsymbol{s}^T\boldsymbol{\omega}^\times\boldsymbol{r}_2 - \frac{k_2}{2}\lambda_1|q_0|\|\boldsymbol{\omega}\|\boldsymbol{s}^T\,\mathbf{sgn}(\boldsymbol{s}) - \boldsymbol{s}^T\boldsymbol{\omega}^\times\boldsymbol{J}\boldsymbol{\omega} \\
&\quad + \frac{k_2}{2}q_0\boldsymbol{s}^T\boldsymbol{J}\boldsymbol{\omega} - \frac{k_2}{2}\boldsymbol{s}^T\boldsymbol{J}\boldsymbol{q}_v^\times\boldsymbol{\omega} \\
&= -k_d\boldsymbol{s}^T\boldsymbol{s} + k_2\boldsymbol{q}_v^T\boldsymbol{\omega}^\times\boldsymbol{r}_1 - \frac{k_2}{2}\lambda_1|q_0|\|\boldsymbol{\omega}\|\boldsymbol{s}^T\,\mathbf{sgn}(\boldsymbol{s}) - k_2\boldsymbol{q}_v^T\boldsymbol{\omega}^\times\boldsymbol{J}\boldsymbol{\omega} \\
&\quad + \frac{k_2}{2}q_0\boldsymbol{s}^T\boldsymbol{J}\boldsymbol{\omega} - \frac{k_2}{2}\boldsymbol{q}_v^T\boldsymbol{\omega}^\times\boldsymbol{J}\boldsymbol{s} \\
&= -k_d\boldsymbol{s}^T\boldsymbol{s} - k_2\lambda_2\left(\|\boldsymbol{\omega}\| + \frac{1}{2}\|\boldsymbol{s}\|\right)\boldsymbol{q}_v^T\boldsymbol{\omega}^\times\,\mathbf{sgn}^T\left(\boldsymbol{q}_v^T\boldsymbol{\omega}^\times\right) \\
&\quad + k_2\boldsymbol{q}_v^T\boldsymbol{\omega}^\times\left(-\widetilde{\boldsymbol{J}}\boldsymbol{\omega} + \frac{1}{2}\widetilde{\boldsymbol{J}}\boldsymbol{s}\right) + \frac{k_2}{2}q_0\boldsymbol{s}^T\boldsymbol{J}\boldsymbol{\omega} - \frac{k_2}{2}\lambda_1|q_0|\|\boldsymbol{\omega}\|\boldsymbol{s}^T\,\mathbf{sgn}(\boldsymbol{s}) \\
&\leq -k_d\boldsymbol{s}^T\boldsymbol{s} \leq 0
\end{aligned} \tag{4.29}$$

At the converge stage ($\beta > \|\boldsymbol{q}_v\|$), calculate the derivative of V_3 it can be got that

$$\begin{aligned}
\dot{V}_3 &= \boldsymbol{s}^T\boldsymbol{J}\dot{\boldsymbol{s}} = \boldsymbol{s}^T\boldsymbol{J}\dot{\boldsymbol{\omega}} + k_2\boldsymbol{s}^T\boldsymbol{J}\dot{\boldsymbol{q}}_v = \boldsymbol{s}^T(\boldsymbol{u} + \boldsymbol{d} - \boldsymbol{\omega}^\times\boldsymbol{J}\boldsymbol{\omega}) + \frac{k_2}{2}\boldsymbol{s}^T\boldsymbol{J}(q_0\boldsymbol{I}_3 + \boldsymbol{q}_v^\times)\boldsymbol{\omega} \\
&\approx \boldsymbol{s}^T\left(-k_d\boldsymbol{\omega} - k_2k_d\boldsymbol{q}_v - k_I\boldsymbol{v} + (c\boldsymbol{\omega} + l_2\boldsymbol{q}_v)^\times\boldsymbol{r}_3 - \frac{k_2}{2}\lambda_1|q_0|\|\omega\|\,\boldsymbol{sgn}(\boldsymbol{s}) - \boldsymbol{\omega}^\times\boldsymbol{J}\boldsymbol{\omega}\right) \\
&\quad + \frac{k_2}{2}q_0\boldsymbol{s}^T\boldsymbol{J}\boldsymbol{\omega} + \frac{k_2}{2}\boldsymbol{s}^T\boldsymbol{J}\boldsymbol{q}_v^\times\boldsymbol{\omega} \\
&= -k_d\boldsymbol{s}^T(\boldsymbol{\omega} + k_2\boldsymbol{q}_v) - k_I\boldsymbol{s}^T\boldsymbol{v} + \boldsymbol{s}^T(c\boldsymbol{\omega} + l_2\boldsymbol{q}_v)^\times\boldsymbol{r}_3 - \frac{k_2}{2}\lambda_1|q_0|\|\omega\|\boldsymbol{s}^T\,\boldsymbol{sgn}(\boldsymbol{s}) \\
&\quad -\boldsymbol{s}^T\boldsymbol{\omega}^\times\boldsymbol{J}\boldsymbol{\omega} + \frac{k_2}{2}q_0\boldsymbol{s}^T\boldsymbol{J}\boldsymbol{\omega} + \frac{k_2}{2}\boldsymbol{s}^T\boldsymbol{J}\boldsymbol{q}_v^\times\boldsymbol{\omega} \\
&= -k_d\boldsymbol{s}^T\boldsymbol{s} - k_I\boldsymbol{s}^T\boldsymbol{v} + c\boldsymbol{s}^T\left(\boldsymbol{s} + \left(\frac{l_2}{c} - k_2\right)\boldsymbol{q}_v\right)^\times r_3 - \frac{k_2}{2}\lambda_1|q_0|\|\omega\|\boldsymbol{s}^T\,\boldsymbol{sgn}(\boldsymbol{s}) \\
&\quad -k_2\boldsymbol{q}_v^T\boldsymbol{\omega}^\times\boldsymbol{J}\boldsymbol{\omega} + \frac{k_2}{2}q_0\boldsymbol{s}^T\boldsymbol{J}\boldsymbol{\omega} + \frac{k_2}{2}\boldsymbol{q}_v^T\boldsymbol{\omega}^\times\boldsymbol{J}\boldsymbol{s} \\
&= -k_d\boldsymbol{s}^T\boldsymbol{s} - k_I\boldsymbol{s}^T\boldsymbol{v} + (ck_2 - l_2)\boldsymbol{q}_v^\times\boldsymbol{r}_3 - \frac{k_2}{2}\lambda_1|q_0|\|\omega\|\boldsymbol{s}^T\,\boldsymbol{sgn}(\boldsymbol{s}) \\
&\quad -k_2\boldsymbol{q}_v^T\boldsymbol{\omega}^\times\boldsymbol{J}\boldsymbol{\omega} + \frac{k_2}{2}q_0\boldsymbol{s}^T\boldsymbol{J}\boldsymbol{\omega} + \frac{k_2}{2}\boldsymbol{q}_v^T\boldsymbol{\omega}^\times\boldsymbol{J}\boldsymbol{s} \\
&\le -k_d\boldsymbol{s}^T\boldsymbol{s} - k_I\boldsymbol{s}^T\boldsymbol{v}
\end{aligned} \tag{4.30}$$

Considering that the globally asymptotic stability of the system, it is reasonable to assume following assumption during the converge process with positive scalar ρ.

$$\|\boldsymbol{s}\| \ge \rho\|\boldsymbol{v}\| \tag{4.31}$$

Select proper parameter to satisfy following equation:

$$k_d \ge \rho k_I \tag{4.32}$$

Thus, $\dot{V}_3 \le 0$ during the converge process.

Based on (4.28)–(4.30), it can be found that $s \to 0$, i.e., the system state reaches the sliding mode. Considering that the convergence rate of sliding mode (4.4) is constant at maneuver stage and exponential at slow down stage and converge stage, the convergence rate of the system could be improved and estimated by selecting proper parameter.

4.3 Angular velocity and control torque constraint

The control torque of satellite is finite, thus it is necessary to guarantee the norm of control torque do not exceed the system upper bound. Moreover, considering that the issue brought by excessive angular velocity, the norm of angular velocity should also be discussed in the design of controller. The goal of this section is to modify the controller presented in the previous section to satisfy the angular velocity and control torque constraint.

The relationship between norm of angular velocity and control parameter is first discussed in this section. Considering that the excessive angular velocity occurs at the initial stage of the system, thus only the maneuver stage is discussed. It is worth noticing that at this stage disturbance torque $\boldsymbol{d}$ and $\overline{d}$ could be ignored. In order to discuss the norm angular velocity, it is necessary to present a lemma.

Lemma 4.4 If the controller governing system (2.1) has the following structure with any three-dimensional vector $\boldsymbol{r}$,

$$\boldsymbol{u} = -k_d\boldsymbol{\omega} - k_p\boldsymbol{q}_v + \boldsymbol{\omega}^{\times}\boldsymbol{r} \tag{4.33}$$

the norm of angular velocity satisfies

$$\|\boldsymbol{\omega}\| \le \frac{k_p}{k_d}\frac{\lambda_{\max}(\boldsymbol{J})}{\lambda_{\min}(\boldsymbol{J})} \tag{4.34}$$

Proof: Define σ as

$$\sigma = \boldsymbol{\omega}^T\boldsymbol{J}\boldsymbol{\omega} \tag{4.35}$$

It is easy to find that

$$\begin{gathered}\lambda_{\min}(\boldsymbol{J})\|\boldsymbol{\omega}\|^2 \le \sigma \le \lambda_{\max}(\boldsymbol{J})\|\boldsymbol{\omega}\|^2 \\ \sqrt{\lambda_{\min}(\boldsymbol{J})}\|\boldsymbol{\omega}\| \le \sigma^{1/2} \le \sqrt{\lambda_{\max}(\boldsymbol{J})}\|\boldsymbol{\omega}\|\end{gathered} \tag{4.36}$$

Calculate the derivative of σ it can be got that

$$\begin{aligned}\dot{\sigma} = 2\boldsymbol{\omega}^T\boldsymbol{J}\dot{\boldsymbol{\omega}} &= 2\boldsymbol{\omega}^T\left(-k_d\boldsymbol{\omega} - k_p\boldsymbol{q}_v + \boldsymbol{\omega}^{\times}\boldsymbol{r} - \boldsymbol{\omega}^{\times}\boldsymbol{J}\boldsymbol{\omega}\right) \\ &= -2k_d\boldsymbol{\omega}^T\boldsymbol{\omega} - 2k_p\boldsymbol{\omega}^T\boldsymbol{q}_v \\ &\le -2k_d\frac{\sigma}{\lambda_{\max}(\boldsymbol{J})} + 2k_p\frac{\sigma^{1/2}}{\sqrt{\lambda_{\min}(\boldsymbol{J})}}\end{aligned} \tag{4.37}$$

Consider system

$$\dot{\sigma} = -2k_d\frac{\sigma}{\lambda_{\max}(\boldsymbol{J})} + 2k_p\frac{\sigma^{1/2}}{\sqrt{\lambda_{\min}(\boldsymbol{J})}} \tag{4.38}$$

The solution of (4.37) and (4.38) satisfies following equation with any initial value satisfies $\sigma_{initial} \le \left(k_d\lambda_{\max}(\boldsymbol{J})/k_p\sqrt{\lambda_{\min}(\boldsymbol{J})}\right)^2$

$$\sigma^{1/2} \le \frac{k_d}{k_p}\frac{\lambda_{\max}(\boldsymbol{J})}{\sqrt{\lambda_{\min}(\boldsymbol{J})}} \tag{4.39}$$

Noticing that

$$\sqrt{\lambda_{\min}(\boldsymbol{J})}\|\boldsymbol{\omega}\| \leq \sigma^{1/2} \tag{4.40}$$

Substitute (4.40) into (4.39), it can be got that

$$\|\boldsymbol{\omega}\| \leq \frac{k_p}{k_d}\frac{\lambda_{\max}(\boldsymbol{J})}{\lambda_{\min}(\boldsymbol{J})} \tag{4.41}$$

Thus Lemma 4.4 has been proved.

Based on Lemma 4.4, it can be found that the norm of angular velocity could be ensured by selecting proper control parameter. It is worth noticing that the method presented in this chapter is conservative that means it leaves much redundancy for the system. Noticing that the derivative term gain factor is locked in controller (4.1), and proportion term gain factor is variable, thus lessen the proportion term gain factor when the angular velocity approaches the system upper bound could satisfy the angular velocity constraint. Noticing that

$$\begin{aligned}\dot{\sigma} &= -2k_d\boldsymbol{\omega}^T\boldsymbol{\omega} - 2k_p\boldsymbol{\omega}^T\boldsymbol{q}_v \\ &\leq -2k_d\|\boldsymbol{\omega}\|^2 + 2k_p\|\boldsymbol{\omega}\|\|\boldsymbol{q}_v\|\end{aligned} \tag{4.42}$$

Thus, if following equation is satisfied, $\dot{\sigma} \leq 0$ could be ensured.

$$\frac{k_p}{k_d} \leq \frac{\|\boldsymbol{\omega}\|}{\|\boldsymbol{q}_v\|} \tag{4.43}$$

Assume that the upper bound of angular velocity norm is $\bar{\omega}$. Substitute the proportional term in (4.43), it can be found that

$$k_1 \leq \bar{\omega} \tag{4.44}$$

If (4.44) is satisfied, $\dot{\sigma} \leq 0$could be ensured. Generally, $\dot{\sigma} \leq 0$could be treated as the decreasing of $\|\boldsymbol{\omega}\|$ except for some extreme conditions.

The next step is to modify controller (4.1) to satisfy the constraint of control torque. First, rewrite controller (4.1) as follows:

$$\boldsymbol{u} = \begin{cases} \rho_1\varsigma_1 + \boldsymbol{\tau}_1 & \|\boldsymbol{q}_v\| \geq \alpha \\ \rho_2\varsigma_2 + \boldsymbol{\tau}_2 & \alpha > \|\boldsymbol{q}_v\| \geq \beta \\ \rho_3\varsigma_3 + \boldsymbol{\tau}_3 & \beta > \|\boldsymbol{q}_v\| \end{cases} \tag{4.45}$$

where

$$\begin{aligned}\boldsymbol{\tau}_1 &= -\bar{d}\boldsymbol{sgn}(\boldsymbol{\omega}) + \boldsymbol{\omega}^\times\boldsymbol{r}_1 \\ \boldsymbol{\tau}_2 &= -\bar{d}\boldsymbol{sgn}(\boldsymbol{\omega}) + \boldsymbol{\omega}^\times\boldsymbol{r}_2 - \frac{k_2}{2}\left(\lambda + \lambda\left(\widehat{\boldsymbol{J}}\right)\right)|q_0|\|\omega\|\,\boldsymbol{sgn}(\boldsymbol{s}) \\ \boldsymbol{\tau}_3 &= -\bar{d}\boldsymbol{sgn}(c\boldsymbol{\omega} + l_2\boldsymbol{q}_v) + (c\boldsymbol{\omega} + l_2\boldsymbol{q}_v)^\times\boldsymbol{r}_3 - \frac{k_2}{2}\left(\lambda + \lambda\left(\widehat{\boldsymbol{J}}\right)\right)|q_0|\|\omega\|\,\boldsymbol{sgn}(\boldsymbol{s})\end{aligned} \tag{4.46}$$

$$\begin{aligned} \varsigma_1 &= -k_d\boldsymbol{\omega} - \gamma_1(k_1k_d/\|\boldsymbol{q}_v\|)\boldsymbol{q}_v, \\ \varsigma_2 &= -k_d\boldsymbol{\omega} - \gamma_2k_2k_d\boldsymbol{q}_v \\ \varsigma_3 &= -k_d\boldsymbol{\omega} - k_2k_d\boldsymbol{q}_v - k_I\boldsymbol{v} \end{aligned} \tag{4.47}$$

and ρ_i, γ_i are gain factors to be determined.

In order to guarantee the norm of angular velocity do not exceed the system upper bound, select γ_i as follows:

$$\gamma_1 = \begin{cases} 1 & \|\boldsymbol{\omega}\| \le \bar{\omega} - \varepsilon \\ \bar{\omega}\dfrac{\lambda_m}{\lambda_M}\dfrac{\|\boldsymbol{q}_v\|}{k_1} & \|\boldsymbol{\omega}\| > \bar{\omega} - \varepsilon \end{cases}, \quad \gamma_2 = \begin{cases} 1 & \|\boldsymbol{\omega}\| \le \bar{\omega} - \varepsilon \\ \bar{\omega}\dfrac{\lambda_m}{\lambda_M}\dfrac{1}{k_2} & \|\boldsymbol{\omega}\| > \bar{\omega} - \varepsilon \end{cases} \tag{4.48}$$

where ε is a small positive scalar and λ_m, λ_M satisfies

$$\lambda_m \le \lambda_{\min}(\boldsymbol{J}), \quad \lambda_M \ge \lambda_{\max}(\boldsymbol{J}) \tag{4.49}$$

Based on (4.48) and Lemma 4.4, the norm of angular velocity could be ensured less than the system upper bound. However, the cost is that the system may deviate the sliding mode (4.4). It is worth noticing that Eq. (4.48) is a strict constraint, and (4.44) could ensure the angular velocity constraint for most common conditions.

The next step is to select proper ρ_i to solve the control torque saturation issue. Assume that the maximum norm of control torque of the system is $\overline{u}$. Select proper parameters to ensure that $\|\boldsymbol{\tau}_i\| \le \overline{u}$. Define ρ_i as follows:

$$\rho_i = \begin{cases} \dfrac{-\varsigma_i^T\boldsymbol{\tau}_i + \sqrt{\left(\varsigma_i^T\boldsymbol{\tau}_i\right)^2 + \overline{u}^2 - \|\boldsymbol{\tau}_i\|^2}}{\|\varsigma_i\|^2} & \|\rho_i\varsigma_i + \boldsymbol{\tau}_i\| \ge \overline{u} \\ 1 & \|\rho_i\varsigma_i + \boldsymbol{\tau}_i\| < \overline{u} \end{cases} \tag{4.50}$$

where η is a small positive scalar.

Based on (4.50), it can be found that the control torque saturation issue is solved by lessen the PID control gain vectors proportionally. The next step is to prove the globally asymptotic stability of the controller developed in this section.

Based on Lemmas 4.1 and 4.2, it is easy to find that PD plus controller is globally asymptotic stable with any positive k_d and k_p. Thus, the maneuver stage and slow down stage are still stable. At converge stage, the parameter k_d'', k_I'' in this situation could be written as follows with positive scalar δ.

$$k_d'' = \delta k_d, \quad k_I'' = \delta k_I \tag{4.51}$$

Select $l_1'' = \delta^4 l_1$, $l_2'' = \delta^2 l_2$ and substitute l_i'', l_2'' and k_d'', k_p'', k_I'' in (4.33) it could be got that

$$cl_1'' - \frac{1}{2}\left(l_2''\right)^2\lambda_{\max}(J) = \delta^4\left(cl_1 - \frac{1}{2}l_2^2\lambda_{\max}(J)\right) > 0 \tag{4.52}$$

$$
\begin{aligned}
& l_2''\left(ck_d''-\frac{1}{2}l_2''\lambda(\boldsymbol{J})\right)\left(k_2k_d''-c_3k_I''\right)-\frac{1}{4}\left(l_2''\right)^2\bar{\omega}^2\lambda^2(\boldsymbol{J}) \\
& =\delta^4\left[l_2\left(ck_d-\frac{1}{2}l_2\lambda(\boldsymbol{J})\right)\left(k_2k_d-c_3k_I\right)-\frac{1}{4}l_2^2\bar{\omega}^2\lambda^2(\boldsymbol{J})\right] \\
& >0
\end{aligned}
\tag{4.53}
$$

Thus, in this situation, $V_2 \geq 0, \dot{V}_2 \leq 0$is still satisfied. Therefore, the system governed by the controller (4.45) is globally asymptotic stable for all conditions considered in this section.

4.4 Simulation results

The controller proposed in this chapter is compared with standard PID controller written as follows:

$$
\begin{aligned}
\boldsymbol{u} &= -k_d\boldsymbol{\omega}-k_p\boldsymbol{q}_v-k_I\boldsymbol{v}-\bar{d}\,\boldsymbol{sgn}\left(c\boldsymbol{\omega}+l_2\boldsymbol{q}_v\right) \\
\dot{\boldsymbol{v}} &= c_1\boldsymbol{\omega}+c_2\boldsymbol{q}_v-\frac{1}{2}c_3\boldsymbol{F}\boldsymbol{\omega}
\end{aligned}
\tag{4.54}
$$

Considering that the constraint on angular velocity of star tracker is at the level of $5-10$ degree/s, and the maneuver rate of agile satellite is at the level of $3-5$ degree/s, set the system parameters as follows:

$$
\begin{aligned}
& \boldsymbol{J}=\operatorname{diag}([25\ \ 20\ \ 15])\mathrm{kg}\cdot\mathrm{m}^2, \widehat{\boldsymbol{J}}=\operatorname{diag}([24\ \ 21\ \ 16])\mathrm{kg}\cdot\mathrm{m}^2, \lambda=2 \\
& \bar{d}=0.001\,\mathrm{N}\cdot\mathrm{m}, \bar{u}=0.5\mathrm{N}\cdot\mathrm{m}, \bar{\omega}=0.1\,\mathrm{rad/s} \\
& \boldsymbol{\omega}_{\text{initial}}=[0.07\ \ -0.05\ \ -0.04]^T\,\mathrm{rad/s}, \boldsymbol{q}_{\text{initial}}=[0.4\ \ 0.2\ \ 0.4\ \ -0.8]^T
\end{aligned}
\tag{4.55}
$$

Based on the loose constraint of angular velocity (4.441), select k_1, k_2, α, β as follows:

$$
\alpha=0.2, \quad \beta=0.05, \quad k_1=0.1, \quad k_2=0.25 \tag{4.56}
$$

Select other parameters as follows:

$$
\begin{aligned}
& k_d=20, \quad k_p=5, \quad k_I=1, \quad c=1, \quad l_2=0.01, \quad c_1=0.25, \quad c_2 \\
& =0.05, \quad c_3=1
\end{aligned}
\tag{4.57}
$$

Based on (4.22), it can be calculated that

$$
l_3=\frac{ck_I}{2c_1}=2, \quad l_1=ck_p+l_2k_d-2c_1c_3l_3=4.2 \tag{4.58}
$$

Based on (4.58), it can be found that

$$
\begin{aligned}
& k_d'=k_d-\frac{k_2}{2}\left(\lambda+\lambda\left(\widehat{\boldsymbol{J}}\right)\right)=16.75>0 \\
& \frac{1}{2}cl_1\lambda_{\min}(\boldsymbol{J})-\frac{1}{4}l_2^2\lambda^2(\boldsymbol{J})=\frac{63}{2}-\frac{1}{16}>0 \\
& \left(ck_d'-\frac{1}{2}l_2\lambda(\boldsymbol{J})\right)\left(k_2k_d'-c_3k_I\right)-\frac{1}{4}l_2\bar{\omega}^2\lambda^2(\boldsymbol{J})\approx 51>0
\end{aligned}
\tag{4.59}
$$

Hence, the constraints on control parameter (4.15) and (4.26) have been satisfied.

Assume that the disturbance torque consists of Gauss white noise and noise of sine function character, i.e.,

$$d_i = 5 \times 10^{-4}(\mathrm{rand}(1) + \sin t) \tag{4.60}$$

The simulation results of standard PID controller (4.54) are given in Figs. 4.1–4.4.

Based on Figs. 4.1–4.4, it can be found that the system converges to zero. However, the converge time is more than 200 s. According to Figs. 4.1 and 4.4, it can be found that the angular velocity at about 5 s is too large and more than the system upper bound. Afterward, the angular velocity decreases severely,

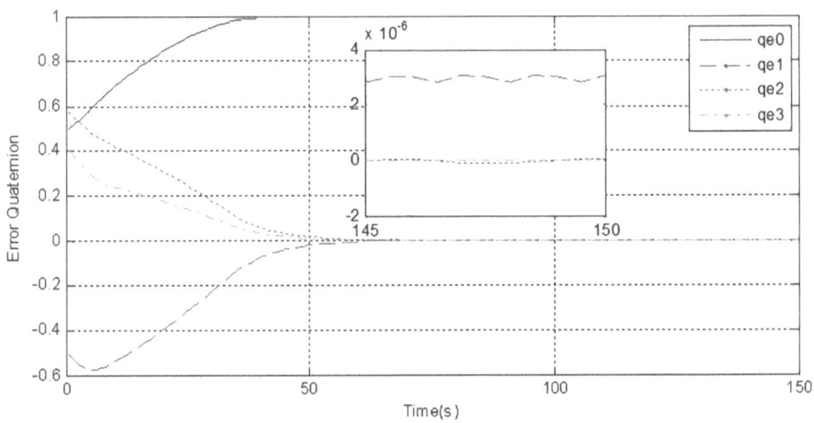

FIG. 4.1 Curve of angular velocity.

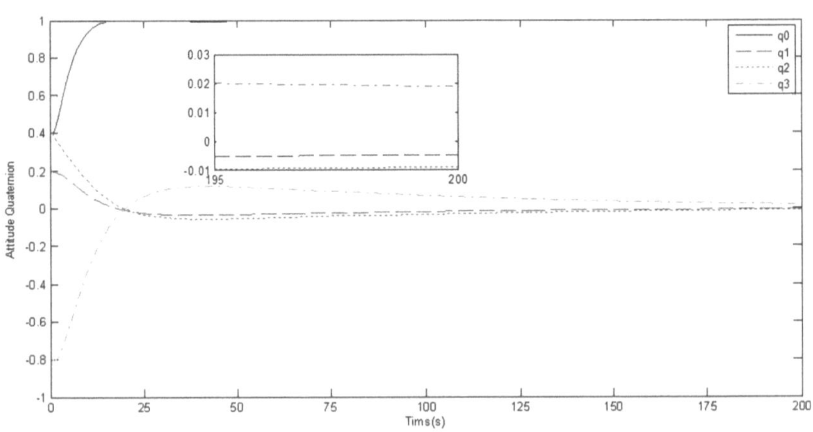

FIG. 4.2 Curve of attitude quaternion.

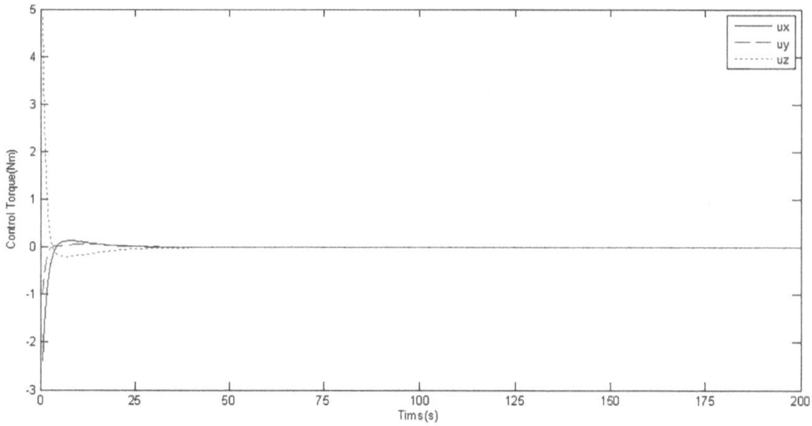

FIG. 4.3 Curve of control torque.

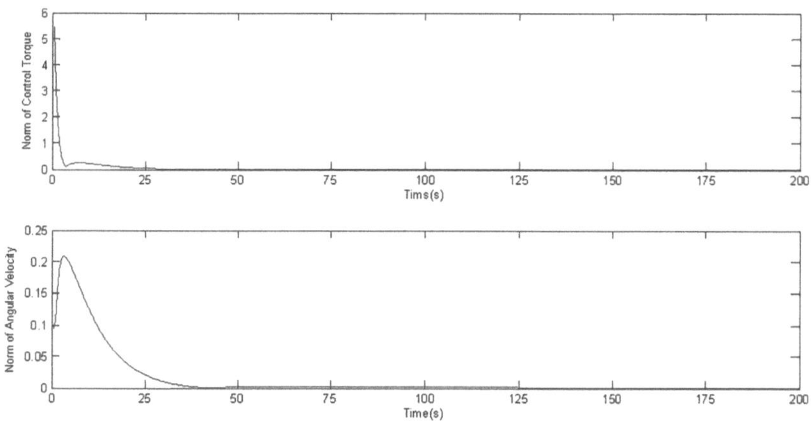

FIG. 4.4 Curve of norm of control torque and angular velocity.

thus the convergence rate of attitude quaternion decreases. This is the main cause of the low convergence rate of standard PID controller. Moreover, based on Fig. 4.4, it can be found that the initial control torque is too large and much more than the system upper bound and later than 5 s the control torque descends severely. This causes the low efficiency of control torque of standard PID controller.

The simulation results of time-efficient PID plus controller (4.1) without angular velocity and control torque constraint are given from Figs. 4.5–4.9.

From Figs. 4.5–4.9, it can be found that the system state converges to zero. The convergence time is about 50 s, which is much shorter than standard PID controller, and the accuracy at the steady stage is maintained at a relatively high level. Based on Fig. 4.8, it can be found that the system state reaches the sliding mode at about 5 s and moves along the sliding mode afterwards. This

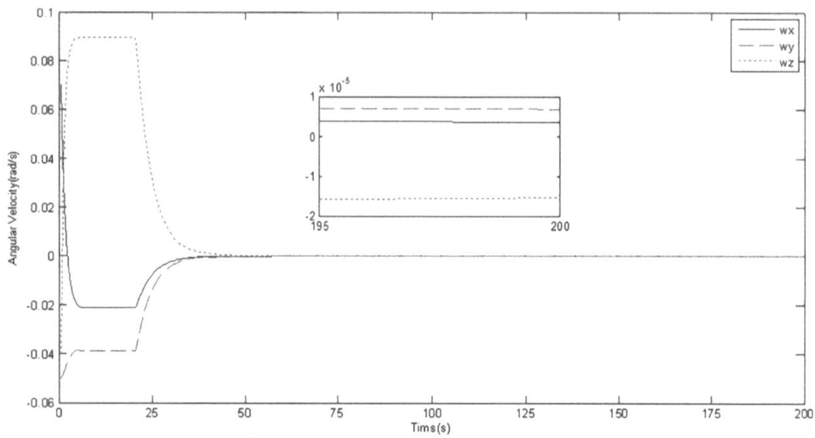

FIG. 4.5 Curve of angular velocity.

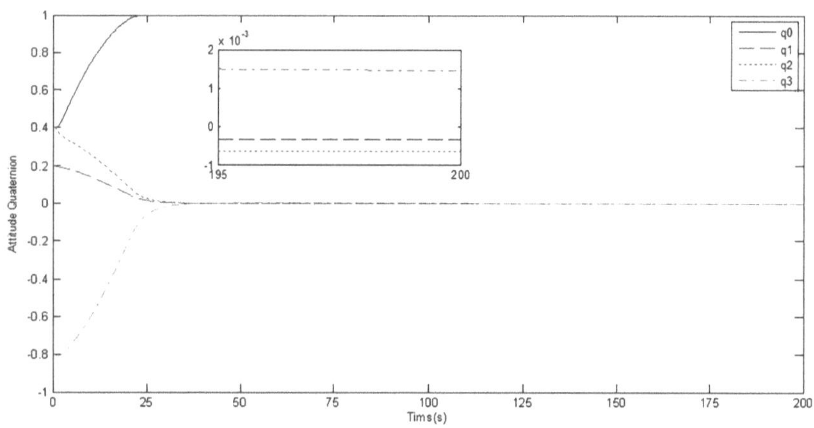

FIG. 4.6 Curve of attitude quaternion.

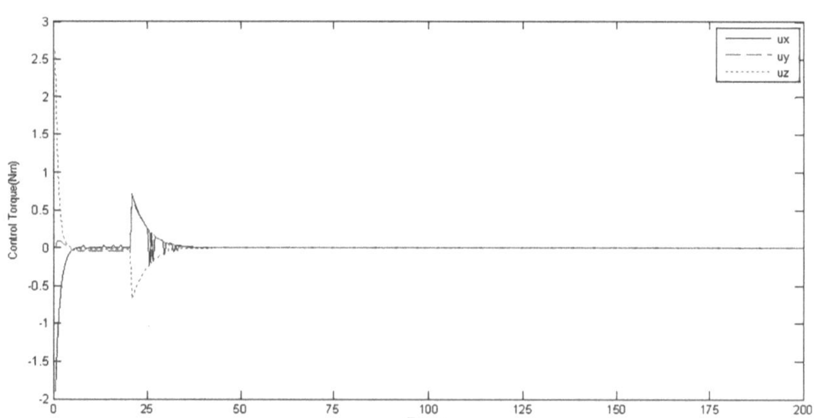

FIG. 4.7 Curve of control torque.

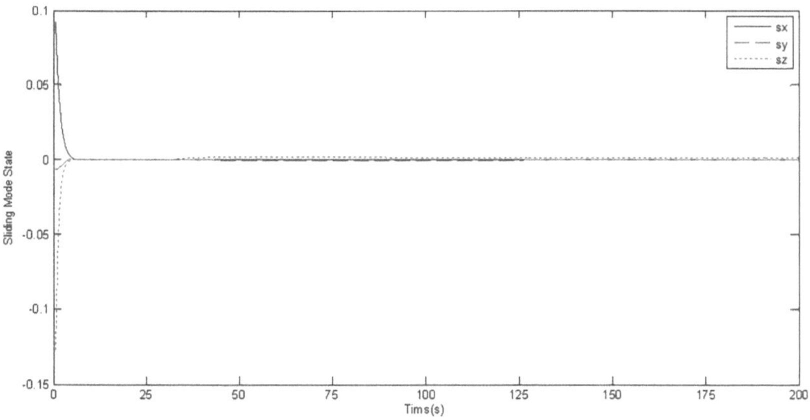

FIG. 4.8 Curve of sliding mode state.

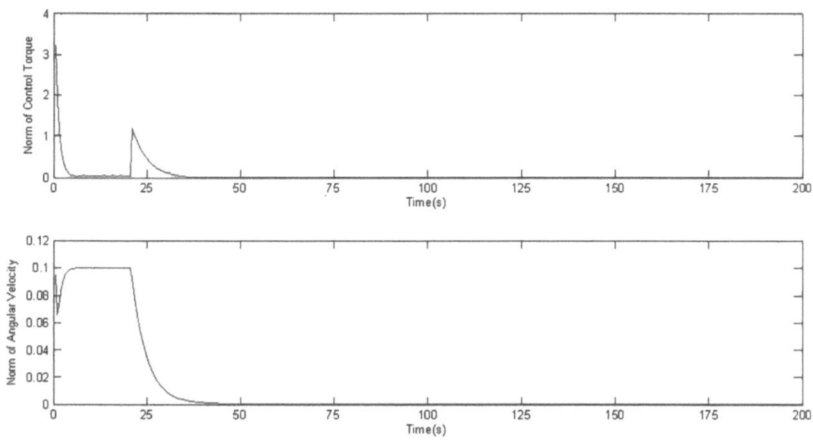

FIG. 4.9 Curve of norm of control torque and angular velocity.

demonstrates that the convergence rate could be controlled and estimated by designing proper sliding mode. Based on Fig. 4.9, it can be found that the norm of angular velocity does not exceed the system upper bound, and this proves that if (4.43), i.e., the loose constraint is satisfied, the system angular velocity could be ensured for most conditions and Eq. (4.48) is a strict constraint for extreme condition. Meanwhile based on Fig. 4.9, it could be found that the norm of control torque does not satisfy the system constraint.

The simulation results of time-efficient PID plus controller (4.44) under angular velocity and control torque constraint are given from Figs. 4.10–4.14.

From Figs. 4.10–4.14, it can be found that the system state converges to zero. The convergence time is also about 50 s, and the system steady accuracy is also maintained. Based on Fig. 4.13, it can be found that the system state

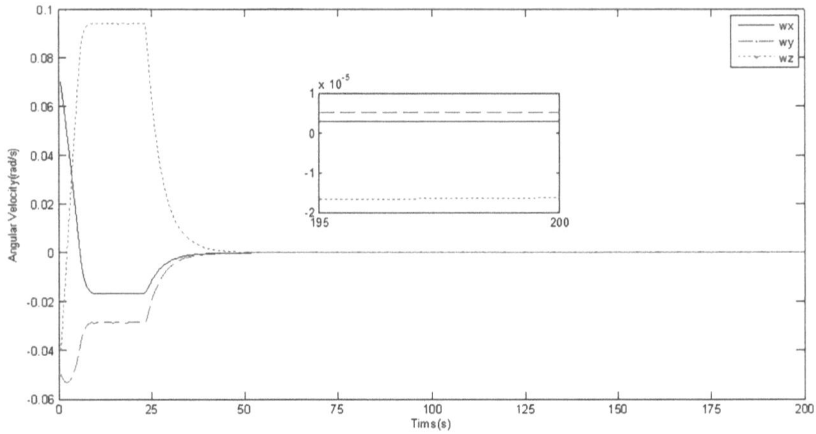

FIG. 4.10 Curve of angular velocity.

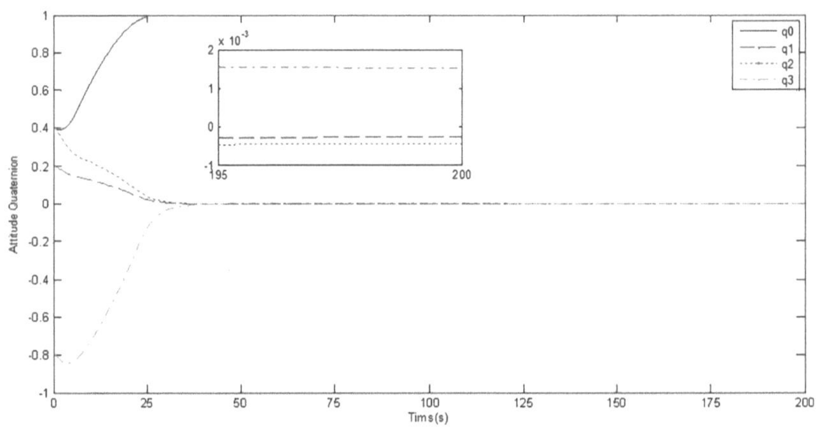

FIG. 4.11 Curve of attitude quaternion.

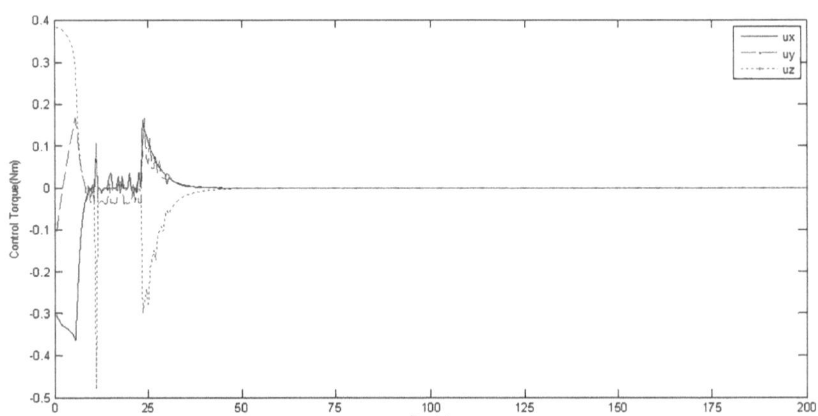

FIG. 4.12 Curve of control torque.

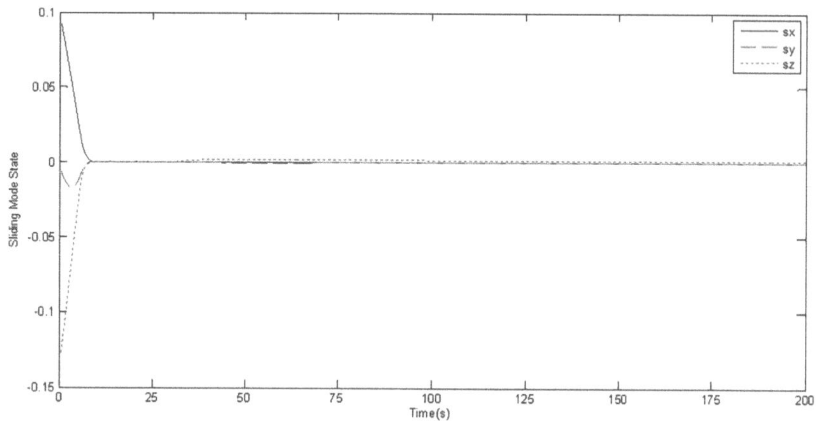

FIG. 4.13 Curve of sliding mode state.

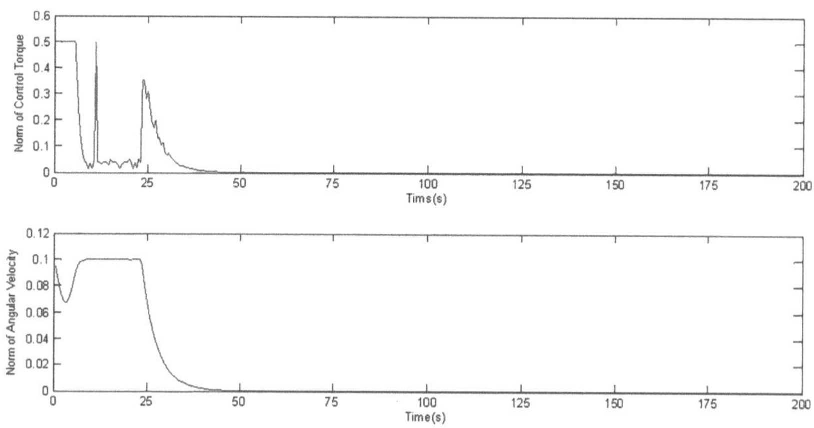

FIG. 4.14 Curve of norm of control torque and angular velocity.

reaches the sliding mode at about 10 s and moves along the sliding mode afterward. Based on Fig. 4.14, it can be found that the norm of angular velocity and control torque does not exceed the system upper bound.

According to Figs. 4.5–4.14, it can be concluded that the controller presented in this chapter has a better convergence rate with a high accuracy under angular velocity and control torque constraint.

4.5 Conclusion

This chapter presents a time-efficient robust PID plus controller for satellite attitude stabilization control with angular velocity and control torque constraint. A sliding mode with two-stage structure is proposed, and the convergence rate is improved by selecting proper sliding mode parameter. Generally, the longer the maneuver stage, the better the convergence rate of the system. The controller presented in this chapter works even if the system inertia matrix is not known accurately; hence, the system inertia matrix uncertainty is fixed. An estimation method for the maximum norm of angular velocity is proposed, and the relationship between the maximum value and system parameter is revealed. The control torque is ensured less than the system upper bound by the gain factor. Simulation results demonstrate the performance of the controller presented in this chapter.

It is worth noticing that although this chapter discusses the issues such as convergence rate, system inertia matrix uncertainty, and constraint of system state, researchers have pointed out that there are still important issues that need to be solved such as the control torque allocation, actuator failure, flexibility appendages, finite time stability, and so on; hence, further work could be done on these fields.

References

[1] E.D. Jin, X.L. Jiang, Z.W. Sun, Robust decentralized attitude coordination control of spacecraft formation, Syst. Control Lett. 57 (7) (2008) 567–577.

[2] E.D. Jin, Z.W. Sun, Robust attitude tracking control of flexible spacecraft for achieving globally asymptotic stability, Int. J. Robust Nonlinear Control 19 (11) (2009) 1201–1223.

[3] S. Kamesh, Nonlinear PID-like control for spacecraft attitude regulation, J. Astronaut. Sci. 52 (2) (2004) 61–74.

[4] F.H. Clarke, Y.S. Ledyaev, E.D. Songtag, Asymptotic controllability implies feedback stabilization, IEEE Trans. Autom. Control 42 (10) (1997) 1394–1407.

[5] B. Wie, D. Bailry, C. Heiberg, Rapid multitarget-target acquisition and pointing control of agile spacecraft, in: AIAA Guidance, Navigation, and Control Conference and Exihibit, 2000.

[6] B. Wie, P.M. Barba, Quaternion feedback for spacecraft large angle maneuvers, J. Guid. Control Dynam. 8 (3) (1985) 360–365.

[7] B. Wie, Q. Liu, J. Sunkel, Robust stabilization of the space station in the presence of the inertia matrix uncertainty, in: American Control Conference, 1993, pp. 2212–2216.

[8] B. Wie, J. Lu, Feedback control logic for spacecraft eigenaxis rotations under slew rate and control constraints, J. Guid. Control Dynam. 18 (6) (1995) 1372–1379.

[9] B. Wie, H. Weiss, A. Arapostathis, Quaternion feedback regulator for spacecraft eigenaxis rotations, J. Guid. Control Dynam. 12 (3) (1989) 375–380.

[10] Y.Y. Lu, Q.L. Hu, G.F. Ma, J.K. Zhou, Decentralized time-delay adaptive sliding mode control for attitude coordination formation under actuator misalignment, Acta Aeronaut. 32 (9) (2011) 1686–1695.

[11] Y.N. Yang, J.Z. Wu, W. Zheng, Attitude control for a station keeping airship using feedback linearization and fuzzy sliding mode control, Int. J. Innov. Comput. 8 (12) (2012) 8299–8310.

[12] J. Yang, X.P. Shi, Fault tolerant control and vibration suppression of satellite attitude control system via higher-order sliding mode control, Electr. Mach. Control 19 (10) (2015) 100–106.

[13] S.N. Wu, Z.G. Wu, G. Radice, R. Wang, Adaptive control for spacecraft relative translation with parametric uncertainty, Aerosp. Sci. Technol. 31 (1) (2012) 53–58.

[14] S.N. Wu, G. Radice, Z.W. Sun, Robust finite-time control for flexible spacecraft attitude maneuver, J. Aerosp. Eng. 27 (1) (2014) 185–190.

[15] H.Z. Liang, Z.W. Sun, J.Y. Wang, Robust decentralized attitude control for spacecraft formations under time-varying topologies, model uncertainties and disturbances, Acta Astronaut. 81 (2) (2012) 445–455.
[16] D. Verbin, V.J. Lappas, Rapid rotational maneuvering of rigid satellites with hybrid actuators configuration, J. Guid. Control Dynam. 36 (2) (2013) 532–547.
[17] D. Verbin, V.J. Lappas, Rapid rotational maneuvering of rigid satellites with reaction wheels, J. Guid. Control Dynam. 36 (5) (2013) 1538–1544.
[18] D. Verbin, Z.B. Joseph, Time efficient closed loop steering laws for rigid satellite large rotational maneuver, in: AIAA Guidance, Navigation, and Control Conference and Exhibit, 2005.
[19] D. Verbin, V.J. Lappas, Z.B. Joseph, Time efficient angular steering laws for rigid satellites, J. Guid. Control Dynam. 34 (3) (2011) 878–892.
[20] K.M. Ma, P. Ma, Robust control method of satellite attitude control, Acta Aeronaut. 25 (3) (2004) 289–292.
[21] L.N. Wang, Z. Li, B. Wang, Robust satellite attitude control, Adv. Intell. Soft Comput. 168 (1) (2012) 649–654.
[22] Q.L. Hu, Robust adaptive backstepping attitude and vibration control with L2 gain performance for flexible spacecraft under angular velocity constraint, J. Sound Vib. 327 (3) (2009) 285–298.
[23] Q.L. Hu, B. Li, Y.M. Zhang, Robust attitude control design for spacecraft under assigned velocity and control constraints, ISA Trans. 52 (4) (2013) 480–493.
[24] Q.L. Hu, L. Li, Anti-unwinding attitude control of spacecraft considering input saturation and angular velocity constraint, Acta Aeronaut. 36 (4) (2015) 1259–1266.
[25] Q. Hu, J.R. Zhang, Attitude control and vibration suppression for flexible spacecraft using control moment gyroscopes, J. Aerosp. Eng. 29 (1) (2016).

Chapter 5

Bang–bang logic-based time-efficient sliding mode control

5.1 Introduction

Sliding mode control is a mature and widely used control algorithm in satellite attitude control. The advantages of sliding mode control could be concluded as simple structure, clear physical meaning, and so on, and these advantages have attract lot of attention of researchers. Xiao et al. [1–3] designed robust sliding mode controllers for satellite attitude control considering perturbation such as system inertia matrix uncertainty, unknown disturbance torque, and actuator failure. The focus of his work is the system robustness and steady accuracy. Wu et al. [4–7] designed sliding mode controllers for satellite attitude stabilization and attitude tracking control. The chattering issue is discussed in his work, and the controller is designed under system uncertainty and disturbance torque. Verbin and Lappas [8,9] and Cao et al. [10] designed "braking curve" of angular velocity for satellite attitude control. The trajectory of angular velocity is optimized to improve system convergence rate. The focus of their work is the improvement of system convergence rate. Lo and Chen [11], Xia et al. [12], Levant and Michael [13], Tian et al. [14], Zong et al. [15], and Ernesto et al. [16] designed integral sliding mode controllers for satellite attitude control, flexible manipulator control, and aircraft attitude control. The focus of these works is the integral sliding mode that could improve system steady accuracy and relieve state chattering issue. Above all, sliding mode controller is a closed-loop controller that is robust to system perturbation when the sliding mode is properly designed.

Bang-bang control is a time-optimal control algorithm, which could fully utilize system capability. Lot of work focused on satellite attitude control has been done based on bang-bang logic. Bai and Junkins [17] proved that bang-bang character trajectory is time-optimal or near time-optimal trajectory for satellite attitude control under sphere control torque assumption (the maximum norm of control torque is fixed and the direction of control torque arbitrary). Nagi et al. [18], Zhang et al. [19], Gong et al. [20] optimized the trajectory of satellite attitude and orbit manoeuver based on bang-bang logic.

Fast Satellite Attitude Maneuver and Control. https://doi.org/10.1016/B978-0-323-95455-6.00006-8

Meier and Bryson [21,22] optimized system convergence rate by bang-bang logic for flexible manipulator control. It is worth noticing that most of these works need the accurate system parameter and the disturbance is not considered. Although bang-bang control algorithm has the advantage of time-optimal or near time-optimal, it is an open-loop control algorithm that means the system robustness to system uncertainty and disturbance is very bad. The controller would fail when system inertia matrix is not accurate known or there exists disturbance torque. Therefore, it is necessary to modify bang-bang control algorithm to improve its robustness.

Considering that the control torque of satellite control system is finite, it is necessary to ensure the norm of control torque does not exceed the system upper bound. Moreover, the excessive angular velocity would cause the descendent of attitude determination accuracy, in some extreme cases, even the failure of attitude determination system. Thus, it is necessary to consider the constraint of control torque and angular velocity when designing controllers.

In this chapter, open-loop bang-bang control algorithm is combined with closed-loop sliding mode control hence the system could both have near time-optimal character and strong robustness. The standard sliding mode is modified, and the sliding mode parameter is designed to be real-time updating so that the system trajectory could have bang-bang character. The convergence process is designed to be two stages: a maneuver stage with fixed angular velocity and a slow-down stage with fixed deceleration. The control torque constraint is discussed, and the constraint on control parameters is given. The system inertia matrix uncertainty and disturbance torque are also considered, and the controller is robust to the perturbation. The system stability is proved by Lyapunov method, and the performance of the controller is demonstrated by simulation results.

5.2 Time-efficient sliding mode controller

Standard sliding mode for satellite attitude control could be written as follows:

$$\boldsymbol{s} = \boldsymbol{\omega} + k\boldsymbol{q}_v, \quad (k > 0) \tag{5.1}$$

Based on the definition of $\boldsymbol{s}$, it could be found that the angular velocity is reversed to attitude quaternion when the system reaches sliding mode (5.1). Considering that

$$\dot{q}_0 = -\frac{1}{2}\boldsymbol{q}_v^T\boldsymbol{\omega} \le \frac{1}{2}\|\boldsymbol{q}_v\|\|\boldsymbol{\omega}\| \tag{5.2}$$

it could be found that the attitude quaternion has the maximum convergence rate with fixed angular velocity when $\boldsymbol{s}=\boldsymbol{0}$. Sliding mode (5.1) could be treated as a structure with the highest efficiency of the use of angular velocity, hence in order to improve the system convergence rate, angular velocity should be reversed to attitude quaternion on the sliding mode to be designed.

Although sliding mode (5.1) has clear physical meaning and the highest efficiency of angular velocity, the convergence rate of sliding mode (5.17) is exponential and need to be improved. The norm of angular velocity descends rapidly hence the convergence rate of attitude quaternion descends rapidly.

Bang-bang control is a time-optimal controller, and the system capability could be used to the utmost extend. However, bang-bang control is an open-loop controller, and the system state could not be feedback in the controller hence the robustness of system is bad. Considering that the essence of bang-bang logic is that the system state should be at the maximum velocity as long as possible and decelerate with maximum control output at proper position, the modified sliding mode based on bang-bang logic could be written as follows:

$$s = \begin{cases} \boldsymbol{\omega} + k_1 \boldsymbol{e} & \|\boldsymbol{q}_v\| \geq \alpha \\ \boldsymbol{\omega} + k_2 \boldsymbol{q}_v & \|\boldsymbol{q}_v\| < \alpha \end{cases} \tag{5.3}$$

$$\begin{cases} k_2(0) = k_1/\alpha \\ \dot{k}_2 = f(\boldsymbol{\omega}, \boldsymbol{q}_v) \end{cases} \tag{5.4}$$

where k_1, α are all positive scalars and $f(\boldsymbol{\omega}, \boldsymbol{q}_v)$ is parameter update law to be designed.

Based on (5.3), it could be found that the angular velocity is reversed to attitude quaternion during the whole converge process hence the advantage of standard sliding mode could be maintained. Sliding mode (5.3) consists of two-stages: at the first stage, the norm of angular velocity is constant hence the convergence rate of quaternion could be maintained at a high level; at the second stage, the system begins to decelerate and by designing proper update law the system could have expected character. It is worth noticing that the initial value of k_2 in (5.4) guarantees angular velocity is continuous at the switching point, and k_1 should be the maximum norm of angular velocity to improve system convergence rate.

Based on the discussion above, the system should decelerate with the maximum control torque at the second stage of (5.3), hence it can be got that

$$\|\boldsymbol{u}\| = \overline{u} \tag{5.5}$$

where $\overline{u}$ is the norm of maximum control torque. When the system converges along sliding mode (5.3), it can be got that

$$\dot{\boldsymbol{q}}_v = \frac{1}{2}\left(q_0 \boldsymbol{I}_3 + \boldsymbol{q}_v^{\times}\right)\boldsymbol{\omega} = -\frac{1}{2}k_2 q_0 \boldsymbol{q}_v \tag{5.6}$$

$$\begin{aligned} \boldsymbol{u} &\approx \boldsymbol{J}\dot{\boldsymbol{\omega}} + \boldsymbol{\omega}^{\times}\boldsymbol{J}\boldsymbol{\omega} \\ &= k_2 \boldsymbol{J}\dot{q}_v + \dot{k}_2 \boldsymbol{J}\boldsymbol{q}_v + k_2^2 \boldsymbol{q}_v^{\times}\boldsymbol{J}\boldsymbol{q}_v \\ &= -\frac{1}{2}k_2^2 q_0 \boldsymbol{J}\boldsymbol{q}_v + k_2^2 \boldsymbol{q}_v^{\times}\boldsymbol{J}\boldsymbol{q}_v + \dot{k}_2 \boldsymbol{J}\boldsymbol{q}_v \end{aligned} \tag{5.7}$$

Hence,

$$
\begin{aligned}
\boldsymbol{u}^T\boldsymbol{u} &= \left(-\frac{1}{2}k_2^2 q_0 \boldsymbol{q}_v^T \boldsymbol{J} - k_2^2 \boldsymbol{q}_v^T \boldsymbol{J} \boldsymbol{q}_v^\times + \dot{k}_2 \boldsymbol{q}_v^T \boldsymbol{J}\right)\left(-\frac{1}{2}k_2^2 q_0 \boldsymbol{J}\boldsymbol{q}_v + k_2^2 \boldsymbol{q}_v^\times \boldsymbol{J}\boldsymbol{q}_v + \dot{k}_2 \boldsymbol{J}\boldsymbol{q}_v\right) \\
&= \frac{1}{4}k_2^4 q_0^2 \boldsymbol{q}_v^T \boldsymbol{J}^2 \boldsymbol{q}_v - \frac{1}{2}k_2^4 q_0 \boldsymbol{q}_v^T \boldsymbol{J} \boldsymbol{q}_v^\times \boldsymbol{J}\boldsymbol{q}_v - \frac{1}{2}k_2^2 q_0 \dot{k}_2 \boldsymbol{q}_v^T \boldsymbol{J}^2 \boldsymbol{q}_v - k_2^4 \boldsymbol{q}_v^T \boldsymbol{J} \boldsymbol{q}_v^\times \boldsymbol{q}_v^\times \boldsymbol{J}\boldsymbol{q}_v \\
&\quad - k_2^2 \dot{k}_2 \boldsymbol{q}_v^T \boldsymbol{J} \boldsymbol{q}_v^\times \boldsymbol{J}\boldsymbol{q}_v - \frac{1}{2}k_2^2 q_0 \dot{k}_2 \boldsymbol{q}_v^T \boldsymbol{J}^2 \boldsymbol{q}_v + k_2^2 \dot{k}_2 \boldsymbol{q}_v^T \boldsymbol{J} \boldsymbol{q}_v^\times \boldsymbol{J}\boldsymbol{q}_v + \left(\dot{k}_2\right)^2 \boldsymbol{q}_v^T \boldsymbol{J}^2 \boldsymbol{q}_v \\
&= A\left(\dot{k}_2\right)^2 - A k_2^2 q_0 \dot{k}_2 + \frac{1}{4}A k_2^4 q_0^2 - \frac{1}{2}B k_2^2 q_0 - k_2^4 C \\
&= \overline{u}^2
\end{aligned}
\tag{5.8}
$$

where

$$
A = \boldsymbol{q}_v^T \boldsymbol{J}^2 \boldsymbol{q}_v, \quad B = \boldsymbol{q}_v^T \boldsymbol{J} \boldsymbol{q}_v^\times \boldsymbol{J}\boldsymbol{q}_v, \quad C = \boldsymbol{q}_v^T \boldsymbol{J} \boldsymbol{q}_v^\times \boldsymbol{q}_v^\times \boldsymbol{J}\boldsymbol{q}_v \tag{5.9}
$$

Based on (24) the update law of k_2 could be got

$$
\dot{k}_2 = \frac{1}{2}k_2^2 q_0 - \frac{\sqrt{2ABk_2^2 q_0 + 4ACk_2^4 + 4A\overline{u}^2}}{2A} \tag{5.10}
$$

It is worth noticing that Eq. (24) has two solutions, and the larger solution is ignored since it corresponds to the situation that the system accelerates with the maximum control torque. By integrating system (2.1), (2.15), (5.10), the switching point α could be calculated.

The update law (5.10) could exploit the system capability to the utmost extend, but the accurate value of inertia matrix is needed, and the disturbance torque is ignored. Moreover, the value of switching point α needs the accurate value of inertia matrix $\boldsymbol{J}$, and the system uncertainty would cause the failure of sliding mode (5.3). Hence, the system robustness is still bad and the defect of open-loop controller is not avoided, and the update law of k_2 in (5.10) still needs improvement.

In order to improve the system robustness, the desired state character could be modified as follows:

$$
d\|\boldsymbol{\omega}\|/dt = -\eta \tag{5.11}
$$

where η is a positive scalar.

The essence of (5.11) is that the system decelerates with constant deceleration. In fact (5.11) is not the time optimal trajectory since the control torque capability is not fully utilized, but it could be treated as a suboptimal trajectory when the parameter is properly selected.

Define σ as follow

$$
\sigma = \boldsymbol{\omega}^T \boldsymbol{\omega} = \|\boldsymbol{\omega}\|^2 \tag{5.12}
$$

Calculate its derivative it can be got that

$$\begin{aligned}\dot{\sigma} &= 2\boldsymbol{\omega}^T\dot{\boldsymbol{\omega}} = 2k_2\boldsymbol{q}_v\left(\dot{k}_2\boldsymbol{q}_v + k_2\dot{\boldsymbol{q}}_v\right) = 2k_2\boldsymbol{q}_v\left(\dot{k}_2\boldsymbol{q}_v - \frac{1}{2}k_2^2 q_0\boldsymbol{q}_v\right) \\ &= 2k_2\|\boldsymbol{q}_v\|^2\dot{k}_2 - \frac{1}{2}k_2^3 q_0\|\boldsymbol{q}_v\|^2\end{aligned} \tag{5.13}$$

Noticing that

$$\dot{\sigma} = 2\|\boldsymbol{\omega}\|\frac{d\|\boldsymbol{\omega}\|}{dt} = -2k_2\|\boldsymbol{q}_v\|\eta \tag{5.14}$$

Hence, the update law under constant deceleration assumption could be written as follows:

$$\dot{k}_2 = \frac{1}{2}k_2^2 q_0 - \frac{\eta}{\|\boldsymbol{q}_v\|} \tag{5.15}$$

Noticing that the norm of angular velocity in the maneuver stage and the deceleration in the slow down stage, the switching point α could be written as follows:

$$\alpha = \sin\left(k_1^2/4\eta\right) \tag{5.16}$$

Based on the discussion above, the modified sliding mode with bang-bang logic character could be written as follows:

$$\begin{aligned} \boldsymbol{s} &= \begin{cases} \boldsymbol{\omega} + k_1\boldsymbol{e} & \|\boldsymbol{q}_v\| \geq \alpha \\ \boldsymbol{\omega} + k_2\boldsymbol{q}_v & \|\boldsymbol{q}_v\| < \alpha \end{cases} \\ \dot{k}_2 &= \begin{cases} 0 & \|\boldsymbol{q}_v\| \geq \alpha \\ \frac{1}{2}k_2^2 q_0 - \frac{\eta}{\|\boldsymbol{q}_v\|} & \beta < \|\boldsymbol{q}_v\| < \alpha \\ 0 & \|\boldsymbol{q}_v\| \leq \beta \end{cases} \\ \alpha &= \sin\left(k_1^2/4\eta\right), \quad k_2(0) = k_1/\alpha \end{aligned} \tag{5.17}$$

where k_1, k_2, η, α, β are all positive scalars.

Based on the structure of (5.17), it could be found that system state constitutes a closed-loop in the sliding mode and the defect of bang-bang control is avoided. The function of parameter β is to stop update k_2 when system state approaches to zero and ensure k_2 is finite.

The next step is to design controllers robust to system inertia matrix and disturbance torque under control torque constraint for satellite attitude stabilization control and attitude tracking control.

5.2.1 Attitude stabilization controller

The attitude stabilization controller based on sliding mode (5.17) could be written as follows:

$$\boldsymbol{u}=\begin{cases}-\rho_1 k_s \boldsymbol{s}+\boldsymbol{\omega}^{\times}\widehat{\boldsymbol{J}}\boldsymbol{\omega}-\dfrac{1}{2}k_1\widehat{\boldsymbol{J}}\boldsymbol{e}^{\times}\left(\boldsymbol{I}_3-\cot\dfrac{\phi}{2}\boldsymbol{e}^{\times}\right)\boldsymbol{\omega}-l_1\,\boldsymbol{sgn}(\boldsymbol{s}) & \|\boldsymbol{q}_v\|\geq\alpha\\ -\rho_2 k_s \boldsymbol{s}+\boldsymbol{\omega}^{\times}\widehat{\boldsymbol{J}}\boldsymbol{\omega}-\dfrac{1}{2}k_2\widehat{\boldsymbol{J}}\boldsymbol{F}\boldsymbol{\omega}-\dfrac{1}{2}k_2^2 q_0\widehat{\boldsymbol{J}}\boldsymbol{q}_v+\eta\widehat{\boldsymbol{J}}\boldsymbol{e}-l_2\,\boldsymbol{sgn}(\boldsymbol{s}) & \beta<\|\boldsymbol{q}_v\|<\alpha\\ -\rho_3 k_s \boldsymbol{s}+\boldsymbol{\omega}^{\times}\widehat{\boldsymbol{J}}\boldsymbol{\omega}-\dfrac{1}{2}k_2\widehat{\boldsymbol{J}}\boldsymbol{F}\boldsymbol{\omega}-l_3\,\boldsymbol{sgn}(\boldsymbol{s}) & \|\boldsymbol{q}_v\|<\beta\end{cases} \tag{5.18}$$

$$\begin{cases}l_1=\lambda\left(\|\boldsymbol{\omega}\|^2+\dfrac{1}{2}k_1\left(1+\cot\dfrac{\varphi}{2}\right)\|\boldsymbol{\omega}\|\right)+\overline{d}\\ l_2=\lambda\left(\|\boldsymbol{\omega}\|^2+\dfrac{1}{2}k_2\|\boldsymbol{\omega}\|+\dfrac{1}{2}k_2^2 q_0\|\boldsymbol{q}_v\|-\eta\right)+\overline{d}\\ l_3=\lambda\left(\|\boldsymbol{\omega}\|^2+\dfrac{1}{2}k_2\|\boldsymbol{\omega}\|\right)+\overline{d}\end{cases} \tag{5.19}$$

where k_s is a positive scalar and λ satisfies $\lambda\geq\lambda_M\left(\widetilde{\boldsymbol{J}}\right)$. Parameter ρ_i is the gain factor to avoid control torque saturation issue and defined as follows:

$$\rho_i=\begin{cases}\dfrac{-k_s\boldsymbol{s}^T\boldsymbol{r}_i+\sqrt{\left(k_s\boldsymbol{s}^T\boldsymbol{r}_i\right)^2+k_s^2\|\boldsymbol{s}\|^2\left(\overline{u}^2-\|\boldsymbol{r}_i\|^2\right)}}{k_s^2\|\boldsymbol{s}\|^2} & \|\boldsymbol{u}\|>\overline{u}\\ 1 & \|\boldsymbol{u}\|\leq\overline{u}\end{cases} \tag{5.20}$$

$$\begin{cases}\boldsymbol{r}_1=\boldsymbol{\omega}^{\times}\widehat{\boldsymbol{J}}\boldsymbol{\omega}-\dfrac{1}{2}k_1\widehat{\boldsymbol{J}}\boldsymbol{e}^{\times}\left(\boldsymbol{I}_3-\cot\dfrac{\varphi}{2}\boldsymbol{e}^{\times}\right)\boldsymbol{\omega}-l_1\,\boldsymbol{sgn}(\boldsymbol{s})\\ \boldsymbol{r}_2=\boldsymbol{\omega}^{\times}\widehat{\boldsymbol{J}}\boldsymbol{\omega}-\dfrac{1}{2}k_2\widehat{\boldsymbol{J}}\boldsymbol{F}\boldsymbol{\omega}-\dfrac{1}{2}k_2^2 q_0\widehat{\boldsymbol{J}}\boldsymbol{q}_v+\eta\widehat{\boldsymbol{J}}\boldsymbol{e}-l_2\,\boldsymbol{sgn}(\boldsymbol{s})\\ \boldsymbol{r}_3=\boldsymbol{\omega}^{\times}\widehat{\boldsymbol{J}}\boldsymbol{\omega}-\dfrac{1}{2}k_2\widehat{\boldsymbol{J}}\boldsymbol{F}\boldsymbol{\omega}-l_3\,\boldsymbol{sgn}(\boldsymbol{s})\end{cases} \tag{5.21}$$

Based on (5.20), it could be found that the function ρ_i is to minify the proportion term $k_s\boldsymbol{s}$ when the norm of control torque exceeds its upper bound. Based on (5.18), it can be found that the controller does not need the accurate value of system inertia matrix hence controller (5.18) is robust to inertia matrix uncertainty.

The next step is to prove the stability of system governed by controller (5.34). Select Lyapunov function as follows:

$$V_s=\frac{1}{2}\boldsymbol{s}^T\boldsymbol{J}\boldsymbol{s} \tag{5.22}$$

Calculate its derivative and substitute the three stages of controller (5.18) in it, it can be got that

$$
\begin{aligned}
\dot{V}_s &= \boldsymbol{s}^T\boldsymbol{J}\dot{\boldsymbol{s}} = \boldsymbol{s}^T\boldsymbol{J}\dot{\boldsymbol{\omega}} + k_1\boldsymbol{s}^T\boldsymbol{J}\dot{\boldsymbol{e}} \\
&= \boldsymbol{s}^T(\boldsymbol{u} + \boldsymbol{d} - \boldsymbol{\omega}^\times\boldsymbol{J}\boldsymbol{\omega}) + \frac{1}{2}k_1\boldsymbol{s}^T\boldsymbol{J}\boldsymbol{e}^\times\left(\boldsymbol{I}_3 - \cot\frac{\varphi}{2}\boldsymbol{e}^\times\right)\boldsymbol{\omega} \\
&= -\rho_1 k_s\boldsymbol{s}^T\boldsymbol{s} + \boldsymbol{s}^T\boldsymbol{\omega}^\times\widehat{\boldsymbol{J}}\boldsymbol{\omega} - \frac{1}{2}k_1\boldsymbol{s}^T\widehat{\boldsymbol{J}}\boldsymbol{e}^\times\left(\boldsymbol{I}_3 - \cot\frac{\varphi}{2}\boldsymbol{e}^\times\right)\boldsymbol{\omega} - l_1\boldsymbol{s}^T\,\boldsymbol{sgn}(\boldsymbol{s}) \\
&\quad + \boldsymbol{s}^T\boldsymbol{d} - \boldsymbol{s}^T\boldsymbol{\omega}^\times\boldsymbol{J}\boldsymbol{\omega} + \frac{1}{2}k_1\boldsymbol{s}^T\boldsymbol{J}\boldsymbol{e}^\times\left(\boldsymbol{I}_3 - \cot\frac{\varphi}{2}\boldsymbol{e}^\times\right)\boldsymbol{\omega} \\
&= -\rho_1 k_s\boldsymbol{s}^T\boldsymbol{s} - \boldsymbol{s}^T\boldsymbol{\omega}^\times\widetilde{\boldsymbol{J}}\boldsymbol{\omega} + \frac{1}{2}k_1\boldsymbol{s}^T\widetilde{\boldsymbol{J}}\boldsymbol{e}^\times\left(\boldsymbol{I}_3 - \cot\frac{\varphi}{2}\boldsymbol{e}^\times\right)\boldsymbol{\omega} + \boldsymbol{s}^T\boldsymbol{d} \\
&\quad -\lambda\boldsymbol{s}^T\,\boldsymbol{sgn}(\boldsymbol{s})\left(\|\boldsymbol{\omega}\|^2 + \frac{1}{2}k_1\left(1 + \cot\frac{\varphi}{2}\right)\|\boldsymbol{\omega}\|\right) - \overline{d}\boldsymbol{s}^T\,\boldsymbol{sgn}(\boldsymbol{s}) \\
&\le -\rho_1 k_s\boldsymbol{s}^T\boldsymbol{s} \le 0
\end{aligned}
\tag{5.23}
$$

$$
\begin{aligned}
\dot{V}_s &= \boldsymbol{s}^T\boldsymbol{J}\dot{\boldsymbol{s}} = \boldsymbol{s}^T\boldsymbol{J}\dot{\boldsymbol{\omega}} + k_2\boldsymbol{s}^T\boldsymbol{J}\dot{\boldsymbol{q}}_v + \dot{k}_2\boldsymbol{s}^T\boldsymbol{J}\boldsymbol{q}_v \\
&= \boldsymbol{s}^T(\boldsymbol{u} + \boldsymbol{d} - \boldsymbol{\omega}^\times\boldsymbol{J}\boldsymbol{\omega}) + \frac{1}{2}k_2\boldsymbol{s}^T\boldsymbol{J}\boldsymbol{F}\boldsymbol{\omega} + \left(\frac{1}{2}k_2^2 q_0 - \frac{\eta}{\|\boldsymbol{q}_v\|}\right)\boldsymbol{s}^T\boldsymbol{J}\boldsymbol{q}_v \\
&= -\rho_2 k_s\boldsymbol{s}^T\boldsymbol{s} + \boldsymbol{s}^T\boldsymbol{\omega}^\times\widehat{\boldsymbol{J}}\boldsymbol{\omega} - \frac{1}{2}k_2\boldsymbol{s}^T\widehat{\boldsymbol{J}}\boldsymbol{F}\boldsymbol{\omega} + \frac{1}{2}k_2^2 q_0\boldsymbol{s}^T\widehat{\boldsymbol{J}}\boldsymbol{q}_v + \eta\boldsymbol{s}^T\widehat{\boldsymbol{J}}\boldsymbol{e} - l_2\boldsymbol{s}^T\,\boldsymbol{sgn}(\boldsymbol{s}) \\
&\quad + \boldsymbol{s}^T\boldsymbol{d} - \boldsymbol{s}^T\boldsymbol{\omega}^\times\boldsymbol{J}\boldsymbol{\omega} + \frac{1}{2}k_2\boldsymbol{s}^T\boldsymbol{J}\boldsymbol{F}\boldsymbol{\omega} + \frac{1}{2}k_2^2 q_0\boldsymbol{s}^T\boldsymbol{J}\boldsymbol{q}_v - \eta\boldsymbol{s}^T\boldsymbol{J}\boldsymbol{e} \\
&= -\rho_2 k_s\boldsymbol{s}^T\boldsymbol{s} - \boldsymbol{s}^T\boldsymbol{\omega}^\times\widetilde{\boldsymbol{J}}\boldsymbol{\omega} + \frac{1}{2}k_2\boldsymbol{s}^T\widehat{\boldsymbol{J}}\boldsymbol{F}\boldsymbol{\omega} + \frac{1}{2}k_2^2 q_0\boldsymbol{s}^T\widetilde{\boldsymbol{J}}\boldsymbol{q}_v - \eta\boldsymbol{s}^T\widetilde{\boldsymbol{J}}\boldsymbol{e} + \boldsymbol{s}^T\boldsymbol{d} \\
&\quad -\lambda\boldsymbol{s}^T\,\boldsymbol{sgn}(\boldsymbol{s})\left(\|\boldsymbol{\omega}\|^2 + \frac{1}{2}k_2\|\boldsymbol{\omega}\| + \frac{1}{2}k_2^2 q_0\|\boldsymbol{q}_v\| - \eta\right) - \overline{d}\boldsymbol{s}^T\,\boldsymbol{sgn}(\boldsymbol{s}) \\
&\le -\rho_2 k_s\boldsymbol{s}^T\boldsymbol{s} \le 0
\end{aligned}
\tag{5.24}
$$

$$
\begin{aligned}
\dot{V}_s &= \boldsymbol{s}^T\boldsymbol{J}\dot{\boldsymbol{s}} = \boldsymbol{s}^T\boldsymbol{J}\dot{\boldsymbol{\omega}} + k_2\boldsymbol{s}^T\boldsymbol{J}\dot{\boldsymbol{q}}_v \\
&= \boldsymbol{s}^T(\boldsymbol{u} + \boldsymbol{d} - \boldsymbol{\omega}^\times\boldsymbol{J}\boldsymbol{\omega}) + \frac{1}{2}k_2\boldsymbol{s}^T\boldsymbol{J}\boldsymbol{F}\boldsymbol{\omega} \\
&= -\rho_3 k_s\boldsymbol{s}^T\boldsymbol{s} + \boldsymbol{s}^T\boldsymbol{\omega}^\times\widehat{\boldsymbol{J}}\boldsymbol{\omega} - \frac{1}{2}k_2\boldsymbol{s}^T\widehat{\boldsymbol{J}}\boldsymbol{F}\boldsymbol{\omega} - l_3\boldsymbol{s}^T\,\boldsymbol{sgn}(\boldsymbol{s}) + \boldsymbol{s}^T\boldsymbol{d} - \boldsymbol{s}^T\boldsymbol{\omega}^\times\boldsymbol{J}\boldsymbol{\omega} + \frac{1}{2}k_2\boldsymbol{s}^T\boldsymbol{J}\boldsymbol{F}\boldsymbol{\omega} \\
&= -\rho_3 k_s\boldsymbol{s}^T\boldsymbol{s} - \boldsymbol{s}^T\boldsymbol{\omega}^\times\widetilde{\boldsymbol{J}}\boldsymbol{\omega} + \frac{1}{2}k_2\boldsymbol{s}^T\widetilde{\boldsymbol{J}}\boldsymbol{F}\boldsymbol{\omega} + \boldsymbol{s}^T\boldsymbol{d} - \lambda\left(\|\boldsymbol{\omega}\|^2 + \frac{1}{2}k_2\|\boldsymbol{\omega}\|\right)\boldsymbol{s}^T\,\boldsymbol{sgn}(\boldsymbol{s}) \\
&\quad -\overline{d}\boldsymbol{s}^T\,\boldsymbol{sgn}(\boldsymbol{s}) \\
&\le -\rho_3 k_s\boldsymbol{s}^T\boldsymbol{s} \le 0
\end{aligned}
\tag{5.25}
$$

Considering that $\dot{V}_s = 0 \Rightarrow \boldsymbol{s} = \boldsymbol{0}$, i.e., $\boldsymbol{\omega} = -k\boldsymbol{q}_v$ and $\boldsymbol{\omega} \to \boldsymbol{0}$, $\boldsymbol{q}_v \to \boldsymbol{0}$, based on the LaSalle invariance principle, system (2.1), (2.15) governed by controller (5.18) is globally asymptotically stable.

In order to improve the system convergence rate, sliding mode parameter k_1 and η should be selected as large as possible so that system capability could be fully utilized. However, large parameters would bring large control torque and when it exceeds its upper bound, the possibility of controller failure increases. Hence, it is necessary to discuss the control torque constraint on sliding mode parameters.

Based on the definition of gain factor ρ_i in (5.18), it can be found the key to solve control torque saturation is to ensure $\|\boldsymbol{r}_i\| \leq \overline{u}$, that is, the norm of $\boldsymbol{r}_i$ does not exceed the system upper bound. When the system converges along the first stage of controller (5.18), angular velocity satisfies

$$\boldsymbol{\omega} = -k_1 \boldsymbol{e} \tag{5.26}$$

Hence,

$$\begin{aligned} \|\boldsymbol{r}_1\| &= \left\| \boldsymbol{\omega}^{\times} \widehat{\boldsymbol{J}} \boldsymbol{\omega} - \frac{1}{2} k_1 \widehat{\boldsymbol{J}} \boldsymbol{e}^{\times} \left(\boldsymbol{I}_3 - \cot \frac{\varphi}{2} \boldsymbol{e}^{\times} \right) \boldsymbol{\omega} - l_1 \, \boldsymbol{sgn} \left(\boldsymbol{s} \right) \right\| \\ &\approx \left\| \boldsymbol{\omega}^{\times} \boldsymbol{J} \boldsymbol{\omega} - \frac{1}{2} k_1 \boldsymbol{J} \boldsymbol{e}^{\times} \left(\boldsymbol{I}_3 - \cot \frac{\varphi}{2} \boldsymbol{e}^{\times} \right) \boldsymbol{\omega} \right\| \\ &= \left\| k_1^2 \boldsymbol{e}^{\times} \boldsymbol{J} \boldsymbol{e} - \frac{1}{2} k_1^2 \boldsymbol{J} \boldsymbol{e}^{\times} \left(\boldsymbol{I}_3 - \cot \frac{\varphi}{2} \boldsymbol{e}^{\times} \right) \boldsymbol{e} \right\| \\ &= \left\| k_1^2 \boldsymbol{e}^{\times} \boldsymbol{J} \boldsymbol{e} \right\| \\ &\leq k_1^2 \lambda_M(\boldsymbol{J}) \end{aligned} \tag{5.27}$$

Therefore if following inequality is satisfies, $\|\boldsymbol{r}_1\| \leq \overline{u}$ could be ensured.

$$k_1^2 \lambda_M(\boldsymbol{J}) \leq \overline{u} \tag{5.28}$$

On the second stage of controller (5.18), angular velocity satisfies

$$\boldsymbol{\omega} = -k_2 \boldsymbol{q}_v \tag{5.29}$$

Hence,

$$\begin{aligned} \|\boldsymbol{r}_2\| &\approx \left\| \boldsymbol{\omega}^{\times} \boldsymbol{J} \boldsymbol{\omega} - \frac{1}{2} k_2 \boldsymbol{J} \boldsymbol{F} \boldsymbol{\omega} - \frac{1}{2} k_2^2 q_0 \boldsymbol{J} \boldsymbol{q}_v + \eta \boldsymbol{J} \boldsymbol{e} \right\| \\ &= \left\| k_2^2 \boldsymbol{q}_v^{\times} \boldsymbol{J} \boldsymbol{q}_v + \frac{1}{2} k_2^2 q_0 \boldsymbol{J} \boldsymbol{q}_v - \frac{1}{2} k_2^2 q_0 \boldsymbol{J} \boldsymbol{q}_v + \eta \boldsymbol{J} \boldsymbol{e} \right\| \\ &\leq \lambda_M(\boldsymbol{J}) k_2^2 \|\boldsymbol{q}_v\|^2 + \lambda_M(\boldsymbol{J}) \eta \end{aligned} \tag{5.30}$$

Define z as follows:

$$z = \lambda_M(\boldsymbol{J}) k_2^2 \|\boldsymbol{q}_v\|^2 + \lambda_M(\boldsymbol{J}) \eta \tag{5.31}$$

It could be found that z is the norm upper bound of $\boldsymbol{r}_2$. Calculate its derivative

$$\begin{aligned} \dot{z} &= 2\lambda_M(\boldsymbol{J}) k_2^2 \|\boldsymbol{q}_v\| \frac{d\|\boldsymbol{q}_v\|}{dt} + 2\lambda_M(\boldsymbol{J}) k_2 \|\boldsymbol{q}_v\|^2 \dot{k}_2 \\ &= -\lambda_M(\boldsymbol{J}) k_2^3 q_0 \|\boldsymbol{q}_v\|^2 + 2\lambda_M(\boldsymbol{J}) k_2 \|\boldsymbol{q}_v\|^2 \left(\frac{1}{2} k_2^2 q_0 - \frac{\eta}{\|\boldsymbol{q}_v\|} \right) \\ &= -\lambda_M(\boldsymbol{J}) k_2 \|\boldsymbol{q}_v\|^2 \left(k_2^2 q_0 - k_2^2 q_0 + \frac{2\eta}{\|\boldsymbol{q}_v\|} \right) \\ &= -2\lambda_M(\boldsymbol{J}) \eta k_2 \|\boldsymbol{q}_v\| \\ &\leq 0 \end{aligned} \tag{5.32}$$

Hence, z is monotonically decreasing and if following inequality is satisfied, $\|\boldsymbol{r}_2\| \le \overline{u}$ could be ensured.

$$\lambda_M(\boldsymbol{J})\left(k_2^2(0)\alpha^2 + \eta\right) \le \overline{u} \tag{5.33}$$

On the third stage of controller (5.18), k_2 no longer increases hence if there is no saturation issue in the second stage, the saturation issue would not occur in the third stage.

The control torque constraint on control parameters, i.e., (5.28) and (5.33), is strict constraint since the worst condition is discussed.

5.2.2 Attitude tracking controller

Similar as discussed in Section 5.2.1, the sliding mode based on bang-bang logic for attitude tracking control could be written as follows:

$$\begin{aligned}
\boldsymbol{s}_e &= \begin{cases} \boldsymbol{\omega}_e + k_1\boldsymbol{e}_e & \|\boldsymbol{q}_{ev}\| \ge \alpha \\ \boldsymbol{\omega}_e + k_2\boldsymbol{q}_{ev} & \|\boldsymbol{q}_{ev}\| < \alpha \end{cases} \\
\dot{k}_2 &= \begin{cases} 0 & \|\boldsymbol{q}_{ev}\| \ge \alpha \\ \dfrac{1}{2}k_2^2 q_{e0} - \dfrac{\eta}{\|\boldsymbol{q}_{ev}\|} & \beta < \|\boldsymbol{q}_{ev}\| < \alpha \\ 0 & \|\boldsymbol{q}_{ev}\| \le \beta \end{cases} \\
\alpha &= \sin\left(k_1^2/4\eta\right), \quad k_2(0) = k_1/\alpha
\end{aligned} \tag{5.34}$$

where k_1, k_2, η, α, β are all positive scalars.

The structure of sliding mode (5.34) is the same as (5.17), and the character of the bang-bang like sliding mode has been discussed in previous section hence the discussion is omitted in this section. Satellite attitude tracking controller based on sliding mode (5.34) could be written as follows:

$$\boldsymbol{u} = \begin{cases}
-\rho_1 k_s \boldsymbol{s}_e - \dfrac{1}{2}k_1\widehat{\boldsymbol{J}}\boldsymbol{e}_e^\times\left(\boldsymbol{I}_3 - \cot\dfrac{\phi_e}{2}\boldsymbol{e}_e^\times\right)\boldsymbol{\omega}_e + \widehat{\boldsymbol{J}}\boldsymbol{R}\boldsymbol{\omega}_d - \widehat{\boldsymbol{J}}\boldsymbol{\omega}_e^\times\boldsymbol{R}\boldsymbol{\omega}_d & \\
+(\boldsymbol{\omega}_e + \boldsymbol{R}\boldsymbol{\omega}_d)^\times\widehat{\boldsymbol{J}}(\boldsymbol{\omega}_e + \boldsymbol{R}\boldsymbol{\omega}_d) - l_1\,\mathbf{sgn}\,(\boldsymbol{s}_e) & \|\boldsymbol{q}_{ev}\| \ge \alpha \\
-\rho_2 k_s \boldsymbol{s}_e - \dfrac{1}{2}k_2\widehat{\boldsymbol{J}}\boldsymbol{F}_e\boldsymbol{\omega}_e + \widehat{\boldsymbol{J}}\boldsymbol{R}\boldsymbol{\omega}_d - \widehat{\boldsymbol{J}}\boldsymbol{\omega}_e^\times\boldsymbol{R}\boldsymbol{\omega}_d - \dfrac{1}{2}k_2^2 q_{e0}\widehat{\boldsymbol{J}}\boldsymbol{q}_{ev} & \\
+\eta\widehat{\boldsymbol{J}}\boldsymbol{e}_e + (\boldsymbol{\omega}_e + \boldsymbol{R}\boldsymbol{\omega}_d)^\times\widehat{\boldsymbol{J}}(\boldsymbol{\omega}_e + \boldsymbol{R}\boldsymbol{\omega}_d) - l_2\,\mathbf{sgn}\,(\boldsymbol{s}_e) & \beta < \|\boldsymbol{q}_{ev}\| < \alpha \\
-\rho_3 k_s \boldsymbol{s}_e - \dfrac{1}{2}k_2\widehat{\boldsymbol{J}}\boldsymbol{F}_e\boldsymbol{\omega}_e + \widehat{\boldsymbol{J}}\boldsymbol{R}\boldsymbol{\omega}_d - \widehat{\boldsymbol{J}}\boldsymbol{\omega}_e^\times\boldsymbol{R}\boldsymbol{\omega}_d & \\
+(\boldsymbol{\omega}_e + \boldsymbol{R}\boldsymbol{\omega}_d)^\times\widehat{\boldsymbol{J}}(\boldsymbol{\omega}_e + \boldsymbol{R}\boldsymbol{\omega}_d) - l_3\,\mathbf{sgn}\,(\boldsymbol{s}_e) & \|\boldsymbol{q}_{ev}\| < \beta
\end{cases} \tag{5.35}$$

$$\begin{cases} l_1 = \lambda\left(\|\omega\|^2 + \dfrac{1}{2}k_1\left(1 + \cot\dfrac{\varphi_e}{2}\right)\|\boldsymbol{\omega}_e\| + \|\dot{\boldsymbol{\omega}}_d\| + \|\boldsymbol{\omega}_e\|\|\boldsymbol{\omega}_d\|\right) + \overline{d} \\ l_2 = \lambda\left(\|\omega\|^2 + \dfrac{1}{2}k_2\|\boldsymbol{\omega}_e\| + \|\dot{\boldsymbol{\omega}}_d\| + \|\boldsymbol{\omega}_e\|\|\boldsymbol{\omega}_d\| + \dfrac{1}{2}k_2^2 q_{e0}\|\boldsymbol{q}_{ev}\| - \eta\right) + \overline{d} \\ l_3 = \lambda\left(\|\omega\|^2 + \dfrac{1}{2}k_2\|\boldsymbol{\omega}_e\| + \|\dot{\boldsymbol{\omega}}_d\| + \|\boldsymbol{\omega}_e\|\|\boldsymbol{\omega}_d\|\right) + \overline{d} \end{cases} \tag{5.36}$$

where k_s is a positive scalar and λ satisfies $\lambda \geq \lambda_M\left(\widetilde{\boldsymbol{J}}\right)$. Parameter ρ_i is the gain factor to avoid control torque saturation issue and defined as follows:

$$\rho_i = \begin{cases} \dfrac{-k_s\boldsymbol{s}_e^T\boldsymbol{r}_i + \sqrt{\left(k_s\boldsymbol{s}_e^T\boldsymbol{r}_i\right)^2 + k_s^2\|\boldsymbol{s}_e\|^2\left(\overline{u}^2 - \|\boldsymbol{r}_i\|^2\right)}}{k_s^2\|\boldsymbol{s}_e\|^2} & \|\boldsymbol{u}\| > \overline{u} \\ 1 & \|\boldsymbol{u}\| \leq \overline{u} \end{cases} \tag{5.37}$$

$$\begin{cases} \boldsymbol{r}_1 = -\dfrac{1}{2}k_1\widehat{\boldsymbol{J}}\boldsymbol{e}_e^{\times}\left(\boldsymbol{I}_3 - \cot\dfrac{\varphi_e}{2}\boldsymbol{e}_e^{\times}\right)\boldsymbol{\omega}_e + \widehat{\boldsymbol{J}}\boldsymbol{R}\dot{\boldsymbol{\omega}}_d - \widehat{\boldsymbol{J}}\boldsymbol{\omega}_e^{\times}\boldsymbol{R}\boldsymbol{\omega}_d - l_1\,\boldsymbol{sgn}(\boldsymbol{s}_e) \\ \quad + (\boldsymbol{\omega}_e + \boldsymbol{R}\boldsymbol{\omega}_d)^{\times}\widehat{\boldsymbol{J}}(\boldsymbol{\omega}_e + \boldsymbol{R}\boldsymbol{\omega}_d) \\ \boldsymbol{r}_2 = -\dfrac{1}{2}k_2\widehat{\boldsymbol{J}}\boldsymbol{F}_e\boldsymbol{\omega}_e + \widehat{\boldsymbol{J}}\boldsymbol{R}\dot{\boldsymbol{\omega}}_d - \widehat{\boldsymbol{J}}\boldsymbol{\omega}_e^{\times}\boldsymbol{R}\boldsymbol{\omega}_d - \dfrac{1}{2}k_2^2 q_{e0}\widehat{\boldsymbol{J}}\boldsymbol{q}_{ev} + \eta\widehat{\boldsymbol{J}}\boldsymbol{e}_e \\ \quad + (\boldsymbol{\omega}_e + \boldsymbol{R}\boldsymbol{\omega}_d)^{\times}\widehat{\boldsymbol{J}}(\boldsymbol{\omega}_e + \boldsymbol{R}\boldsymbol{\omega}_d) - l_2\,\boldsymbol{sgn}(\boldsymbol{s}_e) \\ \boldsymbol{r}_3 = -\rho_3 k_s\boldsymbol{s}_e - \dfrac{1}{2}k_2\widehat{\boldsymbol{J}}\boldsymbol{F}_e\boldsymbol{\omega}_e + \widehat{\boldsymbol{J}}\boldsymbol{R}\dot{\boldsymbol{\omega}}_d - \widehat{\boldsymbol{J}}\boldsymbol{\omega}_e^{\times}\boldsymbol{R}\boldsymbol{\omega}_d \\ \quad + (\boldsymbol{\omega}_e + \boldsymbol{R}\boldsymbol{\omega}_d)^{\times}\widehat{\boldsymbol{J}}(\boldsymbol{\omega}_e + \boldsymbol{R}\boldsymbol{\omega}_d) - l_3\,\boldsymbol{sgn}(\boldsymbol{s}_e) \end{cases} \tag{5.38}$$

The function of ρ_i is to ensure the norm of control torque does not exceed the system upper bound by minifying the proportion term $k_s\boldsymbol{s}$ when control torque saturation occurs.

The next step is to prove the stability of attitude error system governed by controller (5.39). Select Lyapunov function as follows:

$$V_s = \frac{1}{2}\boldsymbol{s}_e^T\boldsymbol{J}\boldsymbol{s}_e \tag{5.39}$$

Calculate its derivative and substitute the three stages of controller (5.34) in it, it can be got that

$$\begin{aligned}
\dot{V}_s &= s_e^T J \dot{s}_e = s_e^T J \dot{\omega}_e + k_1 s_e^T J \dot{e}_e \\
&= s_e^T \left(u + d - JR\dot{\omega}_d + J\omega_e^\times R\omega_d - (\omega_e + R\omega_d)^\times J(\omega_e + R\omega_d)\right) \\
&\quad + \frac{1}{2}k_1 s_e^T J e_e^\times \left(I_3 - \cot\frac{\varphi_e}{2} e_e^\times\right)\omega_e \\
&= -\rho_1 k_s s_e^T s_e + s_e^T d - s_e^T \tilde{J} R\dot{\omega}_d + s_e^T \tilde{J}\omega_e^\times R\omega_d - s_e^T(\omega_e + R\omega_d)^\times \tilde{J}(\omega_e + R\omega_d) \\
&\quad + \frac{1}{2}k_1 s_e^T J e_e^\times \left(I_3 - \cot\frac{\varphi_e}{2} e_e^\times\right)\omega_e - l_1 s_e^T\, sgn(s_e) \\
&\le -\rho_1 k_s s_e^T s_e \le 0
\end{aligned} \tag{5.40}$$

$$\begin{aligned}
\dot{V}_s &= s_e^T J \dot{s}_e = s_e^T J \dot{\omega}_e + k_2 s_e^T J \dot{q}_{ev} + \dot{k}_2 s_e^T J q_{ev} \\
&= s_e^T \left(u + d - JR\dot{\omega}_d + J\omega_e^\times R\omega_d - (\omega_e + R\omega_d)^\times J(\omega_e + R\omega_d)\right) + \frac{1}{2}k_2 s_e^T J F_e \omega_e \\
&\quad + \left(\frac{1}{2}k_2^2 q_{e0} - \frac{\eta}{\|q_{ev}\|}\right) s_e^T J q_{ev} \\
&= -\rho_2 k_s s_e^T s_e + s_e^T d - s_e^T \tilde{J} R\dot{\omega}_d + s_e^T \tilde{J}\omega_e^\times R\omega_d - s_e^T(\omega_e + R\omega_d)^\times \tilde{J}(\omega_e + R\omega_d) \\
&\quad + \frac{1}{2}k_2 s_e^T \tilde{J} F_e \omega_e + \frac{1}{2}k_2^2 q_{e0} \tilde{J} q_{ev} - \eta s_e^T J e_e \\
&\quad - \left(\lambda\left(\|\omega\|^2 + \frac{1}{2}k_2\|\omega_e\| + \|\dot{\omega}_d\| + \|\omega_e\|\|\omega_d\|\right) + \bar{d}\right) s_e^T\, sgn(s_e) \\
&\le -\rho_2 k_s s_e^T s_e \le 0
\end{aligned} \tag{5.41}$$

$$\begin{aligned}
\dot{V}_s &= s_e^T J \dot{s}_e = s_e^T J \dot{\omega}_e + k_2 s_e^T J q_{ev} \\
&= s_e^T \left(u + d - JR\dot{\omega}_d + J\omega_e^\times R\omega_d - (\omega_e + R\omega_d)^\times J(\omega_e + R\omega_d)\right) \\
&\quad + \frac{1}{2}k_1 s_e^T J e_e^\times \left(I_3 - \cot\frac{\varphi_e}{2} e_e^\times\right)\omega_e \\
&= -\rho_1 k_s s_e^T s_e + s_e^T d - s_e^T \tilde{J} R\dot{\omega}_d + s_e^T \tilde{J}\omega_e^\times R\omega_d - s_e^T(\omega_e + R\omega_d)^\times \tilde{J}(\omega_e + R\omega_d) \\
&\quad + \frac{1}{2}k_1 s_e^T \tilde{J} e_e^\times \left(I_3 - \cot\frac{\varphi_e}{2} e_e^\times\right)\omega_e - l_1 s_e^T\, sgn(s_e) \\
&\le -\rho_1 k_s s_e^T s_e \le 0
\end{aligned} \tag{5.42}$$

Considering that $\dot{V}_s = 0 \Rightarrow s_e = \mathbf{0}$, i.e., $\omega_e = -kq_{ev}$, and $\omega_e \to \mathbf{0}$, $q_{ev} \to \mathbf{0}$, based on the LaSalle invariance principle, system (2.21), (2.24) governed by controller (5.35) is globally asymptotically stable.

Similar as discussed in previous section, larger sliding mode parameters would bring better system convergence rate, but the possibility of controller failure caused by control torque saturation also increases. Hence, it is necessary to discuss the control torque constraint on control parameters. Based on the definition of control gain factor ρ_i, it could be found that if $\|r_i\| \le \bar{u}$ could be ensured, the control torque saturation issue could be solved.

On the first stage of controller (5.35), error angular velocity satisfies

$$\boldsymbol{\omega}_e = -k_1 \boldsymbol{e}_e \tag{5.43}$$

Hence

$$\begin{aligned}
\|\boldsymbol{r}_1\| &= \left\| -\frac{1}{2}k_1\widehat{\boldsymbol{J}}\boldsymbol{e}_e^\times\left(\boldsymbol{I}_3 - \cot\frac{\phi_e}{2}\boldsymbol{e}_e^\times\right)\boldsymbol{\omega}_e + \widehat{\boldsymbol{J}}\boldsymbol{R}\dot{\boldsymbol{\omega}}_d - \widehat{\boldsymbol{J}}\boldsymbol{\omega}_e^\times\boldsymbol{R}\boldsymbol{\omega}_d + \boldsymbol{\omega}^\times\widehat{\boldsymbol{J}}\boldsymbol{\omega} - l_1\,\boldsymbol{sgn}\,(s_e) \right\| \\
&\approx \left\| \boldsymbol{\omega}^\times\boldsymbol{J}\boldsymbol{\omega} - \frac{1}{2}k_1\boldsymbol{J}\boldsymbol{e}_e^\times\left(\boldsymbol{I}_3 - \cot\frac{\phi_e}{2}\boldsymbol{e}_e^\times\right)\boldsymbol{\omega}_e + \boldsymbol{J}\boldsymbol{R}\dot{\boldsymbol{\omega}}_d - \boldsymbol{J}\boldsymbol{\omega}_e^\times\boldsymbol{R}\boldsymbol{\omega}_d \right\| \\
&= \left\| (\boldsymbol{\omega}_e + \boldsymbol{R}\boldsymbol{\omega}_d)^\times\boldsymbol{J}(\boldsymbol{\omega}_e + \boldsymbol{R}\boldsymbol{\omega}_d) + \boldsymbol{J}\boldsymbol{R}\dot{\boldsymbol{\omega}}_d - \boldsymbol{J}\boldsymbol{\omega}_e^\times\boldsymbol{R}\boldsymbol{\omega}_d \right\| \\
&\le \lambda_M(\boldsymbol{J})\left(\|\boldsymbol{\omega}_e + \boldsymbol{R}\boldsymbol{\omega}_d\|^2 + \|\dot{\boldsymbol{\omega}}_d\| + \|\boldsymbol{\omega}_e\|\|\boldsymbol{\omega}_d\|\right) \\
&\le \lambda_M(\boldsymbol{J})\left((\|\boldsymbol{\omega}_e\| + \|\boldsymbol{\omega}_d\|)^2 + \|\boldsymbol{\omega}_e\|\|\boldsymbol{\omega}_d\| + \|\dot{\boldsymbol{\omega}}_d\|\right) \\
&= \lambda_M(\boldsymbol{J})\left(\|\boldsymbol{\omega}_e\|^2 + 3\|\boldsymbol{\omega}_e\|\|\boldsymbol{\omega}_d\| + \|\boldsymbol{\omega}_d\|^2 + \|\dot{\boldsymbol{\omega}}_d\|\right) \\
&= \lambda_M(\boldsymbol{J})\left(k_1^2 + 3k_1\|\boldsymbol{\omega}_d\| + \|\boldsymbol{\omega}_d\|^2 + \|\dot{\boldsymbol{\omega}}_d\|\right)
\end{aligned} \tag{5.44}$$

Therefore, if following inequality is satisfied, $\|\boldsymbol{r}_1\| \le \overline{u}$ could be ensured.

$$\lambda_M(\boldsymbol{J})\left(k_1^2 + 3k_1\overline{\omega}_d + \overline{\omega}_d^2 + \dot{\overline{\omega}}_d\right) \le \overline{u} \tag{5.45}$$

where $\overline{\omega}_d$ and $\dot{\overline{\omega}}_d$ is the norm upper bound of $\|\boldsymbol{\omega}_d\|$ and $\|\dot{\boldsymbol{\omega}}_d\|$.

On the second stage of controller (5.35), error angular velocity satisfies

$$\boldsymbol{\omega}_e = -k_2 \boldsymbol{q}_{ev} \tag{5.46}$$

Hence

$$\begin{aligned}
\|\boldsymbol{r}_2\| &\approx \left\| \boldsymbol{\omega}^\times\boldsymbol{J}\boldsymbol{\omega} - \frac{1}{2}k_2\boldsymbol{J}\boldsymbol{F}_e\boldsymbol{\omega}_e + \boldsymbol{J}\boldsymbol{R}\dot{\boldsymbol{\omega}}_d - \boldsymbol{J}\boldsymbol{\omega}_e^\times\boldsymbol{R}\boldsymbol{\omega}_d - \frac{1}{2}k_2^2 q_{e0}\boldsymbol{J}\boldsymbol{q}_{ev} + \eta\boldsymbol{J}\boldsymbol{e}_e \right\| \\
&= \left\| \boldsymbol{\omega}^\times\boldsymbol{J}\boldsymbol{\omega} + \frac{1}{2}k_2^2 q_{e0}\boldsymbol{J}\boldsymbol{q}_{ev} + \boldsymbol{J}\boldsymbol{R}\dot{\boldsymbol{\omega}}_d - \boldsymbol{J}\boldsymbol{\omega}_e^\times\boldsymbol{R}\boldsymbol{\omega}_d - \frac{1}{2}k_2^2 q_{e0}\boldsymbol{J}\boldsymbol{q}_{ev} + \eta\boldsymbol{J}\boldsymbol{e}_e \right\| \\
&= \left\| \boldsymbol{\omega}^\times\boldsymbol{J}\boldsymbol{\omega} + \boldsymbol{J}\boldsymbol{R}\dot{\boldsymbol{\omega}}_d - \boldsymbol{J}\boldsymbol{\omega}_e^\times\boldsymbol{R}\boldsymbol{\omega}_d + \eta\boldsymbol{J}\boldsymbol{e}_e \right\| \\
&\le \lambda_M(\boldsymbol{J})\left(\|\boldsymbol{\omega}_e + \boldsymbol{R}\boldsymbol{\omega}_d\|^2 + \|\dot{\boldsymbol{\omega}}_d\| + \|\boldsymbol{\omega}_e\|\|\boldsymbol{\omega}_d\| + \eta\right) \\
&\le \lambda_M(\boldsymbol{J})\left((\|\boldsymbol{\omega}_e\| + \|\boldsymbol{\omega}_d\|)^2 + \|\dot{\boldsymbol{\omega}}_d\| + \|\boldsymbol{\omega}_e\|\|\boldsymbol{\omega}_d\| + \eta\right) \\
&= \lambda_M(\boldsymbol{J})\left(\|\boldsymbol{\omega}_e\|^2 + 3\|\boldsymbol{\omega}_e\|\|\boldsymbol{\omega}_d\| + \|\boldsymbol{\omega}_d\|^2 + \|\dot{\boldsymbol{\omega}}_d\| + \eta\right) \\
&= \lambda_M(\boldsymbol{J})\left(k_2^2\|\boldsymbol{q}_{ev}\|^2 + 3k_2\|\boldsymbol{q}_{ev}\|\|\boldsymbol{\omega}_d\| + \|\boldsymbol{\omega}_d\|^2 + \|\dot{\boldsymbol{\omega}}_d\| + \eta\right)
\end{aligned} \tag{5.47}$$

Define z as follows:

$$z = \lambda_M(\boldsymbol{J})\left(k_2^2\|\boldsymbol{q}_{ev}\|^2 + 3k_2\|\boldsymbol{q}_{ev}\|\|\boldsymbol{\omega}_d\| + \|\boldsymbol{\omega}_d\|^2 + \|\dot{\boldsymbol{\omega}}_d\| + \eta\right) \tag{5.48}$$

It could be found that z is the norm upper bound of $\|\boldsymbol{r}_2\|$. Calculate its derivative

$$\begin{aligned}
\dot{z} &= \lambda_M(\boldsymbol{J})\left(2k_2^2\|\boldsymbol{q}_{ev}\|\frac{d\|\boldsymbol{q}_{ev}\|}{dt} + 2k_2\|\boldsymbol{q}_{ev}\|^2\dot{k}_2 + 3\|\boldsymbol{q}_{ev}\|\|\boldsymbol{\omega}_d\|\dot{k}_2 + 3k_2\|\boldsymbol{\omega}_d\|\frac{d\|\boldsymbol{q}_{ev}\|}{dt}\right) \\
&= -\frac{1}{2}\lambda_M(\boldsymbol{J})k_2 q_{e0}\|\boldsymbol{q}_{ev}\|\left(2k_2^2\|\boldsymbol{q}_{ev}\| + 3k_2\|\boldsymbol{\omega}_d\|\right) \\
&\quad + \lambda_M(\boldsymbol{J})\left(2k_2\|\boldsymbol{q}_{ev}\|^2 + 3\|\boldsymbol{q}_{ev}\|\|\boldsymbol{\omega}_d\|\right)\left(\frac{1}{2}k_2^2 q_{e0} - \frac{\eta}{\|\boldsymbol{q}_{ev}\|}\right) \\
&= -k_2^3 q_{e0}\|\boldsymbol{q}_{ev}\|^2 - \frac{3}{2}k_2^2 q_{e0}\|\boldsymbol{q}_{ev}\|\|\boldsymbol{\omega}_d\| + k_2^3 q_{e0}\|\boldsymbol{q}_{ev}\|^2 \\
&\quad -2k_2\eta\|\boldsymbol{q}_{ev}\| + \frac{3}{2}k_2^2 q_{e0}\|\boldsymbol{q}_{ev}\|\|\boldsymbol{\omega}_d\| - 3\eta\|\boldsymbol{\omega}_d\| \\
&= -2k_2\eta\|\boldsymbol{q}_{ev}\| - 3\eta\|\boldsymbol{\omega}_d\| \\
&\le 0
\end{aligned} \tag{5.49}$$

Hence z is monotonically decreasing and if following inequality is satisfied, $\|\boldsymbol{r}_2\| \le \overline{u}$ could be ensured.

$$\lambda_M(\boldsymbol{J})\left(k_2^2\alpha^2 + 3k_2\alpha\overline{\omega}_d + \overline{\omega}_d^2 + \dot{\overline{\omega}}_d + \eta\right) \le \overline{u} \tag{5.50}$$

On the third stage of controller (51) k_2 no longer increases hence if there is no saturation issue in the second stage, the saturation issue would not occur in the third stage.

The control torque constraint on control parameters, i.e., (5.45) and (5.50), is strict constraint since the worst condition is discussed.

5.3 Simulation results

5.3.1 Simulation of attitude stabilization controller

In order to demonstrate the superiority of attitude stabilization controller (5.18) presented in this chapter, sliding mode controller (5.51) without inertia matrix uncertainty and control torque constraint constructed as follow is compared.

$$\begin{aligned}
&\boldsymbol{u} = -k\boldsymbol{s} + \boldsymbol{\omega}^{\times}\boldsymbol{J}\boldsymbol{\omega} - \frac{k_1}{2}\left(q_0\boldsymbol{I}_3 + \boldsymbol{q}_v^{\times}\right)\boldsymbol{\omega} - \overline{d}\,\boldsymbol{sgn}(\boldsymbol{s}) \\
&\boldsymbol{s} = \boldsymbol{\omega} + k_1\boldsymbol{q}_v \\
&k_1 = 0.1, \quad k = 5
\end{aligned} \tag{5.51}$$

Set the simulation parameters as follows:

$$\begin{aligned}
&\boldsymbol{J} = \mathrm{diag}([30\ 25\ 20])\mathrm{kg\cdot m^2},\ \hat{\boldsymbol{J}} = \mathrm{diag}([28\ \ 24\ \ 21])\mathrm{kg\cdot m^2},\ \lambda = 2 \\
&\boldsymbol{\omega} = [-0.03\ \ -0.04\ \ 0.05]^T\mathrm{rad/s},\ \boldsymbol{q} = \left[0\ \ \frac{\sqrt{6}}{6}\ \ \frac{\sqrt{3}}{3}\ \ \frac{\sqrt{2}}{2}\right]^T
\end{aligned} \tag{5.52}$$

Assume the disturbance torque consists of Gauss white noise and sinusoidal signal written as follows:

$$d_i = 5 \times 10^{-4}\text{rand}(-1,1) + 5 \times 10^{-4}\sin t \tag{5.53}$$

Hence the norm upper bound of disturbance torque satisfies

$$\overline{d} = 10^{-3} \tag{5.54}$$

The simulation results of standard sliding mode controller (5.51) are given as follows:

Based on Figs. 5.1–5.3, it can be found that system state converges to zero within 200 s. However, the main drawback of controller (5.51) could be concluded as two aspects: (1) system convergence rate descends severely since the angular velocity descends severely and the terminal convergence rate is low. (2) The steady accuracy at 200 s is 2×10^{-5} of angular velocity, 1×10^{-4} of attitude quaternion which needs to be improved. (3) The initial control torque exceeds the system upper bound.

Based on the simulation results, it can be found that although the inertia matrix uncertainty and control torque constraint issue is ignored in controller (5.51), the performance of controller (5.51) still need improvements.

Based on the discussion previously, larger k_1 and smaller η could bring better convergence rate, but the control torque constraint (5.28) and (5.33) may not be satisfied. Considering that the maximum norm of satellite angular velocity is about 0.1 rad/s, set parameter k_1 and η are selected as follows:

$$k_1 = 0.1, \quad \eta = 0.01 \tag{5.55}$$

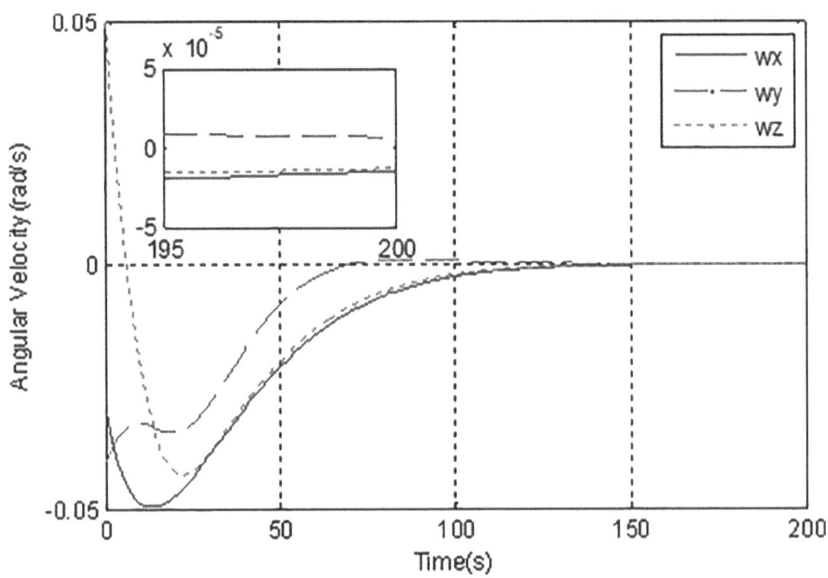

FIG. 5.1 Curve of angular velocity.

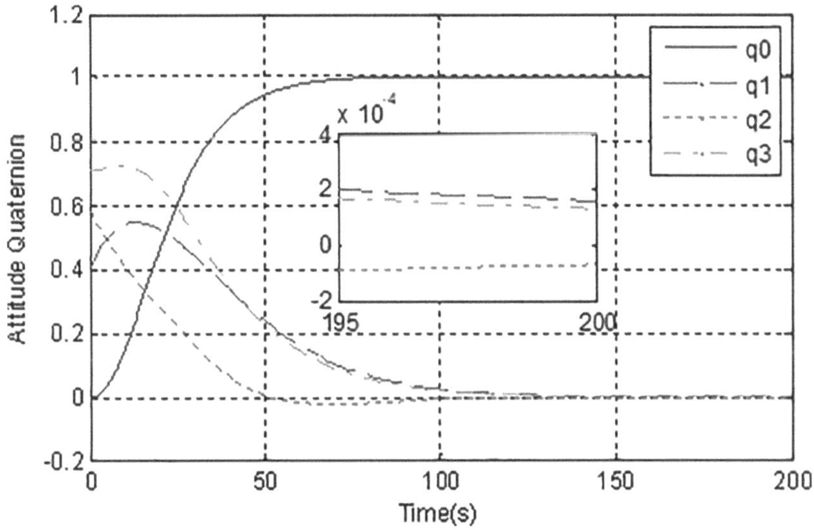

FIG. 5.2 Curve of attitude quaternion.

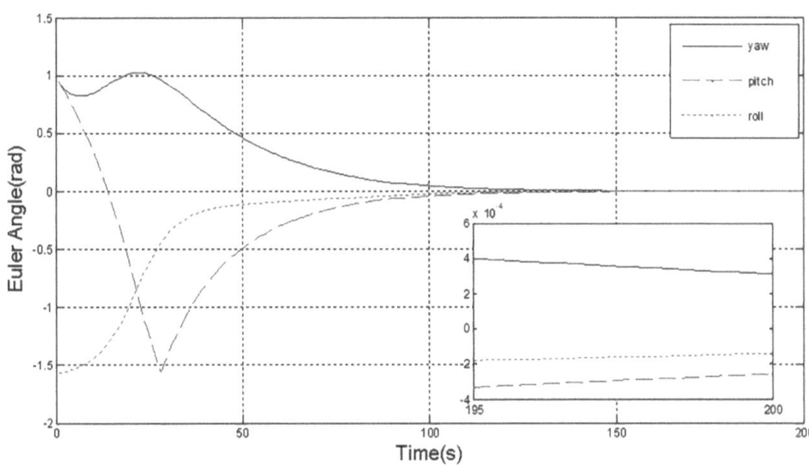

FIG. 5.3 Curve of Euler angle.

Select control parameters as follows:

$$k_s = 2, \quad \beta = 0.005 \tag{5.56}$$

Assume the norm upper bound of control torque satisfies

$$\overline{u} = 0.5 \ \mathrm{N \cdot m} \tag{5.57}$$

The simulation results of controller (5.18) with inertia matrix uncertainty and control torque constraint are given in Figs. 5.4–5.7.

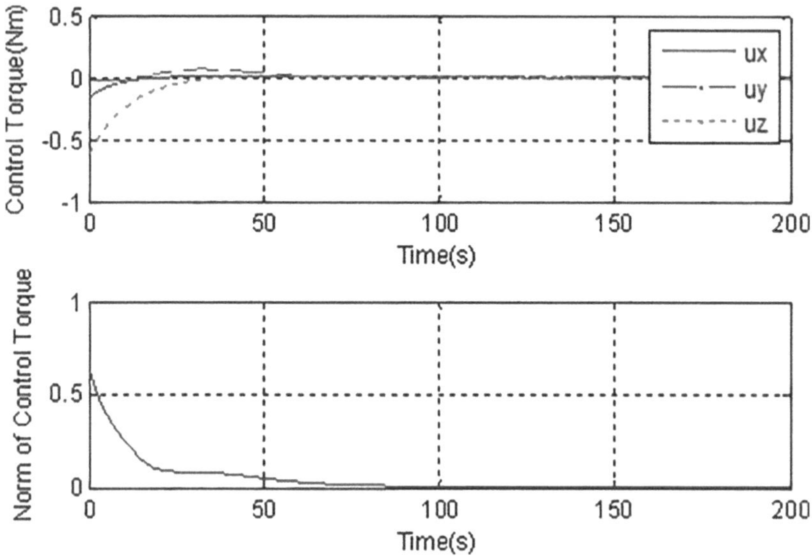

FIG. 5.4 Curve of control torque and its norm.

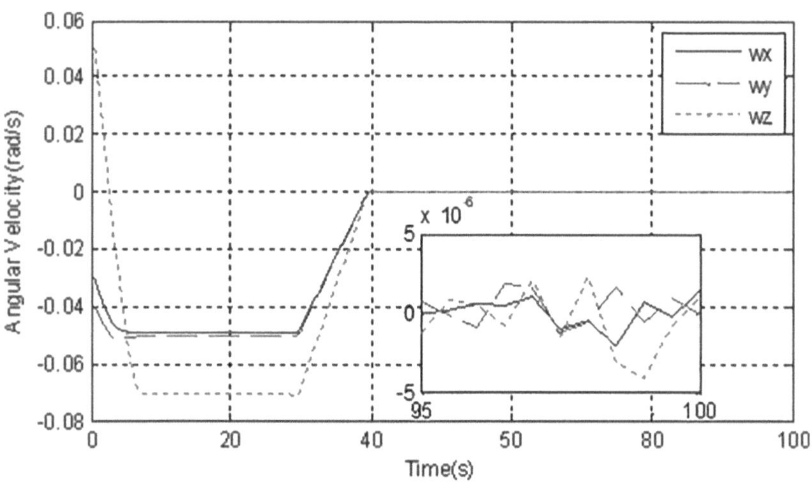

FIG. 5.5 Curve of angular velocity.

Based on Figs. 5.4–5.7, it can be found that system converges to zero at about 40 s. The steady accuracy at 100 s is about 5×10^{-6} of angular velocity and 5×10^{-8} of attitude quaternion. System governed by controller (5.18) has better steady accuracy, while the convergence rate is largely improved. Based on Fig. 5.4, it can be found that the system has about a 20 s maneuver process with constant angular velocity hence the convergence rate of attitude

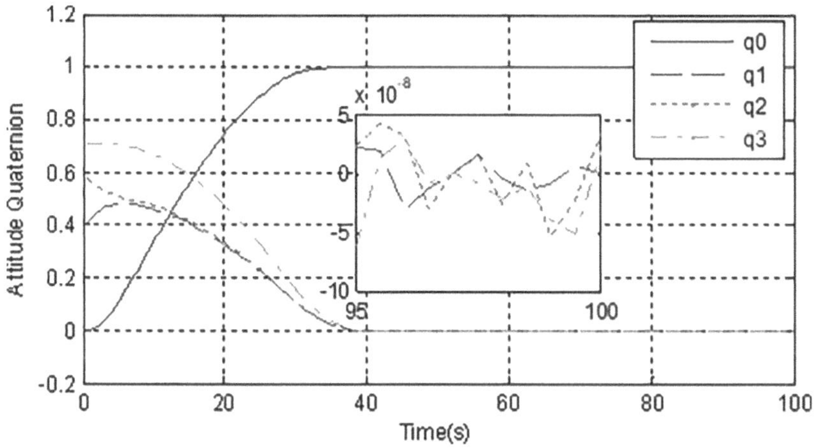

FIG. 5.6 Curve of attitude quaternion.

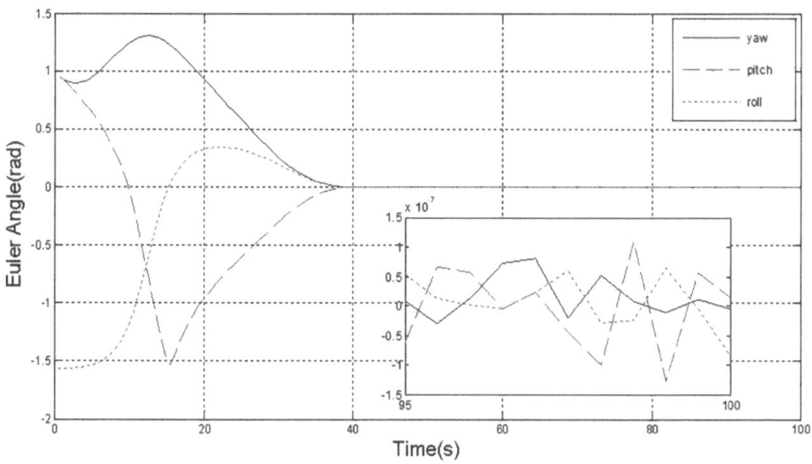

FIG. 5.7 Curve of Euler angle.

quaternion is maintained at a high level. According to Fig. 5.6, it could also be found that during this process the control torque is small; therefore, the maneuver stage with constant angular velocity could also improve the effectiveness of control torque. Moreover, the system begins to slow down with constant deceleration at about 30s, and this bang-bang like trajectory is the key to improve system convergence rate. Based on Fig. 5.7, it could be found that parameter k_2 is enlarged about 60 times and by updating the sliding mode parameter, open-loop bang-bang controller is combined with closed-loop sliding mode controller. Although controller (5.18) has advantage in convergence rate, there still exist two issues: (1) parameter k_2 is enlarged significantly and the sign function term in controller is also enlarged, hence the chattering issue is intensified and (2) based on Fig. 5.6, it can be found that during the deceleration process the norm of control torque does not reach the system upper bound which means the system capability is not fully utilized. These issues need to be solved in later work.

5.3.2 Simulation of attitude tracking controller

In order to demonstrate the superiority of the controller presented in this chapter, sliding mode controller (5.58) without inertia matrix uncertainty and control torque constraint constructed as follow is compared.

$$\begin{aligned}\boldsymbol{u} = & -k\boldsymbol{s} + \boldsymbol{J}\boldsymbol{R}\dot{\boldsymbol{\omega}}_d - \boldsymbol{J}\boldsymbol{\omega}_e^{\times}\boldsymbol{R}\boldsymbol{\omega}_d + (\boldsymbol{\omega}_e + \boldsymbol{R}\boldsymbol{\omega}_d)^{\times}\boldsymbol{J}(\boldsymbol{\omega}_e + \boldsymbol{R}\boldsymbol{\omega}_d) \\ & - \frac{1}{2}k_1\boldsymbol{J}\left(q_{e0}\boldsymbol{I}_3 + \boldsymbol{q}_{ev}^{\times}\right)\boldsymbol{\omega}_e \\ \boldsymbol{s} = & \ \boldsymbol{\omega}_e + k_1\boldsymbol{q}_{ev} \\ k_1 = & \ 0.1, \quad k = 5\end{aligned} \tag{5.58}$$

Considering that the control parameters are basically all the same as previous section, hence the method to select them is omitted. Set simulation parameters as follows:

$$\begin{aligned}&\boldsymbol{J} = \mathrm{diag}([30 \ \ 25 \ \ 20])\mathrm{kg\cdot m^2}, \ \hat{\boldsymbol{J}} = \mathrm{diag}([28 \ \ 24 \ \ 21])\mathrm{kg\cdot m^2}, \ \lambda = 3 \\ &d_i = 5\times10^{-4}\,\mathrm{rand}(-1,1) + 5\times10^{-4}\sin t, \ \ \overline{d} = 0.001 \\ &\boldsymbol{\omega} = [0 \ \ 0 \ \ -0.1]^T\mathrm{rad/s}, \ \ \boldsymbol{q} = \left[1/2 \ \ -1/2 \ \ \sqrt{3}/3 \ \ \sqrt{6}/6\right]^T \\ &\boldsymbol{\omega}_d = [0.01 \ \ -0.005 \ \ 0]^T\mathrm{rad/s}, \ \ \dot{\boldsymbol{\omega}}_d = \boldsymbol{0}_{3\times1}, \boldsymbol{q}_d = [1 \ \ 0 \ \ 0 \ \ 0]^T\end{aligned} \tag{5.59}$$

The simulation results of standard sliding mode controller (5.58) are given in Figs. 5.10–5.13.

Based on Figs. 5.8–5.10, it can be found that the main issue of standard sliding mode controller (5.58) could also be concluded as two aspects: low convergence rate and excessive initial control torque.

Set the control parameter of controller (5.35) as follows:

$$k_1 = 0.1, \quad \eta = 0.01, \quad k_s = 2, \quad \beta = 0.005, \quad \overline{u} = 0.5 \ \mathrm{N\cdot m} \tag{5.60}$$

The simulation results of attitude tracking controller (5.35) presented in this chapter are given as follow.

Based on Figs. 5.14–5.18, it can be found that system state converges to zero at about 30 s. The steady accuracy at 50 s is 2×10^{-6} of error angular velocity and 1×10^{-7} of error quaternion. System convergence rate and steady accuracy are largely improved. Moreover, the norm of control torque does not exceed the system upper bound during whole control process. The reason of this improvement is similar as the analysis of simulation of attitude stabilization control therefore it is omitted in this part.

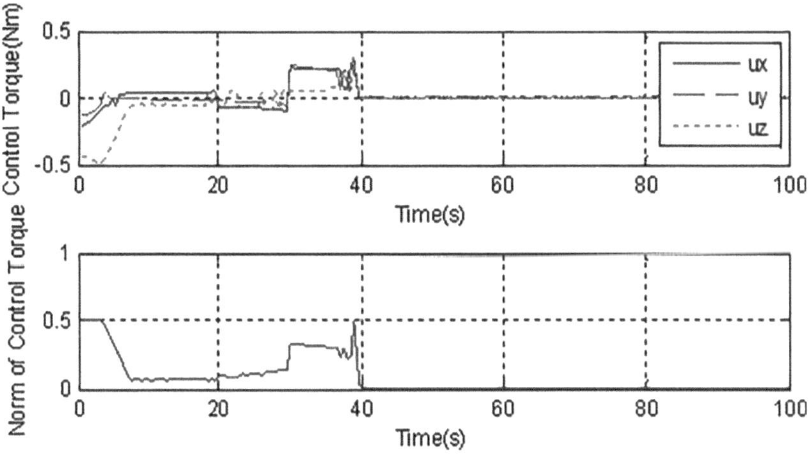

FIG. 5.8 Curve of control torque and its norm.

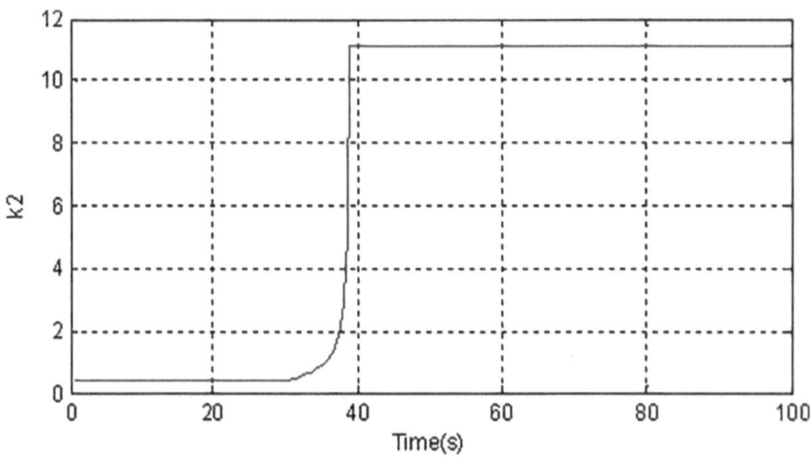

FIG. 5.9 Curve of sliding mode parameter k_2.

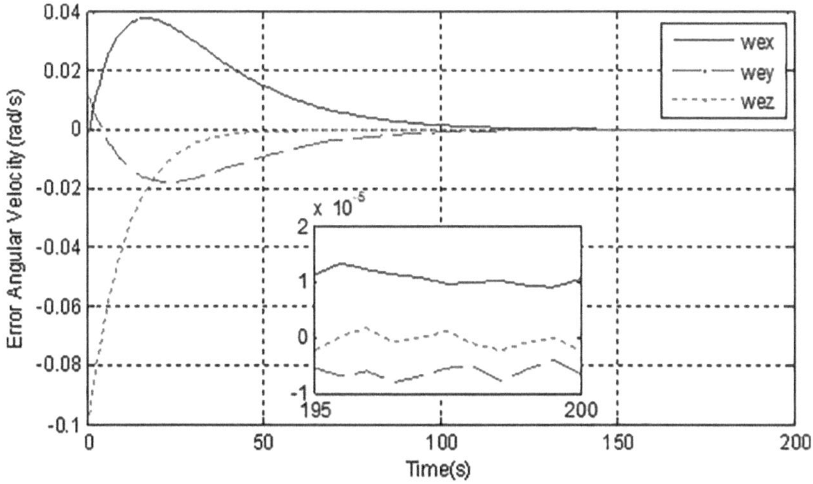

FIG. 5.10 Curve of error angular velocity.

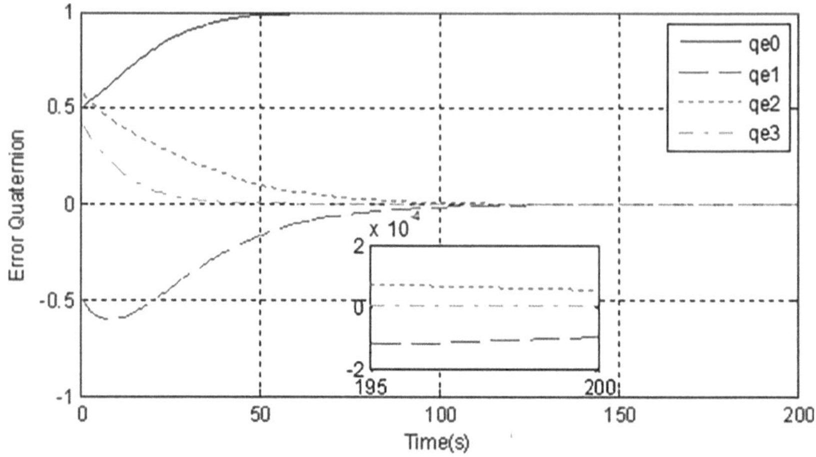

FIG. 5.11 Curve of error quaternion.

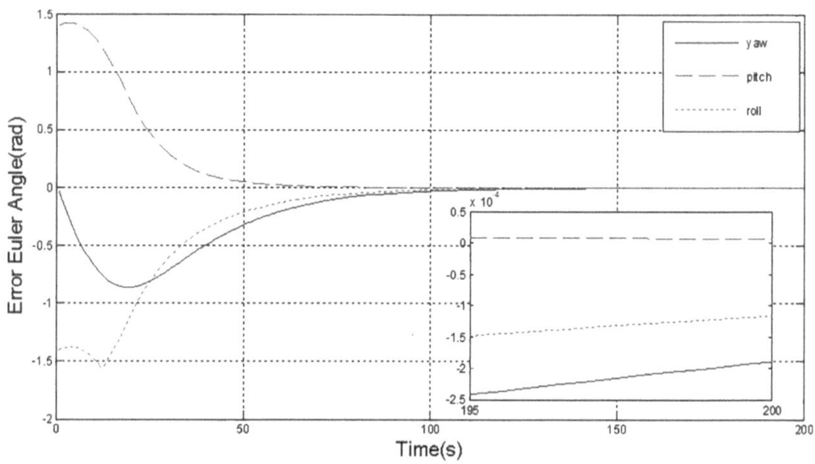

FIG. 5.12 Curve of error Euler angle.

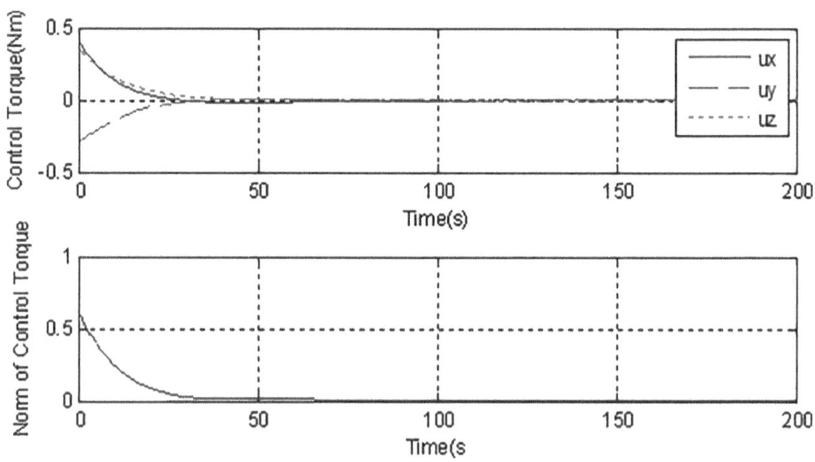

FIG. 5.13 Curve of control torque and its norm.

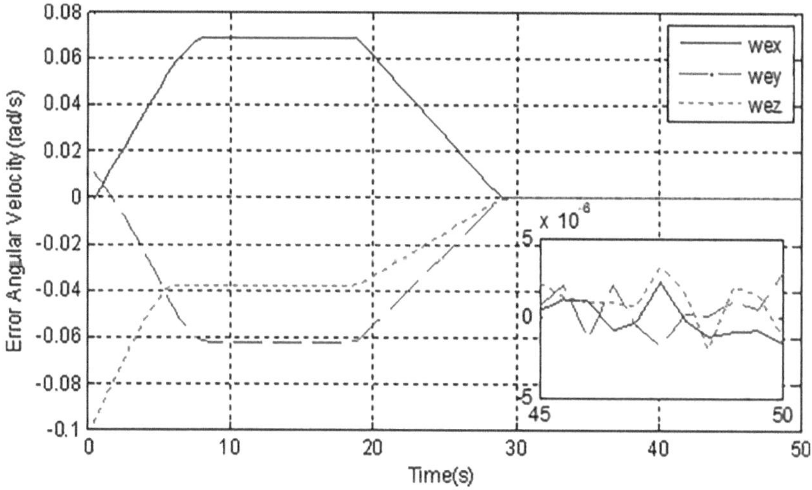

FIG. 5.14 Curve of error angular velocity.

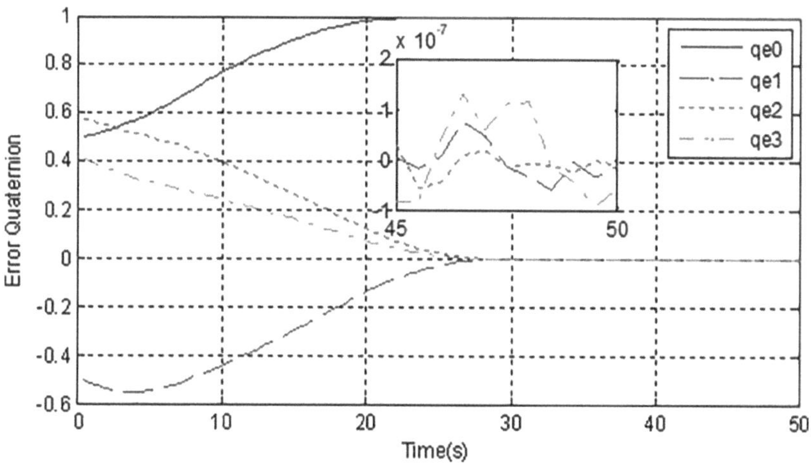

FIG. 5.15 Curve of error quaternion.

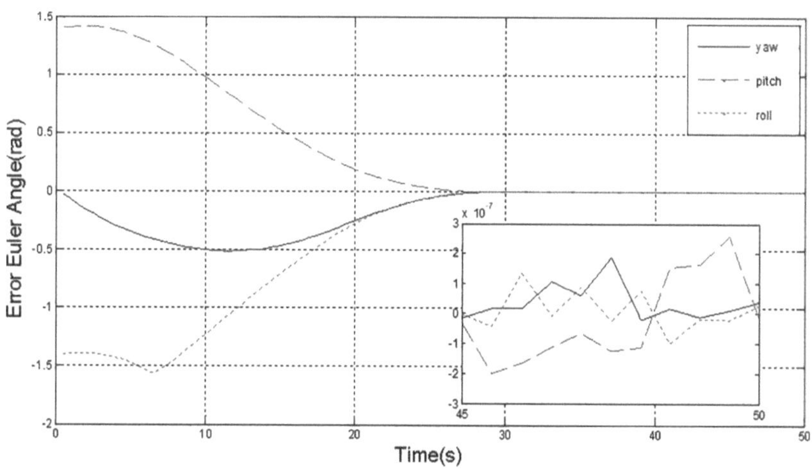

FIG. 5.16 Curve of error Euler angle.

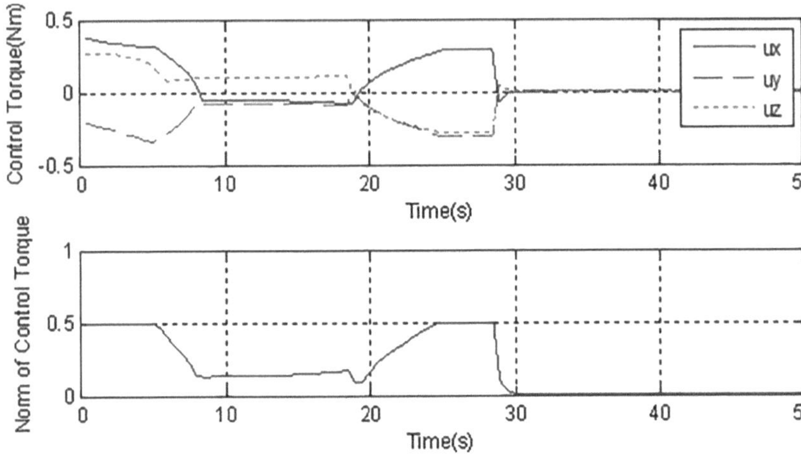

FIG. 5.17 Curve of control torque and its norm.

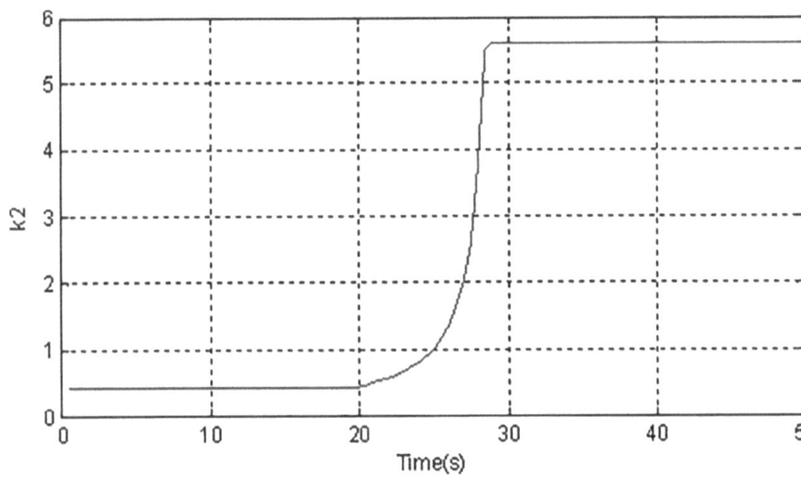

FIG. 5.18 Curve of sliding mode parameter k_2.

5.4 Conclusion

In this chapter, robust sliding mode controllers based on bang-bang logic for satellite attitude control are developed. The open-loop bang-bang control is combined with closed-loop sliding mode control by updating sliding mode parameter. A two-stage structure sliding mode (a maneuver stage with fixed angular velocity and a slow-down stage with fixed deceleration) is proposed, and the system convergence rate is largely improved while the robustness of perturbation is maintained. The system both has time-efficient character of bang-bang logic and strong robustness of sliding mode control. The performance of the controller is demonstrated by numerical simulation results.

In fact, the key to improve system convergence rate could be concluded as following two aspects: (1) the maneuver stage with fixed angular velocity could maintain the convergence rate of quaternion at a high level and (2) updating sliding mode parameter could improve the terminal convergence rate when system state approaches its equilibrium point. However, enlarging the parameter would aggravate the chattering issue, and this should be discussed in later research.

References

[1] B. Xiao, S. Yin, L.G. Wu, A structure simple controller for satellite attitude tracking maneuver, IEEE Trans. Ind. Electron. 64 (2) (2017) 1436–1446.

[2] B. Xiao, S. Yin, H.J. Gao, Tracking control of robotic manipulators with uncertain kinematics and dynamics, IEEE Trans. Ind. Electron. 63 (10) (2016) 6439–6449.

[3] B. Xiao, S. Yin, Velocity-free fault tolerant and uncertainty attenuation control for a class of nonlinear systems, IEEE Trans. Ind. Electron. 63 (7) (2016) 4400–4411.

[4] S.N. Wu, Z.W. Sun, H. Deng, Robust sliding mode controller design for globally fast attitude tracking of target spacecraft, Int. J. Innov. Comput. Inf. Control 7 (5) (2011) 2087–2098.

[5] S.N. Wu, R. Wang, G. Radice, Z.G. Wu, Robust attitude maneuver control of spacecraft with reaction wheel low-speed friction compensation, Aerosp. Sci. Technol. 43 (2015) 213–218.

[6] S.N. Wu, Z.G. Wu, G. Radice, R. Wang, Adaptive control for spacecraft relative reanslation with parametric uncertainty, Aerosp. Sci. Technol. 31 (1) (2013) 53–58.

[7] Y.L. Liu, S.N. Wu, K.M. Zhang, Z.W. Wu, Gravitational orbit-attitude coupling dynamics of a large solar power satellite, Aerosp. Sci. Technol. 62 (2017) 46–54.

[8] D. Verbin, V.J. Lappas, Rapid rotational maneuvering of rigid satellites with hybrid actuators configuration, J. Guid. Control. Dyn. 36 (2) (2013) 532–547.

[9] D. Verbin, V.J. Lappas, Z.B. Joseph, Time efficient angular steering laws for rigid satellites, J. Guid. Control. Dyn. 34 (3) (2011) 878–892.

[10] X.B. Cao, C.F. Yue, M. Liu, Flexible satellite attitude maneuver via constrained torque distribution and active vibration suppression, Aerosp. Sci. Technol. 67 (2017) 387–397.

[11] S.C. Lo, Y.P. Chen, Smooth sliding-mode control for spacecraft attitude tracking maneuvers, J. Guid. Control. Dyn. 18 (6) (1995) 1345–1349.

[12] Y.Q. Xia, Z. Zhu, M.Y. Fu, S. Wang, Attitude tracking of rigid spacecraft with bounded disturbances, IEEE Trans. Ind. Electron. 58 (2) (2011) 647–659.

[13] A. Levant, A. Michael, Adjustment of high-order sliding-mode controllers, Int. J. Robust Nonlinear Control 19 (15) (2009) 1657–1672.

[14] B.L. Tian, Q. Zong, J. Wang, F. Wang, Quasi-continuous high-order sliding mode controller design for reusable launch vehicles in reentry phase, Aerosp. Sci. Technol. 28 (1) (2013) 198–207.

[15] Q. Zong, J. Wang, B.L. Tian, Y. Tao, Quasi-continuous high-order sliding mode controller and observer design for flexible hypersonic vehicle, Aerosp. Sci. Technol. 27 (1) (2013) 127–137.

[16] W. Ernesto, J.L. Duran, P.R. Acosta, Integral sliding-mode active filter control for harmonic distortion compensation, Electr. Power Compon. Syst. 39 (9) (2011) 833–849.

[17] X.L. Bai, J.L. Junkins, New results for time-optimal three-axis reorientation of a rigid spacecraft, J. Guid. Control. Dyn. 32 (4) (2009) 1071–1076.

[18] F. Nagi, A.T. Zulkamain, J. Nagi, Tuning fuzzy bang-bang relay controller for satellite attitude control system, Aerosp. Sci. Technol. 26 (1) (2013) 76–86.
[19] Z.G. Zhang, S.P. Gong, J.F. Liu, The fuel-optimal trajectory for finite-thrust lunar ascent, Aerosp. Sci. Technol. 39 (2014) 675–684.
[20] S.P. Gong, J.F. Liu, F.H. Jiang, Interplanetary trajectory design for a hybrid propulsion system, Aerosp. Sci. Technol. 45 (2015) 104–113.
[21] E.B. Meier, A.E. Bryson, Efficient algorithm for time-optimal control of a 2-link manipulator, J. Guid. Control. Dyn. 13 (5) (1990) 859–866.
[22] R.M. Byers, Minimum time reorientation of an arbitrary rigid body, in: Proceedings of the AAS/AIAA Astrodynamics Conference, 1993, pp. 1889–1903.

Chapter 6

Robust finite-time control based on three-stage structure-sliding mode

6.1 Introduction

Recent years plentiful researches have been done on the field of satellite attitude control. Generally, satellite attitude control issue could be classified into two aspects: attitude stabilization control and attitude tracking control. The latter issue is more complex since the dynamic model is more complex than the former dynamic model. But the control algorithms for these two issues are generally all the same since the kinetic model of these two issues are all the same.

Sliding mode control is a mature and widely used control algorithm in satellite attitude control. The structure of standard sliding mode for satellite attitude control is simple, and it has definite physical meaning. Moreover, sliding mode control is an inherent robust control algorithm. Because of these advantages, sliding mode control algorithm has attracted much attention and research. Boskovi et al. [1] proposed a robust tracking controller under control input saturation. In this paper, a dynamic sliding mode is constructed, and the parameter is updated with the convergence of system state. Lo and Chen [2] presented a smooth sliding mode control algorithm for spacecraft attitude tracking control. In this paper, a class of sliding mode is presented, and a typical Lyapunov function is constructed. Xia et al. [3] presented an adaptive sliding mode controller for satellite attitude tracking control with system inertia matrix uncertainty and bounded disturbances. Leavant [4,5] has done some work focused on high-order sliding mode to improve the system accuracy.

However, the exponential convergence rate is a serious drawback of standard sliding mode control. The convergence rate descends severely, and the system state converges to zero with infinite time. In order to deal with this issue, controllers with finite-time stability are developed. Ryan [6], Wu et al. [7], and Ding and Li [8] presented some methods to analyze finite-time stability such as Lyapunov method and terminal sliding mode method. Amato and Ariola [9]

Fast Satellite Attitude Maneuver and Control. https://doi.org/10.1016/B978-0-323-95455-6.00003-2

designed finite-time controllers for general linear system and presented a method to analyze finite-time stability for discrete systems. Yu et al. [10], Xiao et al. [11] and Meng et al. [12] designed finite controllers for robot manipulators and multiple rigid bodies. Researches about finite-time control have also been done on satellite attitude control. Wu et al. [13] presented a quaternion-based finite-time controller for satellite tracking issue. In this paper, the system model was transformed to Lagrange-like and based on this model, a robust finite-time sliding mode was constructed. The system inertia matrix uncertainty and disturbance torque were taken consideration. In his later work [14], the finite-time controller was modified for the spacecraft flexibility. Liang and Wang [15–17] have done a lot of work focused on the satellite formation finite-time control. A class of finite-time sliding mode was constructed, and the singularity issue was solved in the constructed sliding modes. Xiao et al. [18–20] has done a lot of work focused on the finite-time control with fault tolerant capability. In these papers, the finite-time controllers under fault conditions such as missing angular velocity, actuator misalignment and actuator failure were discussed. Tiwari et al. [21] proposed a higher order sliding mode with finite-time stability for satellite attitude tracking control. The main contribution of this paper is the chattering-free finite-time sliding mode. Generally, there are several issues to be solved in satellite attitude finite-time control: the system inertia matrix uncertainty, the disturbance torque, the control torque constraint and the singularity issue. Most of the above work focused on some of these aspects but an overall discussion of these issues is lacked.

The main contribution of this chapter is the finite-time sliding mode with three stage structure. Considering the convergence rate of attitude quaternion is determined by angular velocity, angular velocity is maintained constant at the initial stage of the sliding mode hence the convergence rate of attitude quaternion could be maintained at a high level. The terminal stage of the sliding mode is also modified based on standard sliding mode hence the convergence time could be finite and the singularity issue is solved. The convergence time and steady accuracy could both be improved by the presented sliding mode. Moreover, the system inertia matrix uncertainty, disturbance torque, and control torque constraint are all taken consideration in this chapter. The finite-time stability for all conditions considered in this chapter is proved, and the performance of the controller is demonstrated by simulation results.

6.2 Finite-time controller for attitude stabilization control

6.2.1 Finite-time controller

The most widely used and mature sliding mode in satellite attitude stabilization control could be written as follows:

$$s = \boldsymbol{\omega} + k\boldsymbol{q}_v, \quad (k > 0) \tag{6.1}$$

Standard sliding mode (6.1) has the advantage such as simple structure, definite physical meaning, thus it is it is convenient for the design of attitude controller. However, a serious defect of sliding mode (6.1) is its convergence rate. The convergence rate of sliding mode (6.1) is exponential and descends severely when system state approaches to zero. Noticing that the physical meaning of sliding mode (6.1) is the reverse of vector $\boldsymbol{\omega}$ and $\boldsymbol{q}_v$, the modified sliding mode presented in this chapter with improved convergence rate could be written as follows:

$$s = \begin{cases} \boldsymbol{\omega} + k_1\boldsymbol{e}, & \|\boldsymbol{q}_v\| > \alpha \\ \boldsymbol{\omega} + k_2\boldsymbol{q}_v, & \beta < \|\boldsymbol{q}_v\| \le \alpha \\ \boldsymbol{\omega} + k_3\|\boldsymbol{q}_v\|^r\boldsymbol{e}, & \|\boldsymbol{q}_v\| \le \beta \end{cases} \tag{6.2}$$

where k_1, k_2, k_2, α, β, r are all positive scalars and r satisfies follow inequality.

$$1/2 < r < 1 \tag{6.3}$$

Noticing that angular velocity should be continuous at the switch points, control parameters should satisfy follow equation.

$$k_1 = \alpha k_2, \quad k_2 = \beta^{r-1}k_3 \tag{6.4}$$

Based on (6.2), it can be found that sliding mode (6.2) consists of three stages: the maneuver stage, i.e., $\boldsymbol{s} = \boldsymbol{\omega} + k_1\boldsymbol{e}$, the slowdown stage, i.e., $\boldsymbol{s} = \boldsymbol{\omega} + k_2\boldsymbol{q}_v$, and the converge stage, i.e., $\boldsymbol{s} = \boldsymbol{\omega} + k_3\|\boldsymbol{q}_v\|^r\boldsymbol{e}$. The angular velocity is constant at the maneuver stage; therefore, the system state should stay at this stage as long as possible to improve system convergence rate, hence smaller parameter α should be selected. Considering that control torque of satellite actuator is upper bounded, α should not be selected too small since the control torque demanded in the slowdown stage would tend to infinite. Moreover, in order to avoid the singularity issue, parameter β should be selected small enough to ensure (6.5) is satisfied before the system reaches the converge stage.

$$\boldsymbol{\omega}^T\boldsymbol{q}_v/(\|\boldsymbol{\omega}\|\|\boldsymbol{q}_v\|) = -1 \tag{6.5}$$

The essence of Eq. (6.5) is that angular velocity and attitude queternion is reversed. This property will be used in the later text.

The next step is to prove the system state converges to zero in finite time along sliding mode (18). Based on the definition of the maneuver stage and the slowdown stage of (18), it can be found that the convergence time t_1 and t_2 of these two stages satisfy

$$t_1 \le \left(\pi - 2\arccos\sqrt{1-\alpha^2}\right)\Big/k_1, t_2 \le (\ln\alpha - \ln\beta)\Big/\left(k_2\sqrt{1-\alpha^2}\right) \tag{6.6}$$

On the converge stage of sliding mode (6.2), select Lyapunov function as follows:

$$V_q = \boldsymbol{q}_v^T\boldsymbol{q}_v \tag{6.7}$$

Calculate its derivative and substitute (6.2) into (6.7), it can be got that

$$\begin{aligned}\dot{V}_q &= 2\boldsymbol{q}_v^T\dot{\boldsymbol{q}}_v = \boldsymbol{q}_v^T\left(q_0\boldsymbol{I}_3 + \boldsymbol{q}_v^\times\right)\boldsymbol{\omega} = -k_3q_0\|\boldsymbol{q}_v\|^r\boldsymbol{q}_v^T\boldsymbol{e} = -k_3q_0\|\boldsymbol{q}_v\|^{r+1}\\ &\le -k_3q_0qV_q^{\frac{r+1}{2}}\end{aligned} \tag{6.8}$$

Since $\sqrt{1-\beta^2} \le q_0 \le 1$ and $1/2 < r < 1$ on this stage, attitude quaternion $\boldsymbol{q}_v$ converges to zero in finite time, therefore angular velocity $\boldsymbol{\omega}$ converges to zero based on the definition of sliding mode (6.2).

The next step is to design satellite attitude stabilization controller to ensure the system state could reach sliding mode (6.2) in finite time. The robust finite controller for attitude stabilization control presented in this chapter could be written as follows:

$$\boldsymbol{u} = \begin{cases} -k\|\boldsymbol{s}\|^p\boldsymbol{s}_e + \boldsymbol{\varphi}_1\boldsymbol{\omega} - l_1\,\text{sgn}\,(\boldsymbol{s}) & \|\boldsymbol{q}_v\| > \alpha \\ -k\|\boldsymbol{s}\|^p\boldsymbol{s}_e + \boldsymbol{\varphi}_2\boldsymbol{\omega} - l_2\,\text{sgn}\,(\boldsymbol{s}) & \beta < \|\boldsymbol{q}_v\| \le \alpha \\ -k\|\boldsymbol{s}\|^p\boldsymbol{s}_e + \boldsymbol{\varphi}_3\boldsymbol{\omega} + \dfrac{k_3^2}{2}rq_0\|\boldsymbol{q}_v\|^{2r-1}\widehat{\boldsymbol{J}}\boldsymbol{e} - l_3\,\text{sgn}\,(\boldsymbol{s}) & \|\boldsymbol{q}_v\| \le \beta \end{cases} \tag{6.9}$$

where

$$\begin{cases} \boldsymbol{\varphi}_1 = \boldsymbol{\omega}^\times\widehat{\boldsymbol{J}} - \dfrac{k_1}{2}\widehat{\boldsymbol{J}}\boldsymbol{e}^\times\left(\boldsymbol{I}_3 + \cot\dfrac{\varphi}{2}\boldsymbol{e}^\times\right) \\ \boldsymbol{\varphi}_2 = \boldsymbol{\omega}^\times\widehat{\boldsymbol{J}} - \dfrac{k_2}{2}\widehat{\boldsymbol{J}}\boldsymbol{F} \\ \boldsymbol{\varphi}_3 = \boldsymbol{\omega}^\times\widehat{\boldsymbol{J}} \end{cases} \tag{6.10}$$

$$\begin{cases} l_1 = \overline{d} + \lambda\|\boldsymbol{\omega}\|^2 + \dfrac{k_1}{2}\lambda\left(1 + \cot\dfrac{\varphi}{2}\right)\|\boldsymbol{\omega}\| \\ l_2 = \overline{d} + \lambda\|\boldsymbol{\omega}\|^2 + \dfrac{k_2}{2}\lambda\|\boldsymbol{\omega}\| \\ l_3 = \overline{d} + \lambda\|\boldsymbol{\omega}\|^2 + \dfrac{k_3^2}{2}r\lambda|q_0|\|\boldsymbol{q}_v\|^{2r-1} \end{cases} \tag{6.11}$$

where p is a positive scalar and satisfies $1 > p > 0$, $\boldsymbol{s}_e$ is unit direction vector of $\boldsymbol{s}$ and satisfies $\boldsymbol{s}_e = \boldsymbol{s}/\|\boldsymbol{s}\|$, λ is a positive scalar and satisfies $\lambda \ge \lambda_M(\widetilde{\boldsymbol{J}})$, sgn($\boldsymbol{r}$) is the sign function of vector $\boldsymbol{r}$.

Noticing that $1/2 < r < 1$, controller (6.9) has no singularity issue.

In order to prove the finite-time stability of controller (6.9), select Lyapunov function as follows:

$$V = \frac{1}{2}\boldsymbol{s}^T\boldsymbol{J}\boldsymbol{s} \tag{6.12}$$

Based on (6.11), it can be found that

$$V \geq \frac{1}{2}\lambda_m(\boldsymbol{J})\|\boldsymbol{s}\|^2 \tag{6.13}$$

Calculate the derivative of V on the maneuver stage ($\|\boldsymbol{q}_v\| > \alpha$) and noticing property (6.13) it can be got that

$$\begin{aligned}
\dot{V} &= \boldsymbol{s}^T\boldsymbol{J}\dot{\boldsymbol{s}} = \boldsymbol{s}^T\boldsymbol{J}\dot{\boldsymbol{\omega}} + k_1\boldsymbol{s}^T\boldsymbol{J}\dot{\boldsymbol{e}} \\
&= \boldsymbol{s}^T\boldsymbol{u} - \boldsymbol{s}^T\boldsymbol{\omega}^\times\boldsymbol{J}\boldsymbol{\omega} + \boldsymbol{s}^T\boldsymbol{d} + \frac{k_1}{2}\boldsymbol{s}^T\boldsymbol{J}\boldsymbol{e}^\times\left(\boldsymbol{I}_3 - \cot\frac{\varphi}{2}\boldsymbol{e}^\times\right)\boldsymbol{\omega} \\
&= -k\|\boldsymbol{s}\|^p\boldsymbol{s}^T\boldsymbol{s}_e - \boldsymbol{s}^T\boldsymbol{\omega}^\times\tilde{\boldsymbol{J}}\boldsymbol{\omega} + \boldsymbol{s}^T\boldsymbol{d} + \frac{k_1}{2}\boldsymbol{s}^T\tilde{\boldsymbol{J}}\boldsymbol{e}^\times\left(\boldsymbol{I}_3 - \cot\frac{\varphi}{2}\boldsymbol{e}^\times\right)\boldsymbol{\omega} \\
&\quad - \left(\bar{d} + \lambda\|\boldsymbol{\omega}\|^2 + \frac{k_1}{2}\lambda\left(1 + \cot\frac{\varphi}{2}\right)\|\boldsymbol{\omega}\|\right)\boldsymbol{s}^T\operatorname{sgn}(\boldsymbol{s}) \\
&\leq -k\|\boldsymbol{s}\|^{p+1} + \bar{d}\|\boldsymbol{s}\| + \lambda_M\left(\tilde{\boldsymbol{J}}\right)\|\boldsymbol{s}\|\|\boldsymbol{\omega}\|^2 + \frac{k_1}{2}\left(1 + \cot\frac{\varphi}{2}\right)\lambda_M\left(\tilde{\boldsymbol{J}}\right)\|\boldsymbol{s}\|\|\boldsymbol{\omega}\| \\
&\quad - \left(\bar{d} + \lambda\|\boldsymbol{\omega}\|^2 + \frac{k_1}{2}\lambda\left(1 + \cot\frac{\varphi}{2}\right)\|\boldsymbol{\omega}\|\right)\|\boldsymbol{s}\| - \bar{d}\|\boldsymbol{s}\| \\
&\leq -k\|\boldsymbol{s}\|^{p+1} \leq -\mu V^{\frac{p+1}{2}}
\end{aligned} \tag{6.14}$$

Calculate the derivative of V on the slowdown stage ($\beta < \|\boldsymbol{q}_v\| \leq \alpha$) and noticing property (6.13) it can be got that

$$\begin{aligned}
\dot{V} &= \boldsymbol{s}^T\boldsymbol{J}\dot{\boldsymbol{s}} = \boldsymbol{s}^T\boldsymbol{J}\dot{\boldsymbol{\omega}} + k_2\boldsymbol{s}^T\boldsymbol{J}\dot{\boldsymbol{q}}_v \\
&= \boldsymbol{s}^T\boldsymbol{u} - \boldsymbol{s}^T\boldsymbol{\omega}^\times\boldsymbol{J}\boldsymbol{\omega} + \boldsymbol{s}^T\boldsymbol{d} + \frac{k_2}{2}\boldsymbol{s}^T\boldsymbol{J}\boldsymbol{F}\boldsymbol{\omega} \\
&= -k\|\boldsymbol{s}\|^p\boldsymbol{s}^T\boldsymbol{s}_e - \boldsymbol{s}^T\boldsymbol{\omega}^\times\tilde{\boldsymbol{J}}\boldsymbol{\omega} + \boldsymbol{s}^T\boldsymbol{d} + \frac{k_2}{2}\boldsymbol{s}^T\tilde{\boldsymbol{J}}\boldsymbol{F}\boldsymbol{\omega} \\
&\quad - \left(\bar{d} + \lambda\|\boldsymbol{\omega}\|^2 + \frac{k_2}{2}\lambda\|\boldsymbol{\omega}\|\right)\boldsymbol{s}^T\operatorname{sgn}(\boldsymbol{s}) \\
&\leq -k\|\boldsymbol{s}\|^{p+1} + \bar{d}\|\boldsymbol{s}\| + \lambda_M\left(\tilde{\boldsymbol{J}}\right)\|\boldsymbol{s}\|\|\boldsymbol{\omega}\|^2 + \frac{k_2}{2}\lambda_M\left(\tilde{\boldsymbol{J}}\right)\|\boldsymbol{s}\|\|\boldsymbol{\omega}\| \\
&\quad - \left(\bar{d} + \lambda\|\boldsymbol{\omega}\|^2 + \frac{k_2}{2}\lambda\|\boldsymbol{\omega}\|\right)\|\boldsymbol{s}\| - \bar{d}\|\boldsymbol{s}\| \\
&\leq -k\|\boldsymbol{s}\|^{p+1} \leq -\mu V^{\frac{p+1}{2}}
\end{aligned} \tag{6.15}$$

Based on (6.14) and (6.15), it can be found that system state would reach the maneuver stage and the slowdown stage of sliding mode (18) in finite time. Noticing the assumption that when the system state reaches the converge stage, angular velocity $\boldsymbol{\omega}$ and attitude quaternion $\boldsymbol{q}_v$ are reversed, therefore $\dot{\boldsymbol{e}} = \boldsymbol{0}$ on this stage. Calculate the derivative of V at this stage ($\|\boldsymbol{q}_v\| \leq \beta$) and noticing property (6.13) it can be got that

$$
\begin{aligned}
\dot{V} &= \boldsymbol{s}^T \boldsymbol{J} \dot{\boldsymbol{s}} = \boldsymbol{s}^T \boldsymbol{J} \dot{\boldsymbol{\omega}} + k_3 \boldsymbol{s}^T \boldsymbol{J} \boldsymbol{e} \frac{d\|\boldsymbol{q}_v\|^r}{dt} + k_3 \|\boldsymbol{q}_v\|^r \boldsymbol{s}^T \boldsymbol{J} \dot{\boldsymbol{e}} \\
&= \boldsymbol{s}^T \boldsymbol{u} - \boldsymbol{s}^T \boldsymbol{\omega}^{\times} \boldsymbol{J} \boldsymbol{\omega} + \boldsymbol{s}^T \boldsymbol{d} - \frac{k_3^2}{2} r q_0 \|\boldsymbol{q}_v\|^{2r-1} \boldsymbol{s}^T \boldsymbol{J} \boldsymbol{e} \\
&= -k\|\boldsymbol{s}\|^p \boldsymbol{s}^T \boldsymbol{s}_e - \boldsymbol{s}^T \boldsymbol{\omega}^{\times} \widetilde{\boldsymbol{J}} \boldsymbol{\omega} - \frac{k_3^2}{2} r q_0 \|\boldsymbol{q}_v\|^{2r-1} \boldsymbol{s}^T \widetilde{\boldsymbol{J}} \boldsymbol{e} \\
&\quad - \left(\overline{d} + \lambda \|\boldsymbol{\omega}\|^2 + \frac{k_3^2}{2} r \lambda |q_0| \|\boldsymbol{q}_v\|^{2r-1} \right) \boldsymbol{s}^T \operatorname{sgn}(\boldsymbol{s}) + \boldsymbol{s}^T \boldsymbol{d} \\
&\le -k\|\boldsymbol{s}\|^{p+1} + \overline{d}\|\boldsymbol{s}\| + \lambda_M\left(\widetilde{\boldsymbol{J}}\right) \|\boldsymbol{s}\| \|\boldsymbol{\omega}\|^2 + \lambda_M\left(\widetilde{\boldsymbol{J}}\right) \frac{k_3^2}{2} r |q_0| \|\boldsymbol{q}_v\|^{2r-1} \|\boldsymbol{s}\| \\
&\quad - \left(\overline{d} + \lambda \|\boldsymbol{\omega}\|^2 + \frac{k_3^2}{2} r \lambda |q_0| \|\boldsymbol{q}_v\|^{2r-1} \right) \|\boldsymbol{s}\| - \overline{d}\|\boldsymbol{s}\| \\
&\le -k\|\boldsymbol{s}\|^{p+1} \le -\mu V^{\frac{p+1}{2}}
\end{aligned}
\tag{6.16}
$$

The unclaimed variable μ in (6.14)–(6.16) is defined as follows:

$$
\mu = k(2/\lambda_m(\boldsymbol{J}))^{\frac{p+1}{2}} \tag{6.17}
$$

Based on (6.14)–(6.16), it can be found that system state reaches sliding mode (6.2) in finite time and the finite-time stability of controller (6.9) has been proved.

6.2.2 Finite-time controller under control torque constraint

The controller presented in Section 6.2.1 does not consider the control torque saturation issue and controller under control torque constraint will be given in this section.

Assume that the norm upper bound of system control torque is $\overline{u}$. The modified controller under control torque constraint could be written as follows:

$$
\boldsymbol{u} = \begin{cases} -\rho_1 \|\boldsymbol{s}\|^p \boldsymbol{s}_e + \boldsymbol{r}_1 & \|\boldsymbol{q}_v\| > \alpha \\ -\rho_2 \|\boldsymbol{s}\|^p \boldsymbol{s}_e + \boldsymbol{r}_2 & \beta < \|\boldsymbol{q}_v\| \le \alpha \\ -\rho_3 \|\boldsymbol{s}\|^p \boldsymbol{s}_e + \boldsymbol{r}_3 & \|\boldsymbol{q}_v\| \le \beta \end{cases} \tag{6.18}
$$

where $\boldsymbol{r}_i$ and ρ_i are defined as follow

$$\begin{cases} \boldsymbol{r}_1 = \boldsymbol{\omega}^{\times}\widehat{\boldsymbol{J}}\boldsymbol{\omega} - \frac{k_1}{2}\widehat{\boldsymbol{J}}\boldsymbol{e}^{\times}\left(\boldsymbol{I}_3 + \cot\frac{\phi}{2}\boldsymbol{e}^{\times}\right)\boldsymbol{\omega} - \left(\overline{d} + \lambda\|\boldsymbol{\omega}\|^2\right. \\ \qquad \left. + \frac{k_1}{2}\lambda\left(1 + \cot\frac{\phi}{2}\right)\|\boldsymbol{\omega}\|\right)\operatorname{sgn}(\boldsymbol{s}) \\ \boldsymbol{r}_2 = \boldsymbol{\omega}^{\times}\widehat{\boldsymbol{J}}\boldsymbol{\omega} - \frac{k_2}{2}\widehat{\boldsymbol{J}}\boldsymbol{F}\boldsymbol{\omega} - \left(\overline{d} + \lambda\|\boldsymbol{\omega}\|^2 + \frac{k_2}{2}\lambda\|\boldsymbol{\omega}\|\right)\operatorname{sgn}(\boldsymbol{s}) \\ \boldsymbol{r}_3 = \boldsymbol{\omega}^{\times}\widehat{\boldsymbol{J}}\boldsymbol{\omega} + \frac{k_3^2}{2}rq_0\|\boldsymbol{q}_v\|^{2r-1}\widehat{\boldsymbol{J}}\boldsymbol{e} - \left(\overline{d} + \lambda\|\boldsymbol{\omega}\|^2\right. \\ \qquad \left. + \frac{k_3^2}{2}r\lambda|q_0|\|\boldsymbol{q}_v\|^{2r-1}\right)\operatorname{sgn}(\boldsymbol{s}) \end{cases} \tag{6.19}$$

$$\rho_i = \begin{cases} k & \|\rho_i\|\boldsymbol{s}\|^p\boldsymbol{s}_e + \boldsymbol{r}_i\| \le \overline{u} \\ \dfrac{-\|\boldsymbol{s}\|^p\boldsymbol{s}_e^T\boldsymbol{r}_i + \sqrt{\|\boldsymbol{s}\|^{2p}\left(\boldsymbol{s}_e^T\boldsymbol{r}_i\right)^2 + \overline{u}^2 - \|\boldsymbol{r}_i\|^2}}{\|\boldsymbol{s}\|^{2p}} & \|\rho_i\|\boldsymbol{s}\|^p\boldsymbol{s}_e + \boldsymbol{r}_i\| > \overline{u} \end{cases} \tag{6.20}$$

Based on (6.18) and (6.20), it can be found that the saturation issue is solved by lessen the proportion term $-k\|\boldsymbol{s}\|^p\boldsymbol{s}_e$ in controller (6.9). When then control torque of controller (6.9) does not exceed the system upper bound, controller (6.18) is totally same as controller (6.9), and when the control torque of controller (6.9) exceeds the system upper bound, the proportion term is lessen to ensure that $\|\boldsymbol{u}\| = \overline{u}$.

It is worth noticing that parameter $k_1, k_2, k_3, \alpha, \beta$ should be proper selected to ensure that $\|\boldsymbol{r}_i\| \le \overline{u}$. Based on (6.18), it can be found that

$$\begin{aligned} \|\boldsymbol{r}_1\| &\le \lambda_M\left(\widehat{\boldsymbol{J}}\right)\overline{\omega}^2 + \frac{1}{2}k_1\lambda_M\left(\widehat{\boldsymbol{J}}\right)\overline{\omega}\left(1 + \sqrt{1-\alpha^2}/\alpha\right) \\ &\quad + \sqrt{3}\left(\overline{d} + \lambda\overline{\omega}^2 + \frac{k_1}{2}\lambda\overline{\omega}\left(1 + \sqrt{1-\alpha^2}/\alpha\right)\right) \\ \|\boldsymbol{r}_2\| &\le \lambda_M\left(\widehat{\boldsymbol{J}}\right)\overline{\omega}^2 + \frac{1}{2}k_2\lambda_M\left(\widehat{\boldsymbol{J}}\right)\overline{\omega} + \sqrt{3}\left(\overline{d} + \lambda\overline{\omega}^2 + \frac{k_2}{2}\lambda\overline{\omega}\right) \\ \|\boldsymbol{r}_3\| &\le \lambda_M\left(\widehat{\boldsymbol{J}}\right)\overline{\omega}^2 + \frac{1}{2}k_3^2r\beta^{2r-1}\lambda_M\left(\widehat{\boldsymbol{J}}\right) + \sqrt{3}\left(\overline{d} + \lambda\overline{\omega}^2 + \frac{k_3^2}{2}r\lambda\beta^{2r-1}\right) \end{aligned} \tag{6.21}$$

where $\overline{\omega}$ is the maximum norm of angular velocity.

Hence, if following equation is satisfied, $\|\boldsymbol{r}_i\| \le \overline{u}$ could be ensured.

$$
\begin{aligned}
&\overline{u} \geq \lambda_M\left(\widehat{\boldsymbol{J}}\right)\overline{\omega}^2 + \frac{1}{2}k_1\lambda_M\left(\widehat{\boldsymbol{J}}\right)\overline{\omega}\left(1+\sqrt{1-\alpha^2}/\alpha\right) \\
&\quad + \sqrt{3}\left(\overline{d} + \lambda\overline{\omega}^2 + \frac{k_1}{2}\lambda\overline{\omega}\left(1+\sqrt{1-\alpha^2}/\alpha\right)\right) \\
&\overline{u} \geq \lambda_M\left(\widehat{\boldsymbol{J}}\right)\overline{\omega}^2 + \frac{1}{2}k_2\lambda_M\left(\widehat{\boldsymbol{J}}\right)\overline{\omega} + \sqrt{3}\left(\overline{d} + \lambda\overline{\omega}^2 + \frac{k_2}{2}\lambda\overline{\omega}\right) \\
&\overline{u} \geq \lambda_M\left(\widehat{\boldsymbol{J}}\right)\overline{\omega}^2 + \frac{1}{2}k_3^2 r\beta^{2r-1}\lambda_M\left(\widehat{\boldsymbol{J}}\right) + \sqrt{3}\left(\overline{d} + \lambda\overline{\omega}^2 + \frac{k_3^2}{2}r\lambda\beta^{2r-1}\right)
\end{aligned} \tag{6.22}
$$

It is worth noticing that (6.22) is a strict constraint of control parameters since the worst condition is considered. The constraint could be loosed properly when it is necessary.

The next step is to prove the finite-time stability of controller (8.18). Select Lyapunov as (6.12) and calculate its derivative, noticing structure of $\boldsymbol{r}_i$ in (6.18) and structure of controller (6.9), it can be found that

$$\dot{V} \leq -\rho_i\|\boldsymbol{s}\|^{p+1} \leq -\mu_i V^{\frac{p+1}{2}} \tag{6.23}$$

where

$$\mu_i = \rho_i(2/\lambda_m(\boldsymbol{J}))^{\frac{p+1}{2}} \tag{6.24}$$

Therefore, controller (6.18) under control torque constraint is also finite-time stable.

6.3 Finite-time controller for attitude tracking control

6.3.1 Finite-time controller

Attitude tracking issue is familiar with attitude stabilization issue since kinetic model (2.15) of attitude quaternion and kinetic model (2.15) of error quaternion has same structure. However, the design of attitude tracking controller is more complex since the dynamic model (2.15) of error angular velocity is much more complex than dynamic model (2.1). Considering the similarities of attitude quaternion and error quaternion, the finite-time sliding mode for error quaternion could be constructed as follows:

$$
\boldsymbol{s} = \begin{cases}
\boldsymbol{\omega}_e + k_1\boldsymbol{e}_e, & \|\boldsymbol{q}_{ev}\| > \alpha \\
\boldsymbol{\omega}_e + k_2\boldsymbol{q}_{ev}, & \beta < \|\boldsymbol{q}_{ev}\| \leq \alpha \\
\boldsymbol{\omega}_e + k_3\|\boldsymbol{q}_{ev}\|^r\boldsymbol{e}_e, & \|\boldsymbol{q}_{ev}\| \leq \beta
\end{cases} \tag{6.25}
$$

where α, β, r, k_1, k_2 are all positive scalars and satisfy following equations.

$$1/2 < r < 1 \tag{6.26}$$

$$k_1 = \alpha k_2, \quad k_2 = \beta^{r-1}k_3 \tag{6.27}$$

Sliding mode (6.25) also has three stages: the maneuver stage, i.e. $\boldsymbol{\omega}_e+k_1\boldsymbol{e}_e$, the slowdown stage, i.e. $\boldsymbol{\omega}_e+k_2\boldsymbol{q}_{ev}$, and the converge stage, i.e. $\boldsymbol{\omega}_e+k_3\|\boldsymbol{q}_{ev}\|^r\boldsymbol{e}_e$. Sliding mode (6.25) has the same property as sliding mode (6.2) that system state should stay on the maneuver stage as long as possible to improve system convergence rate. Moreover, parameter β should be property selected to ensure that when system state reaches the converge stage, error angular velocity is reversed to error quaternion, i.e.

$$\boldsymbol{\omega}_e^T\boldsymbol{q}_{ev}/(\|\boldsymbol{\omega}_e\|\|\boldsymbol{q}_{ev}\|) = -1 \tag{6.28}$$

The next step is to prove the system state converges to zero along sliding mode (41). Similar as the discussion in Section 6.2.1, convergence time t_1, t_2 of the maneuver stage, and the slowdown stage satisfy the following equation:

$$t_1 \le \left(\pi - 2\arccos\sqrt{1-\alpha^2}\right)\Big/k_1, t_2 \le (\ln\alpha - \ln\beta)\Big/\left(k_2\sqrt{1-\alpha^2}\right) \tag{6.29}$$

On the converge stage, select Lyapunov function as follows and calculate its derivative it can be got that

$$V_q = \boldsymbol{q}_{ev}^T\boldsymbol{q}_{ev} \tag{6.30}$$

$$\begin{aligned}\dot{V}_q &= 2\boldsymbol{q}_{ev}^T\dot{\boldsymbol{q}}_{ev} = \boldsymbol{q}_{ev}^T\left(q_{e0}\boldsymbol{I}_3 + \boldsymbol{q}_{ev}^\times\right)\boldsymbol{\omega}_e = -k_3q_{e0}\|\boldsymbol{q}_{ev}\|^r\boldsymbol{q}_{ev}^T\boldsymbol{e}_e = -k_3q_{e0}\|\boldsymbol{q}_{ev}\|^{r+1}\\ &\le -k_3q_{e0}qV_q^{\frac{r+1}{2}}\end{aligned} \tag{6.31}$$

Based on (6.29)–(6.31), it can be found that system state converges to zero in finite time along sliding mode (6.25).

The next step is to design satellite attitude tracking controller to ensure system state could reach sliding mode (6.25) in finite time. The robust finite controller for attitude tracking control presented in this chapter could be written as follows:

$$\boldsymbol{u} = \begin{cases} -k\|\boldsymbol{s}\|^p\boldsymbol{s}_e + \boldsymbol{\varphi}_1 - l_1\,\mathrm{sgn}\,(\boldsymbol{s}) & \|\boldsymbol{q}_{ev}\| > \alpha \\ -k\|\boldsymbol{s}\|^p\boldsymbol{s}_e + \boldsymbol{\varphi}_2 - l_2\,\mathrm{sgn}\,(\boldsymbol{s}) & \beta < \|\boldsymbol{q}_{ev}\| \le \alpha \\ -k\|\boldsymbol{s}\|^p\boldsymbol{s}_e + \boldsymbol{\varphi}_3 - l_3\,\mathrm{sgn}\,(\boldsymbol{s}) & \|\boldsymbol{q}_{ev}\| \le \beta \end{cases} \tag{6.32}$$

where

$$\begin{cases} \boldsymbol{\phi}_1 = \hat{\boldsymbol{J}}\boldsymbol{R}\dot{\boldsymbol{\omega}}_d - \hat{\boldsymbol{J}}\boldsymbol{\omega}_e^\times\boldsymbol{R}\boldsymbol{\omega}_d - \dfrac{1}{2}k_1\hat{\boldsymbol{J}}\boldsymbol{e}_e^\times\left(\boldsymbol{I}_3 - \cot\dfrac{\phi_e}{2}\boldsymbol{e}_e^\times\right)\boldsymbol{\omega}_e \\ \quad +\left(\boldsymbol{\omega}_e + \boldsymbol{R}\boldsymbol{\omega}_d\right)^\times\hat{\boldsymbol{J}}(\boldsymbol{\omega}_e + \boldsymbol{R}\boldsymbol{\omega}_d) \\ \boldsymbol{\phi}_2 = \hat{\boldsymbol{J}}\boldsymbol{R}\dot{\boldsymbol{\omega}}_d - \hat{\boldsymbol{J}}\boldsymbol{\omega}_e^\times\boldsymbol{R}\boldsymbol{\omega}_d - \dfrac{1}{2}k_2\hat{\boldsymbol{J}}\boldsymbol{F}_e\boldsymbol{\omega}_e + (\boldsymbol{\omega}_e + \boldsymbol{R}\boldsymbol{\omega}_d)^\times\hat{\boldsymbol{J}}(\boldsymbol{\omega}_e + \boldsymbol{R}\boldsymbol{\omega}_d) \\ \boldsymbol{\phi}_3 = \hat{\boldsymbol{J}}\boldsymbol{R}\dot{\boldsymbol{\omega}}_d - \hat{\boldsymbol{J}}\boldsymbol{\omega}_e^\times\boldsymbol{R}\boldsymbol{\omega}_d + \dfrac{k_3^2}{2}rq_{e0}\|\boldsymbol{q}_{ev}\|^{2r-1}\hat{\boldsymbol{J}}\boldsymbol{e}_e \\ \quad + (\boldsymbol{\omega}_e + \boldsymbol{R}\boldsymbol{\omega}_d)^\times\hat{\boldsymbol{J}}(\boldsymbol{\omega}_e + \boldsymbol{R}\boldsymbol{\omega}_d) \end{cases} \tag{6.33}$$

$$\begin{cases} l_1 = \overline{d} + \lambda\left(\|\dot{\boldsymbol{\omega}}_d\| + \|\boldsymbol{\omega}_e\|\|\boldsymbol{\omega}_d\| + \|\boldsymbol{\omega}_e + \boldsymbol{R}\boldsymbol{\omega}_d\|^2 + \frac{k_1}{2}\left(1 + \cot\frac{\varphi_e}{2}\right)\|\boldsymbol{\omega}_e\| \right) \\ l_2 = \overline{d} + \lambda\left(\|\dot{\boldsymbol{\omega}}_d\| + \|\boldsymbol{\omega}_e\|\|\boldsymbol{\omega}_d\| + \|\boldsymbol{\omega}_e + \boldsymbol{R}\boldsymbol{\omega}_d\|^2 + \frac{k_2}{2}\|\boldsymbol{\omega}_e\| \right) \\ l_3 = \overline{d} + \lambda\left(\|\dot{\boldsymbol{\omega}}_d\| + \|\boldsymbol{\omega}_e\|\|\boldsymbol{\omega}_d\| + \|\boldsymbol{\omega}_e + \boldsymbol{R}\boldsymbol{\omega}_d\|^2 + \frac{k_3^2}{2} r|q_{e0}|\|\boldsymbol{q}_{ev}\|^{2r-1} \right) \end{cases} \tag{6.34}$$

and p is a positive scalar and satisfies $1>p>0$, $\boldsymbol{s}_e$ is unit direction vector of $\boldsymbol{s}$ and satisfies $\boldsymbol{s}_e = \boldsymbol{s}/\|\boldsymbol{s}\|$, λ is a positive scalar and satisfies $\lambda \geq \lambda_M(\widetilde{\boldsymbol{J}})$, sgn($\boldsymbol{r}$) is the sign function of vector $\boldsymbol{r}$.

It is worth noticing that $1/2<r<1$, therefore controller (6.32) has no singularity issue.

In order to prove the finite-time stability of controller (6.32), select Lyapunov function as follows:

$$V = \frac{1}{2}\boldsymbol{s}^T\boldsymbol{J}\boldsymbol{s} \tag{6.35}$$

Based on (6.35), it can be found that

$$V \geq \frac{1}{2}\lambda_m(\boldsymbol{J})\|\boldsymbol{s}\|^2 \tag{6.36}$$

Calculate the derivative of V on the maneuver stage ($\|\boldsymbol{q}_{ev}\|>\alpha$) and noticing property (6.36) it can be got that

$$\begin{aligned} \dot{V} &= \boldsymbol{s}^T\boldsymbol{J}\dot{\boldsymbol{s}} = \boldsymbol{s}^T\boldsymbol{J}\dot{\boldsymbol{\omega}}_e + k_1\boldsymbol{s}^T\boldsymbol{J}\dot{\boldsymbol{e}}_e \\ &= \boldsymbol{s}^T\boldsymbol{u} + \boldsymbol{s}^T\boldsymbol{d} - \boldsymbol{s}^T\boldsymbol{J}\boldsymbol{R}\dot{\boldsymbol{\omega}}_d + \boldsymbol{s}^T\boldsymbol{J}\boldsymbol{\omega}_e^{\times}\boldsymbol{R}\boldsymbol{\omega}_d + \frac{k_1}{2}\boldsymbol{s}^T\boldsymbol{J}\boldsymbol{e}_e^{\times}\left(\boldsymbol{I}_3 - \cot\frac{\varphi_e}{2}\boldsymbol{e}_e^{\times}\right)\boldsymbol{\omega}_e \\ &\quad -\boldsymbol{s}^T(\boldsymbol{\omega}_e + \boldsymbol{R}\boldsymbol{\omega}_d)^{\times}\boldsymbol{J}(\boldsymbol{\omega}_e + \boldsymbol{R}\boldsymbol{\omega}_d) \\ &= -k\|\boldsymbol{s}\|^p\boldsymbol{s}^T\boldsymbol{s}_e + \boldsymbol{s}^T\boldsymbol{d} - \boldsymbol{s}^T\widetilde{\boldsymbol{J}}\boldsymbol{R}\dot{\boldsymbol{\omega}}_d + \boldsymbol{s}^T\widetilde{\boldsymbol{J}}\boldsymbol{\omega}_e^{\times}\boldsymbol{R}\boldsymbol{\omega}_d + \frac{k_1}{2}\boldsymbol{s}^T\widetilde{\boldsymbol{J}}\boldsymbol{e}_e^{\times}\left(\boldsymbol{I}_3 - \cot\frac{\varphi_e}{2}\boldsymbol{e}_e^{\times}\right)\boldsymbol{\omega}_e \\ &\quad -\boldsymbol{s}^T(\boldsymbol{\omega}_e + \boldsymbol{R}\boldsymbol{\omega}_d)^{\times}\widetilde{\boldsymbol{J}}(\boldsymbol{\omega}_e + \boldsymbol{R}\boldsymbol{\omega}_d) - l_1\boldsymbol{s}^T\,\mathrm{sgn}(\boldsymbol{s}) \\ &\leq -k\|\boldsymbol{s}\|^{p+1} + \overline{d}\|\boldsymbol{s}\| + \lambda_M(\widetilde{\boldsymbol{J}})\|\boldsymbol{s}\|\|\dot{\boldsymbol{\omega}}_d\| + \lambda_M(\widetilde{\boldsymbol{J}})\|\boldsymbol{s}\|\|\boldsymbol{\omega}_e\|\|\boldsymbol{\omega}_d\| \\ &\quad + \lambda_M(\widetilde{\boldsymbol{J}})\|\boldsymbol{s}\|\|\boldsymbol{\omega}_e + \boldsymbol{R}\boldsymbol{\omega}_d\|^2 + \frac{k_1}{2}\left(1 + \cot\frac{\varphi}{2}\right)\lambda_M(\widetilde{\boldsymbol{J}})\|\boldsymbol{s}\|\|\boldsymbol{\omega}_e\| - l_1\boldsymbol{s}^T\,\mathrm{sgn}(\boldsymbol{s}) \\ &\leq -k\|\boldsymbol{s}\|^{p+1} \leq -\mu V^{\frac{p+1}{2}} \end{aligned} \tag{6.37}$$

Calculate the derivative of V on the slowdown stage ($\beta<\|\boldsymbol{q}_{ev}\|\leq\alpha$) and noticing property (6.36) it can be got that

$$\begin{aligned}
\dot{V} &= \boldsymbol{s}^T \boldsymbol{J}\dot{\boldsymbol{s}} = \boldsymbol{s}^T \boldsymbol{J}\dot{\boldsymbol{\omega}}_e + k_2 \boldsymbol{s}^T \boldsymbol{J}\dot{\boldsymbol{q}}_{ev} \\
&= \boldsymbol{s}^T \boldsymbol{u} + \boldsymbol{s}^T \boldsymbol{d} - \boldsymbol{s}^T \boldsymbol{J}\boldsymbol{R}\dot{\boldsymbol{\omega}}_d + \boldsymbol{s}^T \boldsymbol{J}\boldsymbol{\omega}_e^{\times} \boldsymbol{R}\boldsymbol{\omega}_d + \frac{k_2}{2}\boldsymbol{s}^T \boldsymbol{J}\boldsymbol{e}\boldsymbol{F}_e \boldsymbol{\omega}_e \\
&\quad -\boldsymbol{s}^T (\boldsymbol{\omega}_e + \boldsymbol{R}\boldsymbol{\omega}_d)^{\times} \boldsymbol{J}(\boldsymbol{\omega}_e + \boldsymbol{R}\boldsymbol{\omega}_d) \\
&= -k\|\boldsymbol{s}\|^p \boldsymbol{s}^T \boldsymbol{s}_e + \boldsymbol{s}^T \boldsymbol{d} - \boldsymbol{s}^T \tilde{\boldsymbol{J}}\boldsymbol{R}\dot{\boldsymbol{\omega}}_d + \boldsymbol{s}^T \tilde{\boldsymbol{J}}\boldsymbol{\omega}_e^{\times} \boldsymbol{R}\boldsymbol{\omega}_d + \frac{k_2}{2}\boldsymbol{s}^T \tilde{\boldsymbol{J}}\boldsymbol{F}_e \boldsymbol{\omega}_e \\
&\quad -\boldsymbol{s}^T (\boldsymbol{\omega}_e + \boldsymbol{R}\boldsymbol{\omega}_d)^{\times} \tilde{\boldsymbol{J}}(\boldsymbol{\omega}_e + \boldsymbol{R}\boldsymbol{\omega}_d) - l_2 \boldsymbol{s}^T \operatorname{sgn}(\boldsymbol{s}) \\
&\le k\|\boldsymbol{s}\|^{p+1} + \overline{d}\|\boldsymbol{s}\| + \lambda_M(\tilde{\boldsymbol{J}})\|\boldsymbol{s}\|\|\dot{\boldsymbol{\omega}}_d\| + \lambda_M(\tilde{\boldsymbol{J}})\|\boldsymbol{s}\|\|\boldsymbol{\omega}_e\|\|\boldsymbol{\omega}_d\| \\
&\quad + \lambda_M(\tilde{\boldsymbol{J}})\|\boldsymbol{s}\|\|\boldsymbol{\omega}_e + \boldsymbol{R}\boldsymbol{\omega}_d\|^2 + \frac{k_2}{2}\lambda_M(\tilde{\boldsymbol{J}})\|\boldsymbol{s}\|\|\boldsymbol{\omega}_e\| - l_2 \boldsymbol{s}^T \operatorname{sgn}(\boldsymbol{s}) \\
&\le -k\|\boldsymbol{s}\|^{p+1} \le -\mu V^{\frac{p+1}{2}}
\end{aligned} \tag{6.38}$$

Based on the assumption that error angular velocity $\boldsymbol{\omega}_e$ is reversed to error quaternion $\boldsymbol{q}_{ev}$ when system reaches the converge stage of sliding mode (6.25), it can be found that $\dot{\boldsymbol{e}}_e = \boldsymbol{0}$ along the converge stage. Calculate derivative of (6.35) and substitute $\dot{\boldsymbol{e}}_e = \boldsymbol{0}$ in its derivative on the converge stage ($\|\boldsymbol{q}_{ev}\| \le \beta$) it can be got that

$$\begin{aligned}
\dot{V} &= \boldsymbol{s}^T \boldsymbol{J}\dot{\boldsymbol{s}} = \boldsymbol{s}^T \boldsymbol{J}\dot{\boldsymbol{\omega}}_e + k_3 \boldsymbol{s}^T \boldsymbol{J}\boldsymbol{e}_e \frac{d\|\boldsymbol{q}_{ev}\|^r}{dt} + k_3\|\boldsymbol{q}_{ev}\|^r \boldsymbol{s}^T \boldsymbol{J}\dot{\boldsymbol{e}}_e \\
&= \boldsymbol{s}^T \boldsymbol{u} + \boldsymbol{s}^T \boldsymbol{d} - \boldsymbol{s}^T \boldsymbol{J}\boldsymbol{R}\dot{\boldsymbol{\omega}}_d + \boldsymbol{s}^T \boldsymbol{J}\boldsymbol{\omega}_e^{\times} \boldsymbol{R}\boldsymbol{\omega}_d - \frac{k_3^2}{2} r q_{e0}\|\boldsymbol{q}_{ev}\|^{2r-1} \boldsymbol{s}^T \boldsymbol{J}\boldsymbol{e}_e \\
&\quad -\boldsymbol{s}^T (\boldsymbol{\omega}_e + \boldsymbol{R}\boldsymbol{\omega}_d)^{\times} \boldsymbol{J}(\boldsymbol{\omega}_e + \boldsymbol{R}\boldsymbol{\omega}_d) \\
&= -k\|\boldsymbol{s}\|^p \boldsymbol{s}^T \boldsymbol{s}_e + \boldsymbol{s}^T \boldsymbol{d} - \boldsymbol{s}^T \tilde{\boldsymbol{J}}\boldsymbol{R}\dot{\boldsymbol{\omega}}_d + \boldsymbol{s}^T \tilde{\boldsymbol{J}}\boldsymbol{\omega}_e^{\times} \boldsymbol{R}\boldsymbol{\omega}_d - \frac{k_3^2}{2} r q_{e0}\|\boldsymbol{q}_{ev}\|^{2r-1} \boldsymbol{s}^T \tilde{\boldsymbol{J}}\boldsymbol{e}_e \\
&\quad -\boldsymbol{s}^T (\boldsymbol{\omega}_e + \boldsymbol{R}\boldsymbol{\omega}_d)^{\times} \tilde{\boldsymbol{J}}(\boldsymbol{\omega}_e + \boldsymbol{R}\boldsymbol{\omega}_d) - l_3 \boldsymbol{s}^T \operatorname{sgn}(\boldsymbol{s}) \\
&\le k\|\boldsymbol{s}\|^{p+1} + \overline{d}\|\boldsymbol{s}\| + \lambda_M(\tilde{\boldsymbol{J}})\|\boldsymbol{s}\|\|\dot{\boldsymbol{\omega}}_d\| + \lambda_M(\tilde{\boldsymbol{J}})\|\boldsymbol{s}\|\|\boldsymbol{\omega}_e\|\|\boldsymbol{\omega}_d\| \\
&\quad + \lambda_M(\tilde{\boldsymbol{J}})\|\boldsymbol{s}\|\|\boldsymbol{\omega}_e + \boldsymbol{R}\boldsymbol{\omega}_d\|^2 + \lambda_M(\tilde{\boldsymbol{J}})\frac{k_3^2}{2} r|q_{e0}|\|\boldsymbol{q}_{ev}\|^{2r-1}\|\boldsymbol{s}\| - l_3 \boldsymbol{s}^T \operatorname{sgn}(\boldsymbol{s}) \\
&\le -k\|\boldsymbol{s}\|^{p+1} \le -\mu V^{\frac{p+1}{2}}
\end{aligned} \tag{6.39}$$

The unclaimed variable μ in (6.34), (6.38), and (6.39) is defined as follows:

$$\mu = k(2/\lambda_m(\boldsymbol{J}))^{\frac{p+1}{2}} \tag{6.40}$$

Based on (6.34), (6.38), and (6.39), it can be found that system state reaches sliding mode (6.25) in finite time and the finite-time stability of controller (6.32) has been proved.

6.3.2 Finite-time controller under control torque constraint

The controller presented in Section 6.3.1 does not consider the control torque saturation issue and controller under control torque constraint will be given in this section.

Assume the norm upper bound of system control torque is $\overline{u}$. The modified controller under control torque constraint could be written as follows:

$$\boldsymbol{u} = \begin{cases} -\rho_1 \|\boldsymbol{s}\|^p \boldsymbol{s}_e + \boldsymbol{r}_1 & \|\boldsymbol{q}_{ev}\| > \alpha \\ -\rho_2 \|\boldsymbol{s}\|^p \boldsymbol{s}_e + \boldsymbol{r}_2 & \beta < \|\boldsymbol{q}_{ev}\| \le \alpha \\ -\rho_3 \|\boldsymbol{s}\|^p \boldsymbol{s}_e + \boldsymbol{r}_3 & \|\boldsymbol{q}_{ev}\| \le \beta \end{cases} \tag{6.41}$$

where $\boldsymbol{r}_i$ and ρ_i are defined as follows:

$$\begin{cases} \boldsymbol{r}_1 = \widehat{\boldsymbol{J}}\boldsymbol{R}\dot{\boldsymbol{\omega}}_d - \widehat{\boldsymbol{J}}\boldsymbol{\omega}_e^{\times}\boldsymbol{R}\boldsymbol{\omega}_d - \dfrac{1}{2}k_1\widehat{\boldsymbol{J}}\boldsymbol{e}_e^{\times}\left(\boldsymbol{I}_3 - \cot\dfrac{\phi_e}{2}\boldsymbol{e}_e^{\times}\right)\boldsymbol{\omega}_e \\ \quad + (\boldsymbol{\omega}_e + \boldsymbol{R}\boldsymbol{\omega}_d)^{\times}\widehat{\boldsymbol{J}}(\boldsymbol{\omega}_e + \boldsymbol{R}\boldsymbol{\omega}_d) - l_1 \operatorname{sgn}(\boldsymbol{s}) \\ \boldsymbol{r}_2 = \widehat{\boldsymbol{J}}\boldsymbol{R}\dot{\boldsymbol{\omega}}_d - \widehat{\boldsymbol{J}}\boldsymbol{\omega}_e^{\times}\boldsymbol{R}\boldsymbol{\omega}_d - \dfrac{1}{2}k_2\widehat{\boldsymbol{J}}\boldsymbol{F}_e\boldsymbol{\omega}_e + (\boldsymbol{\omega}_e + \boldsymbol{R}\boldsymbol{\omega}_d)^{\times}\widehat{\boldsymbol{J}}(\boldsymbol{\omega}_e + \boldsymbol{R}\boldsymbol{\omega}_d) \\ \quad - l_2 \operatorname{sgn}(\boldsymbol{s}) \\ \boldsymbol{r}_3 = \widehat{\boldsymbol{J}}\boldsymbol{R}\dot{\boldsymbol{\omega}}_d - \widehat{\boldsymbol{J}}\boldsymbol{\omega}_e^{\times}\boldsymbol{R}\boldsymbol{\omega}_d + \dfrac{k_3^2}{2}rq_{e0}\|\boldsymbol{q}_{ev}\|^{2r-1}\widehat{\boldsymbol{J}}\boldsymbol{e}_e \\ \quad + (\boldsymbol{\omega}_e + \boldsymbol{R}\boldsymbol{\omega}_d)^{\times}\widehat{\boldsymbol{J}}(\boldsymbol{\omega}_e + \boldsymbol{R}\boldsymbol{\omega}_d) - l_3 \operatorname{sgn}(\boldsymbol{s}) \end{cases} \tag{6.42}$$

$$\rho_i = \begin{cases} k & \|\rho_i \|\boldsymbol{s}\|^p \boldsymbol{s}_e + \boldsymbol{r}_i\| \le \overline{u} \\ \dfrac{-\|\boldsymbol{s}\|^p \boldsymbol{s}_e^T \boldsymbol{r}_i + \sqrt{\|\boldsymbol{s}\|^{2p}\left(\boldsymbol{s}_e^T\boldsymbol{r}_i\right)^2 + \overline{u}^2 - \|\boldsymbol{r}_i\|^2}}{\|\boldsymbol{s}\|^{2p}} & \|\rho_i \|\boldsymbol{s}\|^p \boldsymbol{s}_e + \boldsymbol{r}_i\| > \overline{u} \end{cases} \tag{6.43}$$

The essence of controller (6.43) is same as the controller in Section 6.3.1. The method used in these two sections to solve control torque constraint is to lessen the proportion term $-k\|\boldsymbol{s}\|^p\boldsymbol{s}_e$ when the saturation issue occurs. When the norm of control torque does not exceed system upper bound, controller (6.41) is totally same as (6.32), and when the norm of control torque exceeds system upper bound, proportion term is lessen to ensure $\|\boldsymbol{u}\| = \overline{u}$.

It is worth noticing that parameter $k_1, k_2, k_3, \alpha, \beta$ should be proper selected to ensure $\|\boldsymbol{r}_i\| \le \overline{u}$. Based on (6.42), it can be found that

$$\begin{aligned}
\|\boldsymbol{r}_1\| &\leq \lambda_M\left(\widehat{\boldsymbol{J}}\right)\|\dot{\boldsymbol{\omega}}_d\| + \lambda_M\left(\widehat{\boldsymbol{J}}\right)\|\boldsymbol{\omega}_e\|\|\boldsymbol{\omega}_d\| + \frac{1}{2}k_1\lambda_M\left(\widehat{\boldsymbol{J}}\right)\left(1+\sqrt{1-\alpha^2}/\alpha\right)\|\boldsymbol{\omega}_e\| \\
&\quad + \lambda_M\left(\widehat{\boldsymbol{J}}\right)\overline{\omega}^2 + \sqrt{3}l_1 \\
\|\boldsymbol{r}_2\| &\leq \lambda_M\left(\widehat{\boldsymbol{J}}\right)\|\dot{\boldsymbol{\omega}}_d\| + \lambda_M\left(\widehat{\boldsymbol{J}}\right)\|\boldsymbol{\omega}_e\|\|\boldsymbol{\omega}_d\| + \frac{1}{2}k_2\lambda_M\left(\widehat{\boldsymbol{J}}\right)\|\boldsymbol{\omega}_e\| + \lambda_M\left(\widehat{\boldsymbol{J}}\right)\overline{\omega}^2 \\
&\quad + \sqrt{3}l_2 \\
\|\boldsymbol{r}_2\| &\leq \lambda_M\left(\widehat{\boldsymbol{J}}\right)\|\dot{\boldsymbol{\omega}}_d\| + \lambda_M\left(\widehat{\boldsymbol{J}}\right)\|\boldsymbol{\omega}_e\|\|\boldsymbol{\omega}_d\| + \frac{1}{2}k_3^2\lambda_M\left(\widehat{\boldsymbol{J}}\right)\beta^{2r-1} \\
&\quad + \lambda_M\left(\widehat{\boldsymbol{J}}\right)\overline{\omega}^2 + \sqrt{3}l_3
\end{aligned} \tag{6.44}$$

where $\overline{\omega}$ is the maximum norm of angular velocity.

Hence, if following equation is satisfied, $\|\boldsymbol{r}_i\| \leq \overline{u}$ could be ensured.

$$\begin{aligned}
\overline{u} &\geq \lambda_M\left(\widehat{\boldsymbol{J}}\right)\|\dot{\boldsymbol{\omega}}_d\| + \lambda_M\left(\widehat{\boldsymbol{J}}\right)\|\boldsymbol{\omega}_e\|\|\boldsymbol{\omega}_d\| + \frac{1}{2}k_1\lambda_M\left(\widehat{\boldsymbol{J}}\right)\left(1+\sqrt{1-\alpha^2}/\alpha\right)\|\boldsymbol{\omega}_e\| \\
&\quad + \lambda_M\left(\widehat{\boldsymbol{J}}\right)\overline{\omega}^2 + \sqrt{3}l_1 \\
\overline{u} &\geq \lambda_M\left(\widehat{\boldsymbol{J}}\right)\|\dot{\boldsymbol{\omega}}_d\| + \lambda_M\left(\widehat{\boldsymbol{J}}\right)\|\boldsymbol{\omega}_e\|\|\boldsymbol{\omega}_d\| + \frac{1}{2}k_2\lambda_M\left(\widehat{\boldsymbol{J}}\right)\|\boldsymbol{\omega}_e\| + \lambda_M\left(\widehat{\boldsymbol{J}}\right)\overline{\omega}^2 \\
&\quad + \sqrt{3}l_2 \\
\overline{u} &\geq \lambda_M\left(\widehat{\boldsymbol{J}}\right)\|\dot{\boldsymbol{\omega}}_d\| + \lambda_M\left(\widehat{\boldsymbol{J}}\right)\|\boldsymbol{\omega}_e\|\|\boldsymbol{\omega}_d\| + \frac{1}{2}k_3^2\lambda_M\left(\widehat{\boldsymbol{J}}\right)\beta^{2r-1} + \lambda_M\left(\widehat{\boldsymbol{J}}\right)\overline{\omega}^2 \\
&\quad + \sqrt{3}l_3
\end{aligned} \tag{6.45}$$

It is worth noticing that (6.45) is a strict constraint of control parameters since the worst condition is considered. The constraint could be loosed properly when it is necessary.

As discussed in Section 6.3.1, the next step is to prove the finite-time stability of controller (6.41). Select Lyapunov as (6.41) and calculate its derivative, noticing structure of $\boldsymbol{r}_i$ in (6.41) and structure of controller (6.32), it can be found that

$$\dot{V} \leq -\rho_i\|\boldsymbol{s}\|^{p+1} \leq -\mu_i V^{\frac{p+1}{2}} \tag{6.46}$$

where

$$\mu_i = \rho_i(2/\lambda_m(\boldsymbol{J}))^{\frac{p+1}{2}} \tag{6.47}$$

Therefore, controller (6.41) under control torque constraint is also finite-time stable.

6.4 Simulation results

6.4.1 Simulation of attitude stabilization controller

In order to demonstrate the superiority of attitude stabilization controller (6.18) presented in this chapter, sliding mode controller (6.48) without inertia matrix uncertainty and control torque constraint constructed as follow is compared.

$$\begin{aligned} \boldsymbol{u} &= -k\boldsymbol{s} + \boldsymbol{\omega}^{\times}\boldsymbol{J}\boldsymbol{\omega} - \frac{k_1}{2}\left(q_0\boldsymbol{I}_3 + \boldsymbol{q}_v^{\times}\right)\boldsymbol{\omega} - \overline{d}\,\boldsymbol{sgn}\,(\boldsymbol{s}) \\ \boldsymbol{s} &= \boldsymbol{\omega} + k_1\boldsymbol{q}_v \end{aligned} \tag{6.48}$$

Set the simulation parameters as follows:

$$\begin{gathered} \boldsymbol{J} = \mathrm{diag}(25, 30, 50)\mathrm{kg}\cdot\mathrm{m}^2,\ \ \hat{\boldsymbol{J}} = \mathrm{diag}(24, 31, 50)\mathrm{kg}\cdot\mathrm{m}^2,\ \ \lambda = 3\mathrm{kg}\cdot\mathrm{m}^2 \\ k_1 = 0.05,\ \ k = 10 \\ \boldsymbol{\omega}_{\mathrm{initial}} = \begin{bmatrix}-0.03 & -0.04 & 0.05\end{bmatrix}^T\mathrm{rad/s},\ \ \boldsymbol{q}_{\mathrm{initial}} = \begin{bmatrix}0 & \sqrt{6}/6 & \sqrt{3}/3 & \sqrt{2}/2\end{bmatrix}^T \end{gathered} \tag{6.49}$$

Assume the disturbance torque consists of Gauss white noise and sinusoidal signal written as follows:

$$d_i = 5 \times 10^{-4}\mathrm{rand}(-1, 1) + 5 \times 10^{-4}\sin t \tag{6.50}$$

Hence, the norm upper bound of disturbance torque satisfies

$$\overline{d} = 10^{-3} \tag{6.51}$$

The simulation results of standard sliding mode controller (6.48) are given as follows:

Based on Figs. 6.1–6.4, it can be found that system state converges to zero. However, the main drawback of controller (6.48) could be concluded as two aspects: (1) convergence rate descends severely and the terminal convergence rate is low. The steady accuracy at 200 s is 6×10^{-4} of angular velocity, 1.2×10^{-2} of attitude quaternion, and 1.5 degree of Euler angle. (The discontinuity of Euler angle at about 5 and 25 s is caused by the mutation of tangent function when the angle passes 90 degree. And in later text this phenomenon is caused by the same reason.) (2) The initial control torque exceeds the system upper bound.

Based on the simulation results, it can be found that although the inertia matrix uncertainty and control torque constraint issue is ignored in controller (6.48), the performance of controller (6.48) still needs improvements.

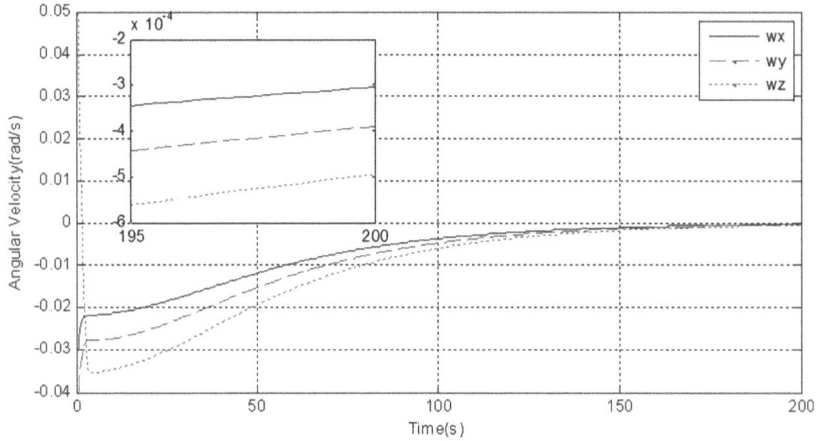

FIG. 6.1 Curve of angular velocity.

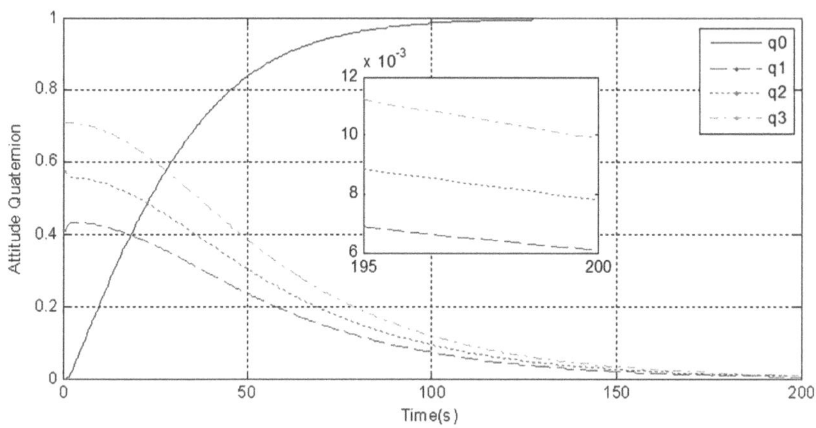

FIG. 6.2 Curve of attitude quaternion.

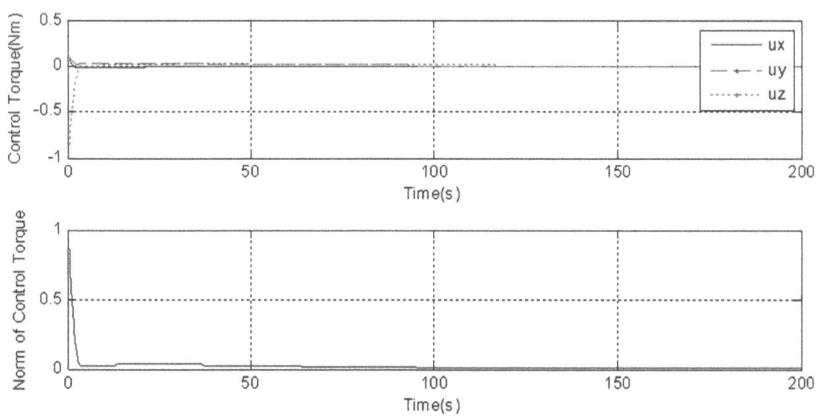

FIG. 6.3 Curve of control torque and its norm.

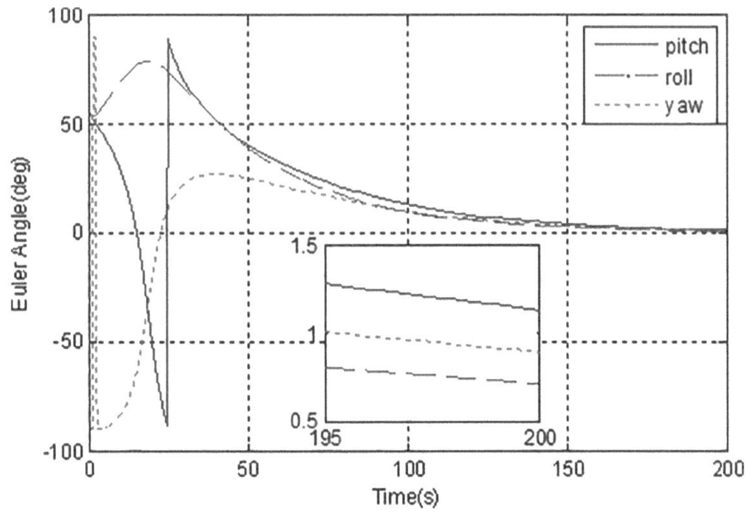

FIG. 6.4 Curve of Euler angle.

Based on the discussion previously, larger β and smaller α could bring better convergence rate, but the control torque constraint (6.22) may not be satisfied and the singularity issue may occur. Hence, parameter α and β are selected as follows:

$$\alpha = 0.2, \quad \beta = 0.02 \tag{6.52}$$

Considering that the maximum norm of angular velocity of satellites is about 5 degree/s, k_i are selected as follows:

$$k_1 = 0.05, \quad k_2 = 0.25, \quad k_3 = 0.0679 \tag{6.53}$$

Generally, the larger k_i, the better convergence rate, but the demanded control torque is enlarged. Assume that the norm upper bound of control torque satisfies

$$\overline{u} = 0.5 \text{ N} \cdot \text{m} \tag{6.54}$$

The simulation results of controller (6.18) with inertia matrix uncertainty and control torque constraint are given as Figs. 6.5–6.8.

Based on Figs. 6.5–6.8, it can be found that system converges to zero at about 100 s. The steady accuracy at 150 s is about 4×10^{-7} of angular velocity, of 4×10^{-8} attitude quaternion and 5×10^{-6} degree of Euler angle. Controller (6.18) has both the advantage of high convergence rate and high steady accuracy. Moreover, the norm of control torque does not exceed system upper bound during the control process and the singularity issue is solved. The improvement of performance is brought by the modified sliding mode. Based on Fig. 6.5, it can be found that angular velocity is constant during 5–60 s, and convergence

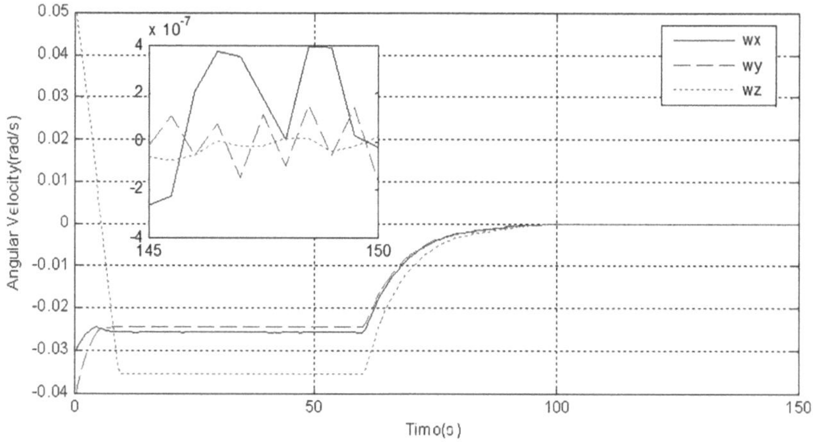

FIG. 6.5 Curve of angular velocity.

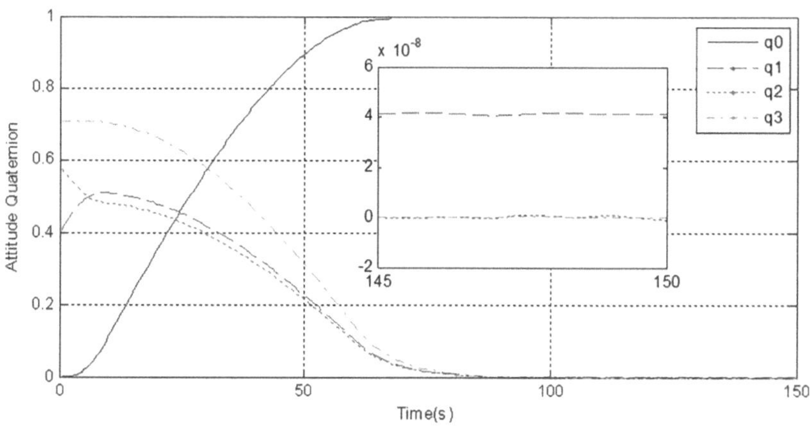

FIG. 6.6 Curve of attitude quaternion.

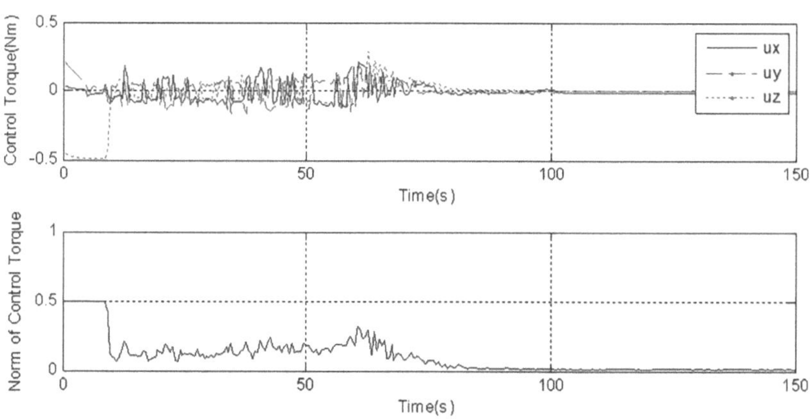

FIG. 6.7 Curve of control torque and its norm.

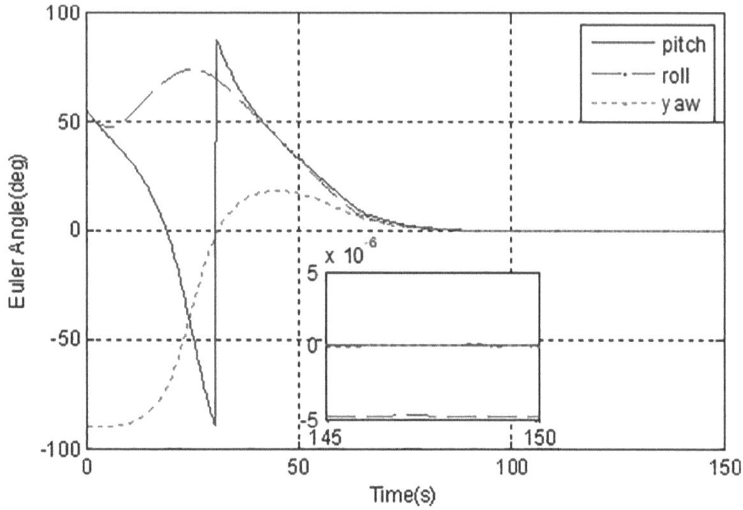

FIG. 6.8 Curve of Euler angle.

rate of quaternion is maintained at a high level during this stage. Moreover, the finite-time stability of controller (6.18) brings a high steady accuracy, which is demonstrated by Figs. 6.5 and 6.6. Comparing Fig. 6.3 with Fig. 6.7, it can be found that the initial norm of control torque of controller (6.48) is huge and descends severely afterward hence the convergence rate descends severely. This reveals the fact that the capability of control torque is not well used in standard sliding mode controller. Based on Figs. 6.3–6.7, it can be found that the norm of control torque of controller (6.18) presented in this chapter is maintained at a moderate level during the convergence process. This reveals that controller (6.18) has a better efficiency on control torque comparing with controller (6.48).

6.4.2 Simulation of attitude tracking controller

In order to demonstrate the superiority of the controller presented in this chapter, sliding mode controller (6.55) without inertia matrix uncertainty and control torque constraint constructed as follows is compared.

$$\begin{aligned} \boldsymbol{u} = & -k\boldsymbol{s} + \boldsymbol{J}\boldsymbol{R}\dot{\boldsymbol{\omega}}_d - \boldsymbol{J}\boldsymbol{\omega}_e^{\times}\boldsymbol{R}\boldsymbol{\omega}_d + (\boldsymbol{\omega}_e + \boldsymbol{R}\boldsymbol{\omega}_d)^{\times}\boldsymbol{J}(\boldsymbol{\omega}_e + \boldsymbol{R}\boldsymbol{\omega}_d) \\ & - \frac{1}{2}k_1\boldsymbol{J}\left(q_{e0}\boldsymbol{I}_3 + \boldsymbol{q}_{ev}^{\times}\right)\boldsymbol{\omega}_e \\ \boldsymbol{s} = & \ \boldsymbol{\omega}_e + k_1\boldsymbol{q}_{ev} \end{aligned} \tag{6.55}$$

Considering that the control parameters are basically all the same as previous section, hence the method to select them is omitted. Set simulation parameters as follows:

$$\begin{gathered}
\boldsymbol{J}=\operatorname{diag}(25,30,50)\mathrm{kg}\cdot\mathrm{m}^2,\ \hat{\boldsymbol{J}}=\operatorname{diag}(24,31,50)\mathrm{kg}\cdot\mathrm{m}^2,\ \lambda=3\mathrm{kg}\cdot\mathrm{m}^2 \\
\overline{u}=0.5\mathrm{N}\cdot\mathrm{m},\ \overline{d}=1e^{-4},\ \alpha=0.2,\ \beta=0.02 \\
r=2/3,\ p=1/2,\ k_1=0.05,\ k_2=0.25,\ k_3=0.0679,\ k=10 \\
\boldsymbol{\omega}_{\text{initial}}=\begin{bmatrix}0 & 0 & -0.1\end{bmatrix}^T\mathrm{rad/s},\ \boldsymbol{q}_{\text{initial}}=\begin{bmatrix}1/2 & -1/2 & \sqrt{3}/3 & \sqrt{6}/6\end{bmatrix}^T \\
\boldsymbol{\omega}_d=\begin{bmatrix}0.01 & -0.005 & 0\end{bmatrix}^T\mathrm{rad/s},\ \dot{\boldsymbol{\omega}}_d=\boldsymbol{0}_{3\times1},\boldsymbol{q}_d=\begin{bmatrix}1 & 0 & 0 & 0\end{bmatrix}^T
\end{gathered} \tag{6.56}$$

The simulation results of standard sliding mode controller (6.55) are given from Figs. 6.9–6.12.

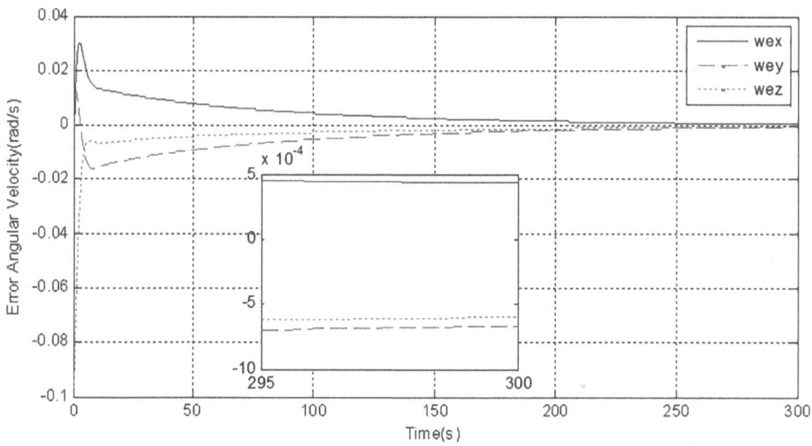

FIG. 6.9 Curve of error angular velocity.

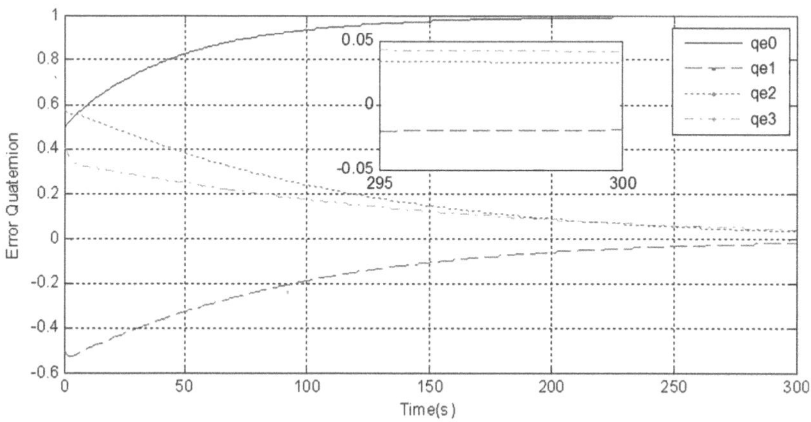

FIG. 6.10 Curve of error quaternion.

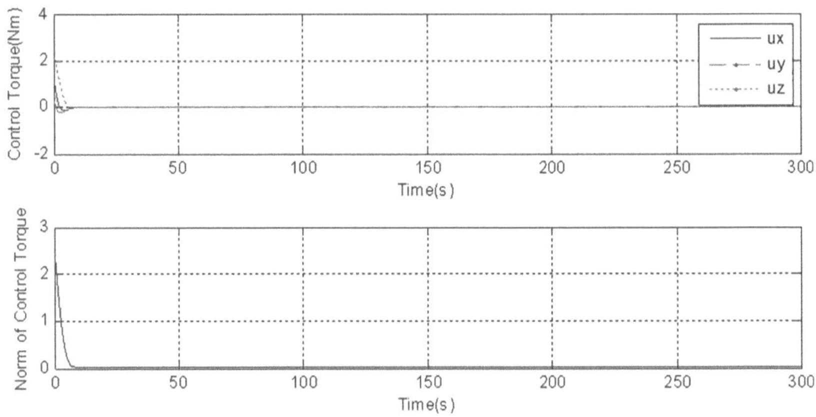

FIG. 6.11 Curve of control torque and its norm.

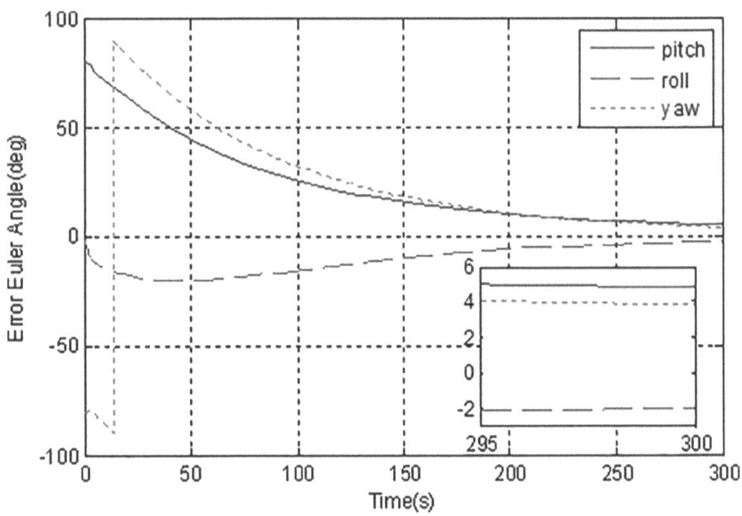

FIG. 6.12 Curve of error Euler angle.

Based on Figs. 6.9–6.12, it can be found that the main issue of standard sliding mode controller (6.55) could also be concluded as two aspects: low convergence rate and excessive initial control torque.

Next, the simulation results of attitude tracking controller (6.41) presented in this chapter are given as follows.

Based on Figs. 6.13–6.16, it can be found that system state converges to zero at about 60 s. The steady accuracy at 150 s is 1×10^{-5} of error angular velocity, 3×10^{-6} of error quaternion, and 4×10^{-4} degree of error Euler angle. System convergence rate and steady accuracy are largely improved. Moreover, the norm of control torque does not exceed the system upper bound during whole

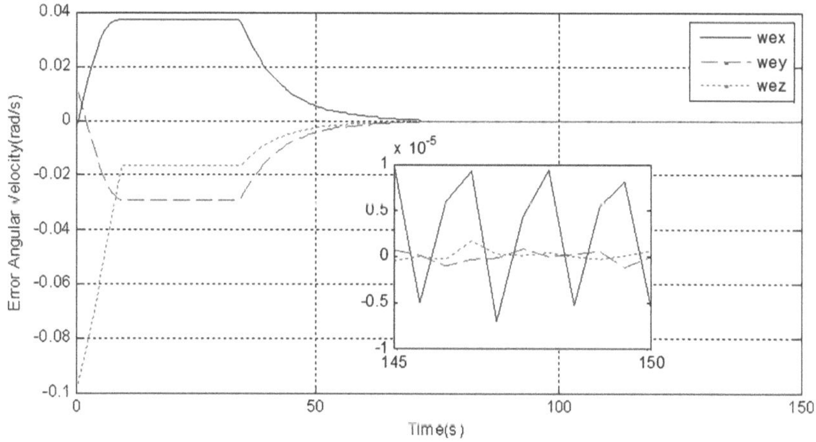

FIG. 6.13 Curve of error angular velocity.

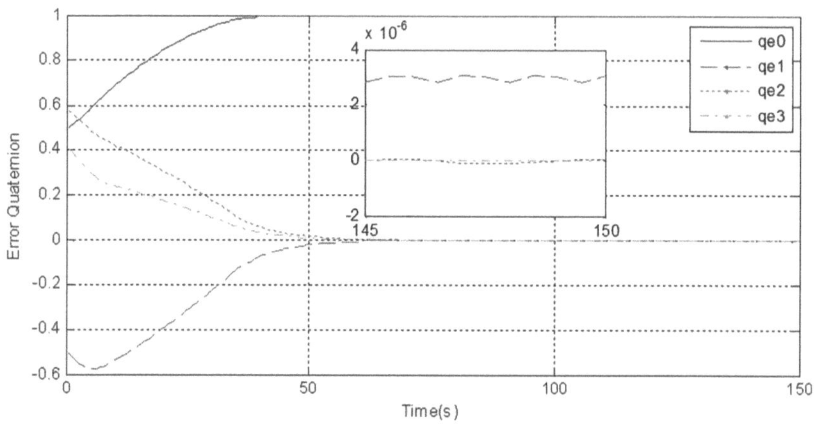

FIG. 6.14 Curve of error quaternion.

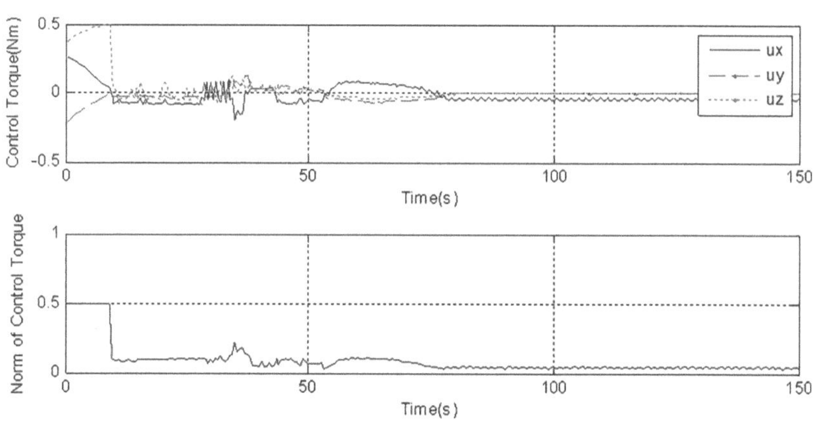

FIG. 6.15 Curve of control torque and its norm.

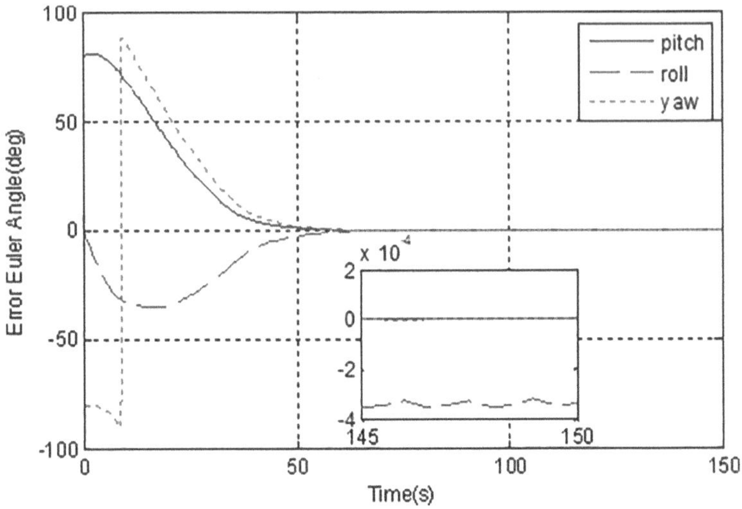

FIG. 6.16 Curve of error Euler angle.

control process and the singularity issue is solved. The reason of this improvement is similar as the analysis of simulation of attitude stabilization control therefore it is omitted in this part.

6.5 Conclusion

In this chapter, a finite-time sliding mode with three-stage structure is constructed. The controller presented in this chapter has both the advantage of high convergence rate and high steady accuracy. The main contribution of the sliding mode is the maneuver stage and the converge stage. Angular velocity in the maneuver stage is constant, hence the convergence rate of attitude quaternion is largely improved. In the converge stage, system state converges to zero in finite time hence the drawback of exponential convergence rate is solved. Moreover, the property that the direction of Euler axis is constant when Euler axis is paralleled to angular velocity is used to solve the singularity issue.

The system inertia matrix uncertainty, unknown disturbance torque with upper bounded norm, is considered in this chapter. Based on the simulation results, the controller could stabilize the system in finite time when the system inertia matrix is not accurately known, and the system is governed by known disturbance torque. Moreover, control torque constraint is taken into consideration, and simulation results demonstrate the effectiveness of the controller.

References

[1] J.D. Boskovi, S. Ming Li, R.K. Mehra, Robust tracking control design for spacecraft under control input saturation, J. Guid. Control. Dyn. 27 (4) (2004) 627–633.

[2] S.C. Lo, Y.P. Chen, Smooth sliding-mode control for spacecraft attitude tracking maneuvers, J. Guid. Control. Dyn. 18 (6) (1995) 1345–1349.

[3] Y. Xia, Z. Zhu, M. Fu, S. Wang, Attitude tracking of rigid spacecraft with bounded disturbances, IEEE Trans. Ind. Electron. 58 (2) (2011) 647–659.

[4] A. Levant, Sliding mode and sliding accuracy in sliding mode control, Int. J. Control. 58 (6) (1993) 1247–1263.

[5] A. Levant, Higher order sliding modes, differentiation and output feedback control, Int. J. Control. 76 (9–10) (2003) 924–941.

[6] E.P. Ryan, Singular optimal controls for second-order saturating systems, Int. J. Control. 30 (4) (1979) 549–564.

[7] Y.Q. Wu, X.H. Yu, Z.H. Man, Terminal sliding mode control design for uncertain dyanmic systems, Syst. Control Lett. 34 (5) (1998) 281–287.

[8] S.H. Ding, S.H. Li, Second-order sliding mode controller design subject to mismatched term, Automatica 77 (2017) 388–392.

[9] F. Amato, M. Ariola, Finite time control of discrete time linear systems, IEEE Trans. Autom. Control 50 (5) (2005) 724–729.

[10] S.H. Yu, X.H. Yu, B. Shirinzadeh, Z.H. Man, Continuous finite time control for robotic manipulators with terminal sliding mode, Automatica 41 (11) (2005) 1957–1964.

[11] F. Xiao, L. Wang, J. Chen, Y.P. Gao, Finite time formation control for multi agent systems, Automatica 45 (11) (2009) 2605–2611.

[12] Z.Y. Meng, W. Ren, Z. You, Distributed finite time attitude containment control for multiple rigid bodies, Automatica 46 (12) (2010) 2092–2099.

[13] S. Wu, G. Radice, Y. Gao, Z. Sun, Quaternion-based finite time control for spacecraft attitude tracking, Acta Astronaut. 69 (1–2) (2011) 48–58.

[14] S. Wu, G. Radice, Z. Sun, Robust finite-time control for flexible spacecraft attitude maneuver, J. Aerosp. Eng. 27 (1) (2014) 185–190.

[15] H. Liang, Z. Sun, J. Wang, Finite-time attitude synchronization controllers design for spacecraft formations via behavior based approach, Proc. Inst. Mech. Eng. G J. Aerospace Eng. 227 (11) (2013) 1737–1753.

[16] J. Wang, H. Liang, Z. Sun, Dual-quaternion-based finite-time control for spacecraft tracking in six degrees of freedom, Proc. Inst. Mech. Eng. G J. Aerospace Eng. 227 (3) (2013) 528–545.

[17] J. Wang, H. Liang, Z. Sun, S. Zhang, M. Liu, Finite-time control for spacecraft formation with dual-number-based description, J. Guid. Control. Dyn. 35 (3) (2012) 950–962.

[18] B. Xiao, S. Yin, L. Wu, A structure simple controller for satellite attitude tracking maneuver, IEEE Trans. Ind. Electron. 64 (2) (2017) 1436–1446.

[19] B. Xiao, S. Yin, H. Gao, Tracking control of robotic manipulators with uncertain kinematics and dynamics, IEEE Trans. Ind. Electron. 63 (10) (2016) 6439–6449.

[20] B. Xiao, S. Yin, Velocity-free fault tolerant and uncertainty attenuation control for a class of nonlinear systems, IEEE Trans. Ind. Electron. 63 (7) (2016) 4400–4411.

[21] P.M. Tiwari, S. Janardhanhan, M. Nabi, Attitude control using higher order sliding mode, Aerosp. Sci. Technol. 64 (2016) 108–113.

Part III

Research on finite time attitude control for fast maneuver rigid-flexible coupling satellites

Chapter 7

Finite-time output feedback attitude control

7.1 Introduction

Flexible appendages are increasingly widely applied in satellites to reduce launch weight, control launch cost, and complete specific on-orbit tasks, such as solar panels on communication satellites, large flexible antennas on remote sensing satellites, and flexible manipulators on-orbit service satellites [1]. However, the attitude maneuver of the satellite causes the vibration of the flexible appendage due to the structural characteristics of the flexible appendage itself (e.g., low modal frequency and weak damping). In addition, the coupling between the vibration of the flexible appendage and the motion of the central rigid body will greatly reduce the attitude control accuracy of the satellite [2]. What is more, the satellite is required to perform the large-angle rapid maneuver in some exceptional cases, and the rigid–flexible coupling characteristic is more obvious [3,4].

The complexity of flexible satellites imposes strict requirements on control schemes. In [5], with the combination of fuzzy mechanism and sliding mode control, an attitude maneuver controller is designed for the three-axis flexible satellite to make the attitude asymptotically stable. An attitude feedback controller is presented based on the Lyapunov method in [6]. The attitude angle and angular velocity can converge asymptotically, but the vibration of the flexible appendage attenuates. A hybrid $H_2/H\infty$ output feedback attitude controller is designed for a flexible satellite when the flexible modal is not measurable [7], which can achieve attitude stability and reduce the flexible vibration concurrently. For the three-axis stabilized flexible satellite with flexible strain gauges installed in flexible appendages, an adaptive sliding mode attitude controller is developed in [8]. A positive position feedback active vibration suppression mechanism is proposed based on linear quadratic regulation. In [9], the angular velocity in attitude maneuver is planned to reduce the excitation to the vibration of flexible appendages. Then a sliding mode control law is derived based on the power reaching law, which can effectively suppress chattering while achieving attitude stability. In [10], an attitude maneuver controller without angular

Fast Satellite Attitude Maneuver and Control. https://doi.org/10.1016/B978-0-323-95455-6.00005-6

velocity is designed, and an extended state observer is used to estimate the angular velocity and the total disturbance of the system. The above attitude controllers can only ensure the asymptotic stability of flexible satellites, and their convergence time is infinite theoretically, so they cannot meet the requirements of rapid maneuver.

Compared to the infinite time stable controller, the finite time one has the advantage of fast stability. Moreover, because fractional power is introduced into the finite-time controller, it also has better robustness and anti-interference ability. Therefore, the finite-time control method has raised wide research interests. Du extends the finite-time attitude tracking controller for a single spacecraft to multiple spacecraft and designs a finite-time attitude cooperative controller [11]. Wu uses the spacecraft dynamics model represented by quaternions to design a finite-time terminal sliding mode (TSM) controller, and an adaptive method is used to estimate the external disturbance [12]. Considering model uncertainties and the unknown upper bound of external disturbances, Tiwari designs a nonsingular fast terminal sliding mode (NFTSM) controller, which keeps the finite-time convergence and has fast convergence speed, and avoids singularities [13]. Lu [14] develops a finite-time NFTSM controller for rigid spacecraft. The adaptive estimation strategy is introduced when the model is uncertain, and the upper bound of external disturbance is unknown. Because the controller is continuous, it weakens the chattering. Moreover, the effectiveness of the controller is verified by the numerical simulation of the Hubble telescope.

However, the above finite-time controllers cannot cope with the impact of flexible appendages on satellite attitude control. Compared to the previously derived controller that only considers the finite-time convergence [15], this chapter studies the finite-time fast attitude control of flexible satellites with large angles and rapid maneuver. Specifically, in the dynamic modeling process, we consider the rigid–flexible coupling effect between the vibration of the flexible appendage and the motion of the rigid body. From this foundation, finite-time fast attitude stabilization control will be studied.

In this chapter, finite-time output attitude control scheme is proposed. Based on the first order approximation rigid–flexible coupling dynamics model, a NFTSM controller (NFTSMC) is designed in the presence of model uncertainties and external disturbances. This controller can provide finite-time fast convergence to the equilibrium with no singularities. Then a modified NFTSM output feedback controller is proposed in the condition that all the flexible vibration measurements are not available, and the dynamic finite-time output feedback controller is based on angle and angular velocity only.

7.2 Main results

In this section, the NFTSM technique is applied to design a finite-time controller to ensure that the system trajectory reaches zero in the finite time. In order to facilitate the controller design, we will start by introducing some definitions and lemmas about finite-time stability.

Definition 7.1 For autonomous system $\dot{\boldsymbol{x}} = f(\boldsymbol{x})$, its equilibrium point is $\boldsymbol{x}=0$. If there exists continuous function $T(x)$, such that $\forall \boldsymbol{x}_0 \in \boldsymbol{D}_0 \subset \boldsymbol{D}$, the solution of the system $\boldsymbol{x}(t, \boldsymbol{x}_0) \in \boldsymbol{D}_0/\{0\}$ when $t < T(\boldsymbol{x}_0)$, and $\boldsymbol{x}(t, \boldsymbol{x}_0)=0$ when $t \geq T(\boldsymbol{x}_0)$. Then $\boldsymbol{x}(t, \boldsymbol{x}_0)$ is finite-time convergent, and the settling time is $T(\boldsymbol{x}_0)$. The zero solution of the system is finite-time stable if it is Lyapunov stable and finite-time convergent.

Lemma 7.1 *Consider autonomous system $\dot{\boldsymbol{x}} = f(\boldsymbol{x})$ and $f(0)=0$ [16]. Suppose that there exists a continuously differentiable function $V:\mathbf{D}\rightarrow \boldsymbol{R}$, which satisfies the following conditions:*

1) V is positive definite;

2) There exist positive constants $\alpha>0$, $1>p>0$ and an open neighborhood $\mathbf{D}_0 \subset \mathbf{D}$, which does not include the origin, such that the function V satisfies

$$\dot{V}(\boldsymbol{x}) < -aV^p(\boldsymbol{x}), \quad \boldsymbol{x} \in \boldsymbol{D}_0$$

Then the system is finite-time stable, and the settling time is

$$T \leq \frac{V_0^{1-p}}{\alpha(1-p)}.$$

Lemma 7.2 *Consider autonomous system $\dot{\boldsymbol{x}} = f(\boldsymbol{x})$ and $f(0)=0$ [16]. Suppose that there exists a continuously differentiable function $V:\mathbf{D}\rightarrow \boldsymbol{R}$, which satisfies the following conditions:*

1) V is positive definite;

2) There exist positive constants $\alpha>0$, $\beta>0$, $1>p>0$ and an open neighborhood $\mathbf{D}_0 \subset \mathbf{D}$, which does not include the origin, such that the function V satisfies

$$\dot{V}(\boldsymbol{x}) < -\beta V(\boldsymbol{x}) - aV^p(\boldsymbol{x}), \quad \boldsymbol{x} \in \boldsymbol{D}_0.$$

Then the system is fast finite-time stable, and the settling time is

$$T \leq \frac{1}{\beta(1-p)} \ln \frac{\alpha + \beta V_0^{1-p}}{\alpha}.$$

Remark 7.1 It should be mentioned that in practical controller design, we do not need system states converge strictly to the equilibrium point in finite time, but only require these states converge to a small neighborhood that satisfies control needs. Such that, some researchers proposed a more practical definition called practical finite-time stable (PFS) [17].

Definition 7.2 Consider a system $\dot{\boldsymbol{x}} = f(\boldsymbol{x}, \boldsymbol{u})$, where $\boldsymbol{x}$ is the state vector, and $\boldsymbol{u}$ is the input vector. The solution is PFS if for all $\boldsymbol{x}(t_0)=\boldsymbol{x}_0$, there exist $\varepsilon>0$ and $T(\varepsilon, \boldsymbol{x}_0)<\infty$, such that $\|\boldsymbol{x}(t)\|<\varepsilon$ for all $t \geq t_0+T$.

We rewrite the error dynamic Eqs. (2.38) and (2.39) in the state space form:

$$J_{\mathrm{mo}}\ddot{\theta}_{\mathrm{e}} + F\dot{\theta}_{\mathrm{e}} + H = T_{\mathrm{c}} + d \tag{7.1}$$

$$\begin{bmatrix} \dot{\boldsymbol{p}} \\ \dot{\boldsymbol{\chi}} \end{bmatrix} = \boldsymbol{A}\begin{bmatrix} \boldsymbol{p} \\ \boldsymbol{\chi} \end{bmatrix} + \boldsymbol{B}\dot{\theta}_{\mathrm{e}} \tag{7.2}$$

where

$$\boldsymbol{A} = \begin{bmatrix} 0 & I \\ -\boldsymbol{M}_{\mathrm{s}}^{-1}\left(\boldsymbol{K}_{\mathrm{s}} + \dot{\theta}_{\mathrm{e}}^{2}\boldsymbol{G}\right) & -\boldsymbol{M}_{\mathrm{s}}^{-1}\boldsymbol{C}_{\mathrm{s}} \end{bmatrix},$$

$$\boldsymbol{B} = \begin{bmatrix} -\boldsymbol{M}_{\mathrm{s}}^{-1}\boldsymbol{U}^{\mathrm{T}} \\ \boldsymbol{M}_{\mathrm{s}}^{-1}\boldsymbol{C}_{\mathrm{s}}\boldsymbol{M}_{\mathrm{s}}^{-1}\boldsymbol{U}^{\mathrm{T}} \end{bmatrix},$$

where $d = J_{\mathrm{mo}}g + J_{\mathrm{mo}}\Delta M$ is the total disturbance of the system including high-order rigid flexible coupling terms, external disturbances, and model uncertainties.

For the first-order approximation, rigid–flexible coupling dynamic model, the linear sliding mode can be designed as

$$s = \dot{\theta}_{\mathrm{e}} + k\theta_{\mathrm{e}} \tag{7.3}$$

where $k > 0$, and the system trajectory can converge to the equilibrium point asymptotically after it reaches the sliding surface.

With the nonlinear term incorporated, TSM can guarantee the finite-time convergence of the trajectory along the sliding surface, and it is designed as

$$s = \dot{\theta}_{\mathrm{e}} + k\mathrm{sig}(\theta_{\mathrm{e}})^{\gamma} \tag{7.4}$$

where $k > 0, 0 < \gamma < 1$, $\mathrm{sig}(x)^{l} = |x|^{l}\mathrm{sign}(x)$, $\mathrm{sign}(\cdot)$ is the sign function. The convergence rate of the conventional TSM is obviously slower than the linear sliding mode when states are far from the equilibrium. In order to increase the convergence rate, the fast TSM is designed by combining these two sliding modes as [18]

$$s = \dot{\theta}_{\mathrm{e}} + k_1\theta_{\mathrm{e}} + k\mathrm{sig}(\theta_{\mathrm{e}})^{\gamma} \tag{7.5}$$

where $k, k_1 > 0, 0 < \gamma < 1$. It can be seen from Eq. (7.5) that the linear term $k_1\theta_e$ can dominate the convergence speed when the system trajectory is far from the equilibrium point, while the nonlinear term $k\mathrm{sig}(\theta_{\mathrm{e}})^{\gamma}$ will guarantee the fast convergence speed when the system trajectory approaches the equilibrium point. Such that the system trajectory can converge fast globally.

Differentiating Eq. (7.5) yields

$$\dot{s} = \ddot{\theta}_{\mathrm{e}} + k_1\dot{\theta}_{\mathrm{e}} + k|\theta_{\mathrm{e}}|^{\gamma-1}\dot{\theta}_{\mathrm{e}}$$

Because $\gamma - 1 < 0$, there exists singularity when $\dot{\theta}_e \neq 0$ and $\theta_e = 0$. To tackle this problem, the NFTSM is proposed as [17]

$$s = \theta_e + a\mathrm{sig}(\theta_e)^{\gamma_1} + b\mathrm{sig}\left(\dot{\theta}_e\right)^{\gamma_2} \tag{7.6}$$

where a, $b > 0$, $2 > \gamma_2 > 1$, and $\gamma_1 > \gamma_2$. Differentiating the Eq. (7.6) yields

$$\dot{s} = \dot{\theta}_e + a\gamma_1 |\theta_e|^{\gamma_1 - 1}\dot{\theta}_e + b\gamma_2 \left|\dot{\theta}_e\right|^{\gamma_2 - 1}\ddot{\theta}_e \tag{7.7}$$

Since $\gamma_1 - 1 > 0$ and $\gamma_2 - 1 > 0$, the singularity can be avoided successfully.

7.2.1 State feedback finite-time attitude control

Substitution of Eq. (7.1) into Eq. (7.7) yields

$$\dot{s} = b\gamma_2 \left|\dot{\theta}_e\right|^{\gamma_2 - 1} J_{mo}^{-1}\left(-F\dot{\theta}_e - H + T_c\right) + \left(1 + a\gamma_1 |\theta_e|^{\gamma_1 - 1}\right)\dot{\theta}_e + b\gamma_2 \left|\dot{\theta}_e\right|^{\gamma_2 - 1} J_{mo}^{-1} d \tag{7.8}$$

Assumption 7.1 Assume that the non-negative function $L(\theta, \boldsymbol{p}, \chi, t)$ satisfies $\|d\| \leq L(\theta, \boldsymbol{p}, \chi, t)$.

Given that the angle and angular velocity are bounded. Due to the high-order rigid–flexible coupling term, external disturbance, and the uncertainty of moment of inertia are minor compared with the control moment. It can be concluded that d is bounded.

Under the condition that both the attitude information and the vibration information of the flexible appendage can be measured, the following state feedback controller is governed by

$$T_c = T_{eq} + T_{vss} \tag{7.9}$$

$$T_{eq} = F\dot{\theta}_e + H - J_{mo}(b\gamma_2)^{-1}\left(1 + a\gamma_1 |\theta_e|^{\gamma_1 - 1}\right) sig\left(\dot{\theta}_e\right)^{2 - \gamma_2} \tag{7.10}$$

$$T_{vss} = -(\lambda + L(\theta, \boldsymbol{p}, \chi, t))\mathrm{sign}(s) \tag{7.11}$$

where the equivalent control law T_{eq} is solved by $\dot{s} = 0$, T_{vss} is the switch control and is robust to the total disturbance of the system, and $\lambda > 0$ is the switching gain.

Theorem 7.1 *For the dynamic systems Eq. (7.1) with disturbances and model uncertainties, with the controller designed as Eq. (7.9), the system trajectory will converge to zero in finite time.*

Proof: Consider a Lyapunov candidate

$$V = \frac{1}{2}s^2 \tag{7.12}$$

Substitution of Eq. (7.9) into the time derivative of Eq. (7.12) yields

$$\begin{aligned}\dot{V}&=s\left(b\gamma_2\left|\dot{\theta}_e\right|^{\gamma_2-1}J_{mo}^{-1}(d-(\lambda+L(\theta,\boldsymbol{p},\boldsymbol{\chi},t))\,\mathrm{sign}(s))\right)\\&\le s\left(-b\gamma_2\left|\dot{\theta}_e\right|^{\gamma_2-1}J_{mo}^{-1}\lambda\,\mathrm{sign}(s)\right)\\&\le-\lambda\varphi\left(\dot{\theta}_e\right)|s|\end{aligned}\tag{7.13}$$

where $\varphi\left(\dot{\theta}_e\right)=\inf\left(b\gamma_2\left|\dot{\theta}_e\right|^{\gamma_2-1}J_{mo}^{-1}\right)$. Let $\alpha=\sqrt{2}\lambda\varphi\left(\dot{\theta}_e\right)$, one has

$$\dot{V}\le-\alpha V^{1/2}\tag{7.14}$$

By using Lemma 7.1, when $\dot{\theta}_e\ne0$ and $\alpha>0$ hold, the trajectory of the system reaches the sliding surface in finite time. According to the literature [18], it can be seen that $\dot{\theta}_e=0$ is not an attractor. There exists a neighborhood $\left|\dot{\theta}_e\right|\le\sigma$, with σ a small positive number. In this neighborhood, $\ddot{\theta}_e\le-J_{mo}^{-1}\lambda$ when $s>0$ and $\ddot{\theta}_e\ge J_{mo}^{-1}\lambda$ when $s<0$. Since the conversion of the system trajectory between the upper and lower bounds of $\left|\dot{\theta}_e\right|\le\sigma$ is in finite time, and the time to converge from $\left|\dot{\theta}_e\right|>\sigma$ to the boundary is also finite. Therefore, the system trajectory can reach the sliding surface in a finite time.

After the trajectory of the system reaches the surface of the sliding mode, we have

$$\theta_e+a\,sig(\theta_e)^{\gamma_1}+b\,sig\left(\dot{\theta}_e\right)^{\gamma_2}=0\tag{7.15}$$

It is clear that $\mathrm{sign}(\theta)=-\mathrm{sign}\left(\dot{\theta}\right)$ and $\left|\dot{\theta}_e\right|=b^{-1/\gamma_2}\left(|\theta_e|+a|\theta_e|^{\gamma_1}\right)^{1/\gamma_2}$. As $|\theta_e+a\theta_e^{\gamma_1}|=|\theta_e|+a|\theta_e|^{\gamma_1}$ holds, Eq. (7.15) can be converted into

$$\dot{\theta}_e=\left|\dot{\theta}_e\right|\mathrm{sign}\left(\dot{\theta}_e\right)=-b^{-1/\gamma_2}sig\left(\theta_e+a\theta_e^{\gamma_1}\right)^{1/\gamma_2}\tag{7.16}$$

By a similar approach as in [17], the system trajectory can converge to zero in finite time, that is, θ_e converges to zero in finite time. Then it is obtained that $\dot{\theta}_e$ will converge to zero in finite time. Here completes the proof.

7.2.2 Dynamic output feedback finite-time attitude control

For on-orbit missions, it is not easy to measure the vibration information of flexible appendages in real time. Therefore, it is not feasible to feedback control the node coordinate information of the flexible appendage as the system state. In this section, an output feedback controller is designed based on the angle and angular velocity information only to achieve the finite-time stability of the system.

Define $\widehat{\boldsymbol{\chi}},\widehat{\boldsymbol{p}}$ as the estimated value of node coordinate information. The estimation error is then denoted by $\boldsymbol{e}_p=\boldsymbol{p}-\widehat{\boldsymbol{p}}$, $\quad\boldsymbol{e}_\chi=\boldsymbol{\chi}-\widehat{\boldsymbol{\chi}}$.

The dynamic output feedback controller is proposed as follows:

$$\begin{aligned} T_c &= F\dot{\theta}_e + \widehat{H} - (\lambda + L(\theta, \boldsymbol{p}, \boldsymbol{\chi}, t))\text{sign}(s) - \lambda_1 s \\ &-J_{mo}(b\gamma_2)^{-1}\left(1 + a\gamma_1|\theta_e|^{\gamma_1 - 1}\right)sig\left(\dot{\theta}_e\right)^{2-\gamma_2} \end{aligned} \tag{7.17}$$

$$\begin{bmatrix} \dot{\widehat{\boldsymbol{p}}} \\ \dot{\widehat{\boldsymbol{\chi}}} \end{bmatrix} = \boldsymbol{A}\begin{bmatrix} \widehat{\boldsymbol{p}} \\ \widehat{\boldsymbol{\chi}} \end{bmatrix} + \boldsymbol{B}\dot{\theta}_e + \boldsymbol{\Gamma}^{-1}\boldsymbol{C} \tag{7.18}$$

with

$\widehat{H} = -\boldsymbol{UM}_s^{-1}\boldsymbol{C}_s\widehat{\boldsymbol{\chi}} - \boldsymbol{UM}_s^{-1}\left(\boldsymbol{K}_s - \dot{\theta}_e^2\boldsymbol{G}\right)\widehat{\boldsymbol{p}},$

$\boldsymbol{C} = \begin{bmatrix} \left(sb\gamma_2|\dot{\theta}_e|^{\gamma_2-1}J_{mo}^{-1}\boldsymbol{UM}_s^{-1}\left(\boldsymbol{K}_s - \dot{\theta}_e^2\boldsymbol{G}\right)\right)^T \\ \left(sb\gamma_2|\dot{\theta}_e|^{\gamma_2-1}J_{mo}^{-1}\boldsymbol{UM}_s^{-1}\boldsymbol{C}_s\right)^T \end{bmatrix}$, where $\lambda > 0$ and $\lambda_1 > 1$ are constants. $\boldsymbol{\Gamma}$ is the positive definite diagonal matrix. It can be seen that $\boldsymbol{\Gamma A} < 0$ by using the assumption that the flexible appendage has non-negative inertia damping.

Theorem 7.2 *For the dynamic system (7.1) with disturbances and model uncertainties, the controller is designed as Eq. (7.17), then* s, $\mathbf{e}_p$, *and* $\mathbf{e}_\chi$ *are uniformly ultimately bounded.*

Proof: Consider a Lyapunov candidate

$$V_1(t) = \frac{1}{2}s^2 + \frac{1}{2}\begin{bmatrix} \boldsymbol{e}_p^T & \boldsymbol{e}_\chi^T \end{bmatrix}\boldsymbol{\Gamma}\begin{bmatrix} \boldsymbol{e}_p \\ \boldsymbol{e}_\chi \end{bmatrix} \tag{7.19}$$

Substitution of the controller (25) into the time derivative of Eq. (27) yields

$$\begin{aligned} \dot{V}_1 &= \begin{bmatrix} \boldsymbol{e}_p^T & \boldsymbol{e}_\chi^T \end{bmatrix}\boldsymbol{\Gamma A}\begin{bmatrix} \boldsymbol{e}_p \\ \boldsymbol{e}_\chi \end{bmatrix} \\ &+ s\left(b\gamma_2|\dot{\theta}_e|^{\gamma_2-1}J_{mo}^{-1}(d - (\lambda + L(\theta, \boldsymbol{p}, \boldsymbol{\chi}, t))\text{sign}(s) - \lambda_1 s)\right) \\ &\le s\left(-b\gamma_2|\dot{\theta}_e|^{\gamma_2-1}J_{mo}^{-1}\lambda\text{sign}(s) - \lambda_1 s\right) + \begin{bmatrix} \boldsymbol{e}_p^T & \boldsymbol{e}_\chi^T \end{bmatrix}\boldsymbol{\Gamma A}\begin{bmatrix} \boldsymbol{e}_p \\ \boldsymbol{e}_\chi \end{bmatrix} \\ &= -b\gamma_2|\dot{\theta}_e|^{\gamma_2-1}J_{mo}^{-1}\left(\lambda|s| + \lambda_1 s^2\right) + \begin{bmatrix} \boldsymbol{e}_p^T & \boldsymbol{e}_\chi^T \end{bmatrix}\boldsymbol{\Gamma A}\begin{bmatrix} \boldsymbol{e}_p \\ \boldsymbol{e}_\chi \end{bmatrix} \end{aligned} \tag{7.20}$$

Since $\boldsymbol{\Gamma A} < 0$, it yields $\dot{V}_1 < 0$.

Here Theorem 7.2 is proved.

However, the above theorem cannot prove that the satellite attitude can converge in finite time [19]. Let

$$u_1 = \widehat{H} - H \tag{7.21}$$

Given that $\boldsymbol{e}_p$ and $\boldsymbol{e}_\chi$ are bounded. It can be seen that u_1 has an upper bound and satisfies $|u_1| \le u_{\max}$, where $u_{\max} > 0$ is a constant.

Next, we prove the finite-time convergence of satellite attitude.

Theorem 7.3 *For the dynamic systems (7.1) with disturbances and model uncertainties, the controller is designed as Eq. (7.17), when the following condition holds*

$$s^2(0) + \left[\boldsymbol{e}_p^T(0)\ \boldsymbol{e}_\chi^T(0)\right]\boldsymbol{\Gamma}\begin{bmatrix}\boldsymbol{e}_p(0)\\ \boldsymbol{e}_\chi(0)\end{bmatrix} < V_{max} \tag{7.22}$$

where $V_{max} > 0$. *The sliding surface* s *will converge to*

$$|s| \le \Delta \tag{7.23}$$

$$\Delta = min\left\{\sqrt{\frac{c}{\lambda_1 - 1}}, \frac{c}{\lambda}\right\} \tag{7.24}$$

in finite time, where $c = u_{max}^2/4 > 0$. *Meanwhile, the attitude angle error and attitude angular velocity error will converge to the following neighborhood in finite time, given by*

$$|\theta_e| \le \Delta_1 = min\left\{\Delta, \left(\frac{\Delta}{a}\right)^{1/\gamma_1}\right\} \tag{7.25}$$

$$\left|\dot{\theta}_e\right| \le \Delta_2 = \left(\frac{\Delta_1 + a\Delta_1^{\gamma_1} + \Delta}{b}\right)^{1/\gamma_2} \tag{7.26}$$

Proof: Consider a Lyapunov candidate

$$V = \frac{1}{2}s^2 \tag{7.27}$$

Substituting controller (7.17) into the time derivative of the above equation yields

$$\begin{aligned}\dot{V} &= s\dot{s}\\ &= s\left(b\gamma_2\left|\dot{\theta}_e\right|^{\gamma_2 - 1}J_{mo}^{-1}(d + u_1 - (\lambda + L(\theta, \boldsymbol{p}, \boldsymbol{\chi}, t))\text{sign}(s) - \lambda_1 s)\right)\\ &\le b\gamma_2\left|\dot{\theta}_e\right|^{\gamma_2 - 1}J_{mo}^{-1}\left(-\lambda_1 s^2 - \lambda s + s^2 + \frac{u_{max}^2}{4}\right)\\ &\le -\varphi b_1 V - \varphi b_2 V^{1/2} + \varphi c\end{aligned} \tag{7.28}$$

Let $\varphi = \inf\left(b\gamma_2\left|\dot{\theta}_e\right|^{\gamma_2 - 1}J_{mo}^{-1}\right)$ as in Eq. (7.13), $b_1 = 2(\lambda_1 - 1) > 0$, and $b_2 = \sqrt{2}\lambda > 0$. Eq. (7.28) has two expressions of

$$\dot{V} \le -\left(\varphi b_1 - \frac{\varphi c}{V}\right)V - \varphi b_2 V^{1/2} \tag{7.29}$$

$$\dot{V} \le -\varphi b_1 V - \left(\varphi b_2 - \frac{\varphi c}{V^{1/2}}\right)V^{1/2} \tag{7.30}$$

When $b_1 - c/V > 0$ holds, by using Lemma 7.2, s will converge to

$$|s| \le \sqrt{\frac{c}{\lambda_1 - 1}} \tag{7.31}$$

And when $b_2 - c/V^{1/2} > 0$ holds, s will converge to

$$|s| \le \frac{c}{\lambda} \tag{7.32}$$

By combining Eqs. (7.31) and (7.32), Eq. (7.24) can be derived. Moreover, from Eq. (7.6), it holds that

$$\theta_e + a sig(\theta_e)^{\gamma_1} + b sig\left(\dot{\theta}_e\right)^{\gamma_2} = \phi \tag{7.33}$$

Similarly, Eq. (7.33) also has two expressions of

$$\left(1 - \frac{\phi}{|\theta_e|}\right)|\theta_e| + a|\theta_e|^{\gamma_1} + b\left|\dot{\theta}_e\right|^{\gamma_2} = 0 \tag{7.34}$$

$$|\theta_e| + \left(a - \frac{\phi}{|\theta_e|^{\gamma_1}}\right)|\theta_e|^{\gamma_1} + b\left|\dot{\theta}_e\right|^{\gamma_2} = 0 \tag{7.35}$$

When $1 - \phi/|\theta_e| > 0$ holds, θ_e will converge to

$$|\theta_e| \le |\phi| \le \Delta \tag{7.36}$$

When $a - \phi/|\theta_e|^{\gamma_1} > 0$ holds, θ_e will converge to

$$|\theta_e| \le (|\phi|/a)^{1/\gamma_1} \le (\Delta/a)^{1/\gamma_1} \tag{7.37}$$

By combining Eqs. (7.36) and (7.37), Eq. (7.25) can also be derived. Furthermore, $\dot{\theta}_e$ will converge to

$$\left|\dot{\theta}_e\right| \le \left(\frac{\Delta_1 + a\Delta_1^{\gamma_1} + \Delta}{b}\right)^{1/\gamma_2} = \Delta_2 \tag{7.38}$$

This completes the proof.

7.3 Simulation results

In this section, numerical simulations are set to verify the effectiveness of the proposed control scheme for rigid–flexible spacecraft. In order to save the limited space, we mainly investigate the most challenging case, in which the vibration information of flexible appendages is not available. The parameters of the

TABLE 7.1 Model parameters of the rigid–flexible coupling spacecraft.

Parameters	Value	Parameters	Value
The moment of inertial of the hub	$128\ \text{kg}\cdot\text{m}^2$	Volume density of the beam	$2.7667\times10^3\ \text{kg/m}^3$
The radius of the hub	$2r_c=1.6\ \text{m}$	Young's modulus of elasticity of the beam	$6.8952\times10^{10}\ \text{N/m}^2$
The length of the beam	$L=20\ \text{m}$	Section moment of inertia of the beam	$I=8.2189\times10^{-9}\ \text{m}^4$
Sectional area of the beam	$A=7.2986\times10^{-5}\ \text{m}^2$	The weight of the tip mass	0.3 kg

rigid–flexible spacecraft are listed in Table 7.1. The initial angle is $\theta(t_0)=0$ rad, the initial angular velocity is $\dot{\theta}(t_0)=0$ rad/s, and the desired attitude angle is $\theta_d=0.5$ rad. It is assumed that $T_d=0.4\dot{\theta}\sin(0.1t)\text{N}\cdot\text{m}$ and $\Delta J_{mo}/J_{mo}=20\%$. The amplitude of the control moment is limited to 5 Nm. Parameters of the controller (7.17) and (7.18) are $a=0.2$, $b=28$, $\gamma_1=7/5$, $\gamma_2=2.3$, $\Gamma=I_{40}$, and $\lambda=30$.

From the analysis of the simulation results in Figs. 7.1 and 7.2, we can see that in the condition when the vibrations of appendages cannot be measured, the angle error and angular velocity error can still converge rapidly in finite time

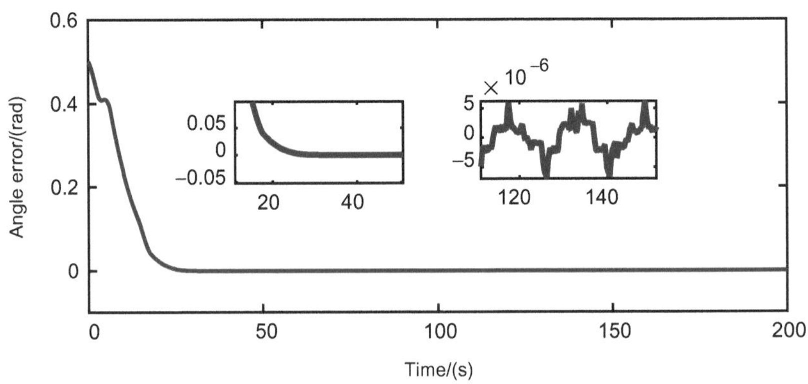

FIG. 7.1 Angle error.

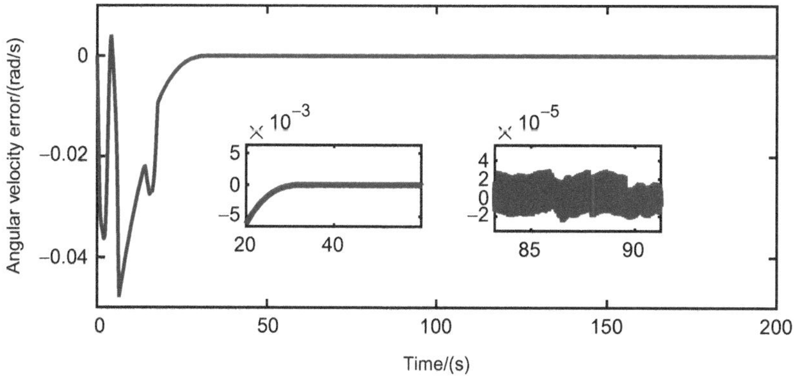

FIG. 7.2 Angular velocity error.

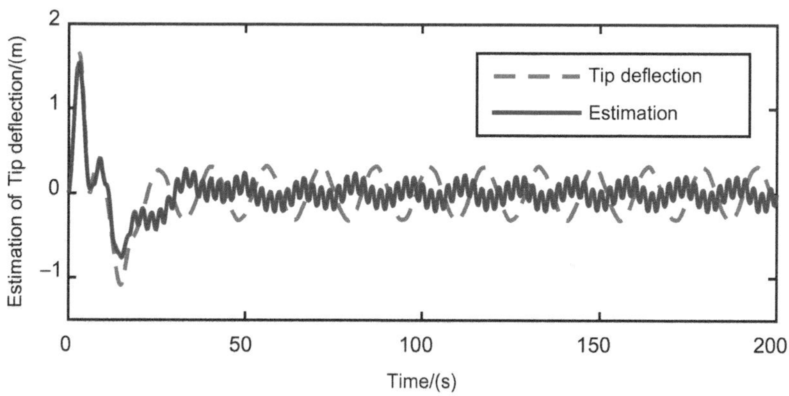

FIG. 7.3 Estimation of tip deflection.

and the steady-state accuracy can still meet the control requirements with the dynamic output feedback controller proposed in this chapter.

Since the sliding mode variable is introduced into the estimation process of flexible vibration of the appendages, the estimation has a larger vibration frequency compared to the real flexible vibration as can be seen in Fig. 7.3. The estimation error in Fig. 7.4 can soon damp to a small value, so it can still meet the requirement of the controller design. In addition, as can be seen from Fig. 7.5, the chattering of the control torque is relatively obvious, but it maintains within a small range. The convergence of the sliding mode surface can be achieved in about 50 s as can be seen in Fig. 7.6, which as well proves the effectiveness of the proposed control scheme.

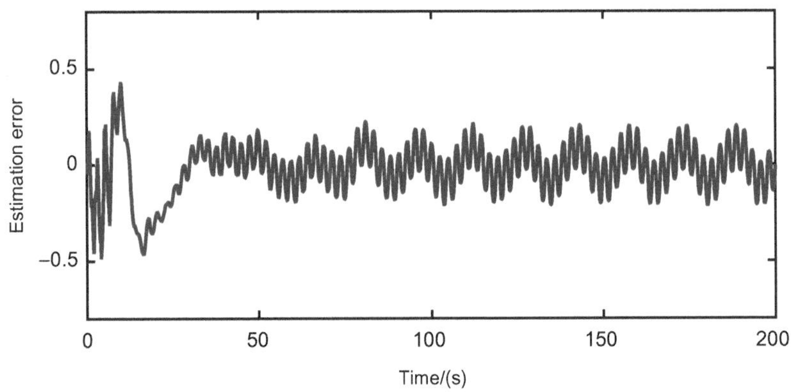

FIG. 7.4 Estimation error.

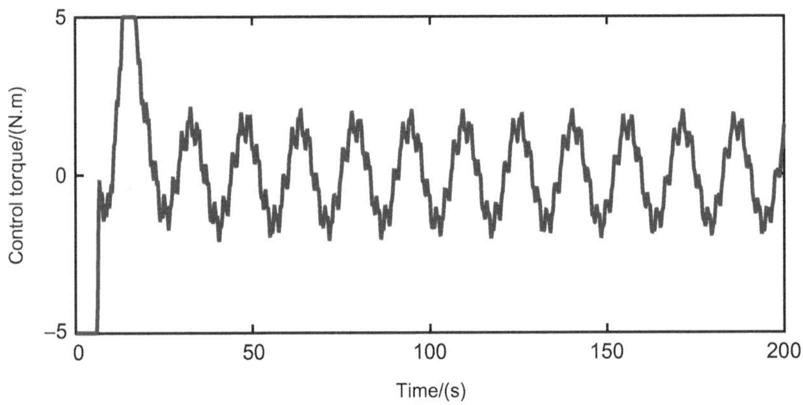

FIG. 7.5 Control torque.

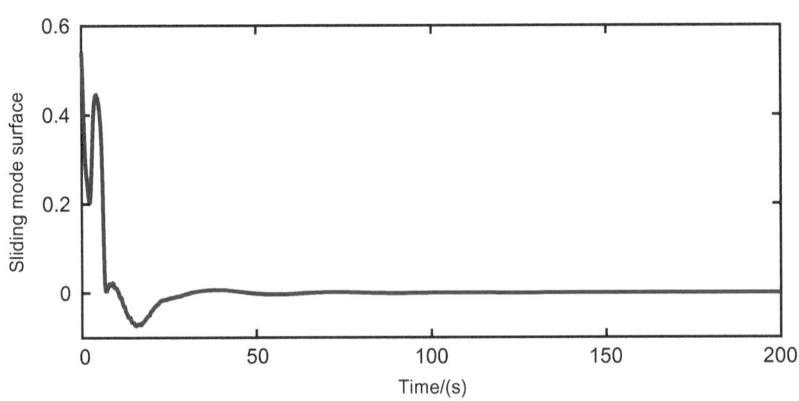

FIG. 7.6 Sliding mode surface.

7.4 Conclusion

In this chapter, a NFTSM controller based on the TSM theory is designed for the improved first-order approximate rigid–flexible coupling dynamic model of flexible satellite, which makes the error angle and error angular velocity converge quickly in finite time, and improves the steady-state accuracy. In the condition that the flexible vibration of the flexible appendage is unavailable, a dynamic output feedback controller is designed based on NFTSM. The controller only needs the information of attitude and angular velocity to realize the rapid convergence of attitude in finite time. In view of the unmeasurable angular velocity information caused by the failure of sensors, the TSM observer is further introduced to reconstruct the attitude and angular velocity information in finite time. Numerical simulation shows that the NFTSM controller based on the observed information can provide fast attitude maneuver control in finite time.

References

[1] J.Y.L. Ho, Direct path method for flexible multibody spacecraft dynamics, J. Spacecr. Rocket. 14 (2) (2012) 102–110.

[2] J.B. Yang, L.J. Jiang, D.C. Chen, Dynamic modelling and control of a rotating Euler–Bernoulli beam, J. Sound Vib. 274 (3–5) (2004) 863–875.

[3] T.R. Kane, R.R. Ryan, A.K. Banerjee, Dynamics of a cantilever beam attached to a moving base, J. Guid. Control. Dyn. 10 (2) (1987) 139–151.

[4] J.P. Jiang, D.X. Li, Modeling and simulation for the rigid-flexible coupling dynamics of the spacecraft with flexible appendages, J. Astronaut. 26 (3) (2005) 270–274.

[5] H. Bang, C.K. Ha, H.K. Jin, Flexible spacecraft attitude maneuver by application of sliding mode control, Acta Astronaut. 57 (11) (2005) 841–850.

[6] Y. Xiao, D. Ye, Z.X. Yang, et al., Research on attitude adjustment control for large angle maneuver of rigid-flexible coupling spacecraft, in: The 8th International Conference on Intelligent Robotics and Applications, Portsmouth, UK, August 24–27, 2015.

[7] H. Liu, Y.M. Zhang, Z.H. Yang, Mixed H2/H∞ output feedback attitude control for flexible spacecraft, J. Astronaut. 33 (9) (2012) 1255–1261.

[8] Q.L. Hu, Robust adaptive sliding mode attitude maneuvering and vibration damping of three-axis-stabilized flexible spacecraft with actuator saturation limits, Nonlinear Dyn. 55 (4) (2009) 301–321.

[9] C.X. Zhong, G. Yu, Y. Zhen, et al., Vibration suppression and sliding mode attitude control of flexible spacecraft with unknown disturbance and uncertainty, in: The 33rd Chinese Control Conference, Nanjing, China, July 28–30, 2014.

[10] C.X. Zhong, A.F. Lai, G. Yu, et al., On attitude maneuver control of flexible spacecraft without angular velocity sensors, in: 2013 6th IEEE/SICE International Symposium on System Integration (SII), Kobe, Japan, December 15–17, 2013.

[11] H.B. Du, S.H. Li, C.J. Qian, Finite-time attitude tracking control of spacecraft with application to attitude synchronization, IEEE Trans. Autom. Control 56 (11) (2011) 2711–2717.

[12] S.N. Wu, G.Q. Wu, S.J. Tan, et al., Quaternion-based adaptive terminal sliding mode control for spacecraft attitude tracking, IEEE Int. Conf. Control Autom. 45 (5) (2013) 913–917.

[13] P.M. Tiwari, S. Janardhanan, M.U. Nabi, Rigid spacecraft attitude control using adaptive non-singular fast terminal sliding mode, J. Control Autom. Electr. Syst. 26 (2) (2015) 115–124.

[14] D. Ye, Y.Y. Tu, Z.W. Sun, Attitude tracking sliding mode control using extended state observer for nonparallel ground track imaging, J. Astronaut. 37 (6) (2016) 720–728.

[15] K.F. Lu, Y.Q. Xia, Adaptive attitude tracking control for rigid spacecraft with finite time convergence, Automatica 49 (12) (2013) 3591–3599.

[16] S.P. Bhat, D.S. Bernstein, Continuous finite-time stabilization of the translational and rotational double integrators, IEEE Trans. Autom. Control 43 (5) (1998) 678–682.

[17] L. Yang, J.Y. Yang, Nonsingular fast terminal sliding-mode control for nonlinear dynamical systems, Int. J. Robust Nonlinear Control 21 (16) (2011) 1865–1879.

[18] Y. Feng, X.H. Yu, Z.H. Man, Non-singular terminal sliding mode control of rigid manipulators, Automatica 38 (12) (2002) 2159–2167.

[19] A.M. Zou, K.D. Kumar, Z.G. Hou, et al., Finite-time attitude tracking control for spacecraft using terminal sliding mode and chebyshev neural network, IEEE Trans. Syst. Man Cybern. Part B Cybern. 41 (4) (2011) 950–963.

Chapter 8

Observer-based continuous finite-time attitude control

8.1 Introduction

Focusing on solving the attitude control problem of flexible satellites, some researchers have done much work. An attitude regulation controller is proposed in [1], which combines fuzzy logic and sliding mode, to guarantee the asymptotically stability of flexible satellites. Based on Lyapunov method, an output feedback attitude regulation controller is designed in [2]. With considering external disturbances, the attitude can converge to the equilibrium point asymptotically and the vibration of the flexible appendages can be attenuated. For the flexible satellite, of which on–off thrusters are attitude controllers and embedded smart material is vibration controller attached to the flexible appendages, an adaptive sliding mode controller is first designed in [3] to guarantee the convergence of the attitude, and then a linear quadratic regulator (LQR) based positive position feedback control method is presented for actively suppressing the induced vibration. A sliding mode controller is presented in [4] for flexible satellites. In this control scheme, an angular velocity planning is applied first to reduce the irritation of the vibration of flexible appendages, then a sliding mode controller is designed based on power reaching law; finally, the attitude asymptotic stability can be achieved while the vibration is attenuated efficiently. In the condition when the angular velocity is unavailable, an observer based attitude controller is designed in [5]. In this literature, the extended state observer (ESO) is introduced to rebuild the angular velocity information, and then the attitude asymptotic stability is achieved with a feedback controller. However, all controllers mentioned above can only guarantee that the satellite attitude is asymptotically stable, which means that the convergence time is theoretically infinite, so the attitude control demand for rapid maneuver cannot be met.

Compared to controllers providing infinite convergence time, the finite-time controller attracts more interests with its fast convergence speed. In addition, because of the fractional power term, finite-time controller has attractive advantages of robustness and disturbance rejection. A nonsingular terminal sliding mode (NTSM) is designed in Ref. [6], and it can avoid the singularity while

Fast Satellite Attitude Maneuver and Control. https://doi.org/10.1016/B978-0-323-95455-6.00009-3

guarantee the finite-time stability. Based on FTSM and NTSM, nonsingular fast terminal sliding mode (NFTSM) is proposed in Ref. [7], which has the same property as NTSM but provides a faster convergence speed. In view of the advantages of TSM, many finite-time attitude control schemes for satellites based on TSM are presented lately. Wu et al. [8] designs an adaptive terminal sliding mode controller for rigid satellites with considering the external disturbance. A NFTSM controller is designed for rigid satellites in Ref. [9] with the bound of the total disturbance unknown, and the adaptive algorithm is introduced to estimate uncertainties and disturbances, then the fast finite-time stability is proved by the simulation based on the model of Hubble Space Telescope. An observer-based finite-time attitude controller is presented in Ref. [10] based on NFTSM, in which a finite-time observer is designed to estimate the total disturbance, and the finite-time stability of the closed-loop system is proved by Lyapunov theory. In the condition when the control input is limited, a more practical NFTSM controller for rigid satellite is presented in Ref. [11], and adaptive method is introduced to estimate external disturbances and model uncertainties. However, the above-mentioned papers mainly deal with rigid spacecraft, which means most of these finite-time controllers are designed without considering the coupling effect between the vibration of flexible appendages and the spatial movement of the central hub.

When using the terminal sliding mode control scheme to achieve the attitude control, the chattering caused by the discontinuous term in the controller is another problem which will seriously degrade the attitude control accuracy. What is worse, compared to the rigid satellite, the rigid–flexible coupling satellite not only suffers bigger uncertainties of the moment of inertia but also is more sensitive to the external disturbances [12,13]. In addition, in order to increase the convergence speed and improve the control accuracy of the attitude angle and angular velocity, we need to increase the switch gain to reject the perturbation. However, higher gain will result in serious chattering and thus deteriorate the control performance.

As we mentioned above, discontinuous switch control term and high control gain will cause the chattering when we use sliding mode control, especially for the control of the flexible satellite with large appendages like the one in this chapter. Since satellites with large flexible appendages always have relatively low-frequency band, which makes the flexible appendages are easier to be motivated by the chattering of the sliding mode control, so the chattering phenomenon must be attenuated during the control design for the rigid–flexible coupling satellite with large flexible appendages.

For the attenuation of the chattering, various approaches have be studied. A time-varying sliding mode for attitude control of flexible satellites is designed in literature [14], then boundary layer approach is introduced to make the controller continuous, while the fuzzy scheme is applied to adjust the thickness of the boundary layer and thus to achieve the chattering attenuation. However, boundary layer approach will decrease the control accuracy for it is not a rigorous sliding mode control. Zhu et al. [15] proposes an adaptive sliding mode control with which disturbances and uncertainties can be estimated, thus the control gain is decreased, and the simulation in the end proves the chattering attenuation. Being analyzed for addressing the problem of the chattering and enhancing the control effect, an ESO is proposed in this chapter to estimate the total disturbance of the system, thus to attenuate the chattering by compensating for the total disturbance. Then based on the high-order characteristic of the fast power reaching rule, a continuous terminal sliding mode control is proposed to further eliminate the chattering and suppress the vibration of the flexible appendages.

Comparing to the exiting attitude controllers for rigid–flexible satellites, the attitude controller presented in this chapter has several advantages:

(1) The satellite model considered in this chapter can fully reflect the coupling effect between the vibration of flexible appendages and the spatial movement of the central hub, and thus is more accurate in the rapid maneuver mode.
(2) A fast finite-time output feedback attitude controller for rigid–flexible coupling satellites is designed combining ESO, NFTSM, and fast power reaching law.
(3) The proposed controller is robust to the external disturbance and uncertainty of moment of inertia without knowing the bound of them.
(4) The proposed controller is continuous and have fast convergence speed with high accuracy, no singularity, less chattering, and vibration.
(5) The proposed control scheme is verified through simulations and experiment both.

8.2 Continuous terminal sliding mode control based on ESO

In the endeavor toward achieving the finite-time control for the first-order approximation rigid–flexible coupling dynamic model, a finite-time attitude control method will be present. More specifically, there exists finite time T, such that $\theta_e=0$, $\dot{\theta}_e=0$ when $t>T$. First, we will design the ESO to estimate the total disturbance. Then the fast power reaching law will be introduced for its second-order sliding mode characteristic, which is used to eliminate the chattering. At last, combining these two methods with NFTSM theory, we will design the continuous finite-time controller and prove the finite stability of the system.

8.2.1 The extended state observer design

In this section, we use the ESO to reconstruct the total perturbation of the system dynamics and the total perturbation including external disturbances, model uncertainties, and flexible coupling terms.

$$x_1 = \dot{\theta}_e + k\theta_e \tag{8.1}$$

$$d_1 = g + \Delta M + J_{mo}^{-1}\boldsymbol{H} \tag{8.2}$$

$$f = -J_{mo}^{-1}\boldsymbol{F}\dot{\theta}_e + k\dot{\theta}_e \tag{8.3}$$

where d_1 is the total perturbation of the system dynamics that incorporates high-order coupling terms of the system, and f is the nominal part of the system dynamics. The error dynamic equation of the system can be formulated into state space form as

$$\dot{x}_1 = f + J_{mo}^{-1}\boldsymbol{T}_c + d_1 \tag{8.4}$$

Then let $x_2 = d_1$ and add the total perturbation to the state space as the new state. The extended state form can be expressed as

$$\begin{cases} \dot{x}_1 = f + J_{mo}^{-1}\boldsymbol{T}_c + x_2 \\ \dot{x}_2 = g(t) \end{cases} \tag{8.5}$$

where $g(t)$ is the first-order derivative of the total perturbation of the system, and it is unknown to the controller design. Based on the method of the ESO in Ref. [16], a second-order ESO can be designed as

$$\begin{cases} E_1 = Z_1 - x_1 \\ \dot{Z}_1 = Z_2 + f + J_{mo}^{-1}\boldsymbol{T} - \beta_1 f_{e1}(E_1) \\ \dot{Z}_2 = -\beta_2 f_{e2}(E_1) \end{cases} \tag{8.6}$$

Correction function $f_{e1} = E_1$, and f_{e2} can be defined by *fal* function, where *fal* and $g(t)$ can be designed as

$$fal(E_1, \varepsilon, \delta) = \begin{cases} |E_1|^{\varepsilon} sign(E_1), & |E_{1i}| > \delta \\ E_1/\delta^{1-\varepsilon}, & |E_{1i}| \le \delta \end{cases}$$
$$g(\cdot) = -\beta_2 fal(E_1, \varepsilon, \delta) \tag{8.7}$$

where $0 < \varepsilon < 1, \delta > 0$. By choosing proper parameters $\beta_1, \beta_2, \varepsilon, \delta$, outputs Z_1 and Z_2 can approach x_1 and d_1 respectively.

Then we prove the stability of the ESO. With observation error $E_1 = Z_1 - x_1$, $E_2 = Z_2 - x_2 = Z_2 - d_1$, the observer error dynamics can be expressed as

$$\begin{cases} \dot{E}_1 = E_2 - \beta_1 E_1 \\ \dot{E}_2 = -\beta_2 fal(E_1, \alpha, \delta) - g(t) \end{cases} \tag{8.8}$$

It is proved in Refs. [17,18] that the observer stability can be reached by choosing proper parameters β_1, β_2, ε, δ. When the observer stability is reached, the error differential vector become zero, and the observation error can be expressed as

$$\begin{cases} E_l = -fal^{-1}(g(t)/\beta_2) \\ E_2 = -\beta_1 fal^{-1}(g(t)/\beta_2) \end{cases} \tag{8.9}$$

when $|E_1| > \delta$, the observation error becomes

$$\begin{cases} |E_1| = |g(t)/\beta_2|^{1/\varepsilon} \\ |E_2| = \beta_1 |g(t)/\beta_2|^{1/\varepsilon} \end{cases} \tag{8.10}$$

when $|E_1| \leq \delta$, the observation error becomes

$$\begin{cases} |E_1| = \left|g(t)\delta^{1-\varepsilon}\right|/\beta_2 \\ |E_2| = \beta_1 \left|g(t)\delta^{1-\varepsilon}\right|/\beta_2 \end{cases} \tag{8.11}$$

It can be noted from Eqs. (8.10) and (8.11) that the observation error is related to parameters β_1, β_2, ε, δ and these parameters are chosen to satisfy $\beta_1 > 0$, $\beta_2 > 0$, $0 < \varepsilon < 1$, $\delta > 0$. In addition, when β_2 is chosen big enough, $|g(t)/\beta_2|$ can be very small even if $g(t)$ is unknown. Certainly, β_1 should be small enough to make sure that the observation error E_2 is as small as possible. Hence, observation errors E_1 and E_2 can become small enough, which means Z_1 and Z_2 can converge to neighborhoods of real states x_1 and d_1, respectively [16].

8.2.2 Fast power reaching law

In order to eliminate the chattering, which appears in the sliding mode control, and increase the convergence speed of the system, fast power reaching law will be introduced to solve this problem.

Fast power reaching law can be expressed as [19]

$$\dot{s} = -k_1 s - k_2 sig(s)^\rho \tag{8.12}$$

where $sig(x)^y = |x|^y \cdot sign(x)$, $sign(\cdot)$ is the sign function, $k_1, k_2 > 0$, $0 < \rho < 1$. The fast power reaching law is the combination of the traditional power reaching law $\dot{s} = -k_2 sig(s)^\rho$ and the exponential reaching law $\dot{s} = -k_1 s$. The traditional reaching law is continuous, and the exponential reaching law is added to increase the convergence rate, so both these two advantages are gained by the fast power reaching law. Thus, it can eliminate the chattering and increase the convergence speed of the system simultaneously. Theoretically speaking, this characteristic is mainly because that fast power reaching law has the ability as the second-order sliding mode, which means the sliding mode s and its first derivative $\dot{s}$ can converge to the equilibrium point in finite time T. In other words, the fast power reaching law can be treated as a second-order sliding mode algorithm, which can be used when the system relative order is one. The second-order sliding mode character of the fast power reaching law is proved in Refs. [20–22] in detail.

8.2.3 Controller design

Rewrite the dynamic Eq. (2.40) of the flexible satellite as

$$\ddot{\theta}_e = -J_{mo}^{-1}\boldsymbol{F}\dot{\theta}_e + J_{mo}^{-1}T_c + d_1 \tag{8.13}$$

where d_1 is defined in Eq. (8.2). Since the flexible deformation and its first derivative are both small values, external disturbances, and uncertainties of the moment of inertia are bounded, the total disturbance of the system satisfies $|d_1| < D$, with D a positive constant. Then based on the ESO and the fast power reaching law, we begin to design the finite-time controller, such that attitude angle θ can converge to the desired angle θ_d in finite time. There are three steps to build this controller.

The first step is to design the sliding surface. According to Ref. [23], the nonsingular fast terminal sliding mode, in the form of first-order differential equation, is designed as

$$s = \theta_e + \alpha sig(\theta_e)^{\gamma_1} + \beta sig\left(\dot{\theta}_e\right)^{\gamma_2} \tag{8.14}$$

where $\alpha > 0$, $\beta > 0$, $1 < \gamma_2 < 2$, $\gamma_1 > \gamma_2$.

The second step is to calculate the continuous finite-time controller. We first insert NFTSM (8.14) into the fast power reaching law (8.12) as

$$\dot{s} = \dot{\theta}_e + \alpha\gamma_1|\theta_e|^{\gamma_1-1}\dot{\theta}_e + \beta\gamma_2\left|\dot{\theta}_e\right|^{\gamma_2-1}\ddot{\theta}_e = -k_1 s - k_2 sig(s)^{\rho} \tag{8.15}$$

Then, inserting the dynamic Eq. (8.13) gives

$$T_c = F\dot{\theta}_e - J_{mo}(k_1 s + k_2 sig(s)^{\rho}) - J_{mo}\beta^{-1}\gamma_2^{-1}\left(1 + \alpha\gamma_1|\theta_e|^{\gamma_1-1}\right)sig\left(\dot{\theta}_e\right)^{2-\gamma_2} \tag{8.16}$$

The last step is to compensate the total disturbance d_1 with its estimation value Z_2 in Eq. (8.19), thus the observer-based continuous finite-time control law can be modified as

$$\begin{aligned} T_c = {} & F\dot{\theta}_e - J_{mo}(k_1 s + k_2 sig(s)^{\rho}) - J_{mo}Z_2 \\ & -J_{mo}\beta^{-1}\gamma_2^{-1}\left(1 + \alpha\gamma_1|\theta_e|^{\gamma_1-1}\right)sig\left(\dot{\theta}_e\right)^{2-\gamma_2} \end{aligned} \tag{8.17}$$

Remark 8.1 Since $2-\gamma_2 > 0$, $\gamma_1 - 1 > 0$, and $\gamma_2 - 1 > 0$, there are no singular terms in both Eqs. (8.15) and (8.17). Thus, NFTSM can guarantee a fast convergence rate while avoiding singularity [24].

Theorem 8.1 *Consider the flexible satellite system (8.13). With application of the NFTSM (8.14), the fast power reaching law (8.12), and the continuous* *terminal sliding mode control law (8.17), there exist following conclusions. In the presence of the total disturbance, the system trajectory will converge fast to the neighborhood of the terminal sliding mode* s = 0

$$|s| \le \Delta = min\{\Delta_1, \Delta_2\} \tag{8.18}$$

$$\Delta_1 = \frac{\left|J_{mo}{}^{-1}(d_1 - Z_2)\right|}{k_1} \tag{8.19}$$

$$\Delta_2 = \left(\frac{\left|J_{mo}{}^{-1}(d_1 - Z_2)\right|}{k_2}\right)^{1/\rho} \tag{8.20}$$

In addition, the attitude tracking error θ_e *and angular velocity tracking error* $\dot{\theta}_e$ *can converge to following neighborhoods in finite time, respectively:*

$$|\theta_e| \le \Delta_{\theta_e} = min\left\{\Delta, \left(\frac{\Delta}{\alpha}\right)^{1/\gamma_1}\right\} \tag{8.21}$$

$$\left|\dot{\theta}_e\right| \le \Delta_{\dot{\theta}_e} = \left(\frac{\Delta + \Delta_{\theta_e} + \alpha\Delta_{\theta_e}^{\gamma_1}}{\beta}\right)^{1/\gamma_2} \tag{8.22}$$

Proof: Consider Lyapunov function $V = |s|^2/2$ and calculate the time derivative of the Lyapunov function

$$\dot{V} = s\dot{s} = s\left(\dot{\theta}_e + \alpha\gamma_1|\theta_e|^{\gamma_1 - 1}\dot{\theta}_e + \beta\gamma_2\left|\dot{\theta}_e\right|^{\gamma_2 - 1}\ddot{\theta}_e\right) \tag{8.23}$$

Inserting the satellite dynamic Eq. (8.13) and the control torque (8.17) yields

$$\dot{V} = s\dot{s} = s\beta\gamma_2\left|\dot{\theta}_e\right|^{\gamma_2 - 1}J_{mo}{}^{-1}(T_1 + d_1 - Z_2) \tag{8.24}$$

$$T_1 = -J_{mo}(k_1 s + k_2 sig(s)^{\rho}) \tag{8.25}$$

We first assume that $d_1 = 0$, then we have

$$\begin{aligned}\dot{V} &= s\beta\gamma\left|\dot{\theta}_e\right|^{\gamma-1}J_{mo}{}^{-1}\left[-J_{mo}(k_1 s + k_2 sig(s)^{\rho})\right]\\ &= -s\beta\gamma\left|\dot{\theta}_e\right|^{\gamma-1}(k_1 s + k_2 sig(s)^{\rho})\\ &= -\bar{k}_1 s^2 - \bar{k}_2|s|^{\rho+1}\end{aligned} \tag{8.26}$$

where $\bar{k}_1 = \beta\gamma_2\left|\dot{\theta}_e\right|^{\gamma-1}k_1$, $\bar{k}_2 = \beta\gamma_2\left|\dot{\theta}_e\right|^{\gamma-1}k_2$.

When $\dot{\theta}_e \ne 0$, $\dot{V} \le -2\bar{k}_1 V - 2^{(\rho+1)/2}\bar{k}_2 V^{(\rho+1)/2}$, with $1/2 < (\rho+1)/2 < 1$, satisfies the form of the Lyapunov finite-time stability [25], then the system trajectory can reach the sliding mode surface in finite time T, where

$$T \le \ln\left(\left(\bar{k}_1 V^{(\rho+1)/2} + 2^{(\rho+1)/2}\bar{k}_2\right)/2^{(\rho+1)/2}\bar{k}_2\right)/\left(\bar{k}_2(1-\rho)\right) \tag{8.27}$$

Then the attitude tracking error will converge to the equilibrium point along the sliding mode surface in finite time.

When $\dot{\theta}_e = 0$, the satellite dynamic equation becomes

$$\ddot{\theta}_e = -\beta^{-1}\gamma_2^{-1}\left(1 + \alpha\gamma_1|\theta_e|^{\gamma_1-1}\right)sig\left(\dot{\theta}_e\right)^{2-\gamma_2} - \left(k_1 s + k_2 sig(s)^{\rho}\right) \quad (8.28)$$

It can be noted that $\ddot{\theta}_e \neq 0$, when $\dot{\theta}_e = 0$, and $s \neq 0$, which means $\dot{\theta}_e = 0$ is not an attractor during the reaching phase [24]. Then the system trajectory can reach the sliding surface in finite time and converge to the equilibrium point along the sliding surface in finite time. Hence, in the absence of the total disturbance, the attitude tracking error θ_e can converge to the equilibrium point in finite time.

Then we deal with the situation when $d_1 \neq 0$. We have $\dot{V} = -s\beta\gamma_2\left|\dot{\theta}_e\right|^{\gamma_2-1}\left[k_1 s + k_2 sig(s)^{\rho} - J_{mo}{}^{-1}(d_1 - Z_2)\right]$, and V can be formulated into the following two forms:

$$\dot{V} = -s\beta\gamma_2\left|\dot{\theta}_e\right|^{\gamma_2-1}\left[k_1 - J_{mo}{}^{-1}(d_1 - Z_2)s^{-1}\right]s - s\beta\gamma_2\left|\dot{\theta}_e\right|^{\gamma_2-1}k_2 sig(s)^{\rho} \quad (8.29)$$

$$\begin{aligned}\dot{V} &= -s\beta\gamma_2\left|\dot{\theta}_e\right|^{\gamma_2-1}k_1 s \\ &\quad -s\beta\gamma_2\left|\dot{\theta}_e\right|^{\gamma_2-1}\left[k_2 - J_{mo}{}^{-1}(d_1 - Z_2)\left(sig(s)^{\rho}\right)^{-1}\right]sig(s)^{\rho}\end{aligned} \quad (8.30)$$

For the first form, if $k_1 - J_{mo}{}^{-1}(d_1 - Z_2)s^{-1} > 0$, then the expression of V accords with the form of Lyapunov finite-time stability [25], thus the system trajectory can converge to the neighborhood Δ_1 in finite time

$$|s| \leq \frac{\left|J_{mo}{}^{-1}(d_1 - Z_2)\right|}{k_1} = \Delta_1 \quad (8.31)$$

Similarly, for the second form, if $k_2 - J_{mo}{}^{-1}(d_1 - Z_2)\left(sig(s)^{\rho}\right)^{-1} > 0$, then V accords with the form of Lyapunov finite-time stability, and the system trajectory can converge to the neighborhood Δ_2 in finite time

$$|s| \leq \left(\frac{\left|J_{mo}{}^{-1}(d_1 - Z_2)\right|}{k_2}\right)^{1/\rho} = \Delta_2 \quad (8.32)$$

Hence, when $d_1 \neq 0$ and $\dot{\theta}_e \neq 0$, the system trajectory can converge to the neighborhood $\Delta = \min\{\Delta_1, \Delta_2\}$ in finite time.

For the condition when $d_1 \neq 0$ and $\dot{\theta}_e = 0$, inserting the control torque (8.17) into the satellite dynamic Eq. (8.13) yields

$$\begin{aligned}\ddot{\theta}_e &= -\beta^{-1}\gamma_2^{-1}\left(1 + \alpha\gamma_1|\theta_e|^{\gamma_1-1}\right)sig\left(\dot{\theta}_e\right)^{2-\gamma_2} \\ &\quad -\left(k_1 s + k_2 sig(s)^{\rho}\right) + J_{mo}{}^{-1}(d_1 - Z_2)\end{aligned} \quad (8.33)$$

It can be seen from the above equation that when $\dot{\theta}_e = 0$, the system trajectory is out of the region Δ_1, and we have

$$\ddot{\theta}_e = -\left(k_1 - \frac{\left|J_{mo}{}^{-1}(d_1 - Z_2)\right|}{s}\right)s - k_2|s|^{\rho} sign(s) \neq 0 \tag{8.34}$$

Thus, when $d_1 \neq 0$ and $\dot{\theta}_e \neq 0$, the system trajectory can converge to the neighborhood $\Delta = \min\{\Delta_1, \Delta_2\}$ in finite time.

Since $|s| \leq \Delta$, based on the definition of the sliding mode (8.14), there exists $|\phi| \leq \Delta$, such that

$$\theta_e + \alpha sig(\theta_e)^{\gamma_1} + \beta sig(\dot{\theta}_e)^{\gamma_2} = \phi \tag{8.35}$$

Rewrite the above equation in two forms

$$\left(1 - \frac{\phi}{\theta_e}\right)\theta_e + \alpha sig(\theta_e)^{\gamma_1} + \beta sig(\dot{\theta}_e)^{\gamma_2} = 0 \tag{8.36}$$

$$\theta_e + \left(\alpha - \frac{\phi}{sig(\theta_e)^{\gamma_1}}\right) sig(\theta_e)^{\gamma_1} + \beta sig(\dot{\theta}_e)^{\gamma_2} = 0 \tag{8.37}$$

It can be seen from Eq. (8.36) that if $1 - \phi/\theta_e > 0$, Eq. (8.35) accords with the form of NFTSM, and it is obvious that θ_e can converge to a certain neighborhood in finite time

$$|\theta_e| \leq |\phi| \leq \Delta \tag{8.38}$$

Similarly for Eq. (8.37), if $\alpha - \phi/sig(\theta_e)^{\gamma_1} > 0$, θ_e can also converge to a certain neighborhood in finite time

$$|\theta_e| \leq \left(\frac{|\phi|}{\alpha}\right)^{1/\gamma_1} \leq \left(\frac{\Delta}{\alpha}\right)^{1/\gamma_1} \tag{8.39}$$

Thus, θ_e can converge to $|\theta_e| \leq \Delta_{\theta_e} = \min\{\Delta, (\Delta/\alpha)^{1/\gamma_1}\}$ in finite time.

Based on the definition of the NTSM (8.14), it can be seen that $\dot{\theta}_e$ can converge to the neighborhood $\Delta_{\dot{\theta}_e}$ in finite time

$$\left|\dot{\theta}_e\right| \leq \left(\frac{\Delta + \Delta_{\theta_e} + \alpha\Delta_{\theta_e}^{\gamma_1}}{\beta}\right)^{1/\gamma_2} = \Delta_{\dot{\theta}_e} \tag{8.40}$$

This thereby completes the proof.

Remark 8.2 Because of the compensation effect of the ESO on the total perturbation consisted of coupling terms, external disturbances and model uncertainties, the unknown total perturbation that the controller counteracts decreases from d_1 to $E_2 = d_1 - Z_2$. Since E_2 is asymptotically convergent, the control gain is reduced effectively, and the chattering is attenuated significantly. What is more, the existence of the fractional power of the sign function in the fast power reaching law makes the controller continuous and thus further restricts the chattering. In addition, there is no negative power parameter appeared in the controller, so that the controller is singularity free as well.

Remark 8.3 The controller proposed in this section can guarantee the finite-time convergence to the neighborhood of the equilibrium point and continuity simultaneously. Moreover, values of the convergence time and the control accuracy can be governed to be as small as possible by tuning related gains.

Remark 8.4 The nonlinear term $\left|\dot{\theta}_e\right|^{\gamma-1}$ can increase the convergence speed of the system trajectory to the terminal sliding manifold in the neighborhood of $\dot{\theta}_e = 0$. And this is because when $\dot{\theta}_e < 1$, functional power $0 < \gamma - 1 < 1$ magnifies the weight of $\dot{\theta}_e$. What is more, this magnifying character can avoid the stagnation when the system trajectory reach the neighborhood of $\dot{\theta}_e = 0$ during the reaching phase and drive the system trajectory further approach the equilibrium point during the sliding phase [19].

8.3 Simulation results

8.3.1 Numerical simulations

In this section, simulations are conducted for demonstrating the performances of the proposed ESO (8.6) and the continuous finite-time terminal sliding mode controller (8.17) based on ESO in attitude adjustment. Based on Refs. [26, 27], the physical parameters of the rigid–flexible coupling satellites are chosen as in Table 7.1. During the attitude adjustment, the initial attitude information of the rigid–flexible coupling satellite is $\theta(t_0)=0$ rad, $\dot{\theta}(t_0) = 0$ rad/s. The desired attitude angle is $\theta_d = -0.5$ rad.Without loss of generality, external disturbance is assigned as $T_d = 0.4\dot{\theta}\sin(0.1t)$N · m, and uncertainty of the moment of inertia is given as $\Delta J_{mo}/J_{mo}=20\%$. Parameters of the ESO are chosen as $\beta_1=5$, $\beta_2=300$, $\varepsilon=0.25$, $\delta=0.2$, $k=1$, while parameters of the controller are $k_1=15$, $k_2=30$, $\alpha=0.17$, $\beta=28$, $\gamma_1=2.6$, $\gamma_2=7/5$, $\rho=1/5$.

The estimation of the total perturbation and the estimation error are shown in Fig. 8.1, and it can be seen that ESO can achieve available observation. Although initial observation error is relatively large, it has fast convergence rate and can reach into a very small neighborhood of zero in about 6 s. Based on the estimation value of the ESO, the attitude angle error and angular velocity error are shown as full lines in Fig. 8.2A, B. Under the control of the proposed continues finite-time controller, the attitude of the system can converge to a small neighborhood of the desired attitude within 19 s, and the attitude point accuracy is 1×10^{-4} rad. The angular velocity can also converge to a small neighborhood of zero in 17 s, and the attitude stability is about 5×10^{-5} rad/s. Simulation results of the attitude angle and angular velocity show the fast finite time convergence of the proposed controller with high accuracy. Fig. 8.3A, B reveals that sliding mode s and its first derivative $\dot{s}$ both converge to zero in finite time, which is in accordance with the analysis of the second-order sliding mode character of the fast power reaching law in this chapter. The time response of the control torque result in Fig. 8.4A shows the obvious attenuation of the chattering, which proves that the ESO and fast power reaching law used in the chapter

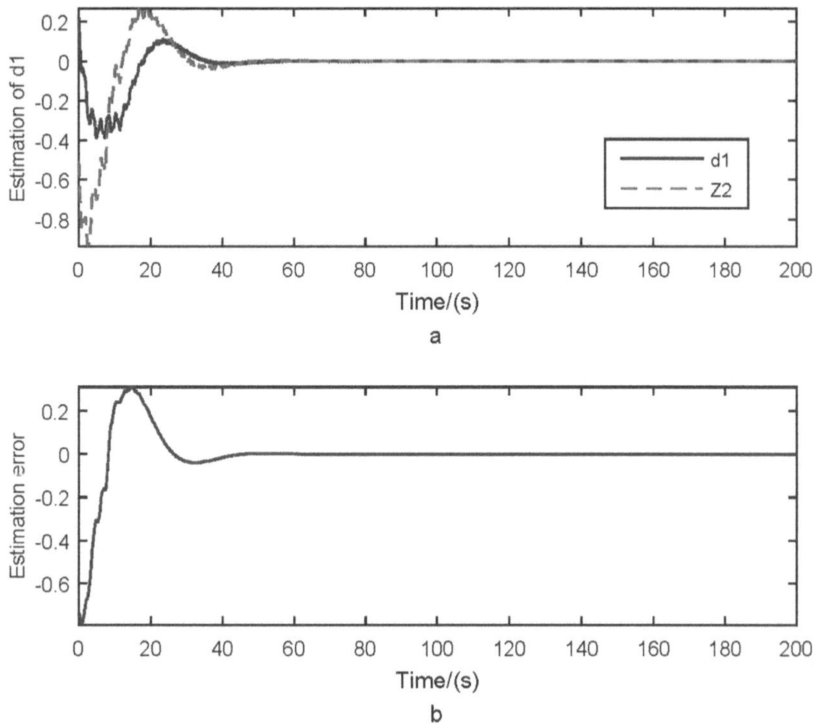

FIG. 8.1 The observe effect of ESO: (A) the estimation of total perturbation; (B) the estimation error of total perturbation.

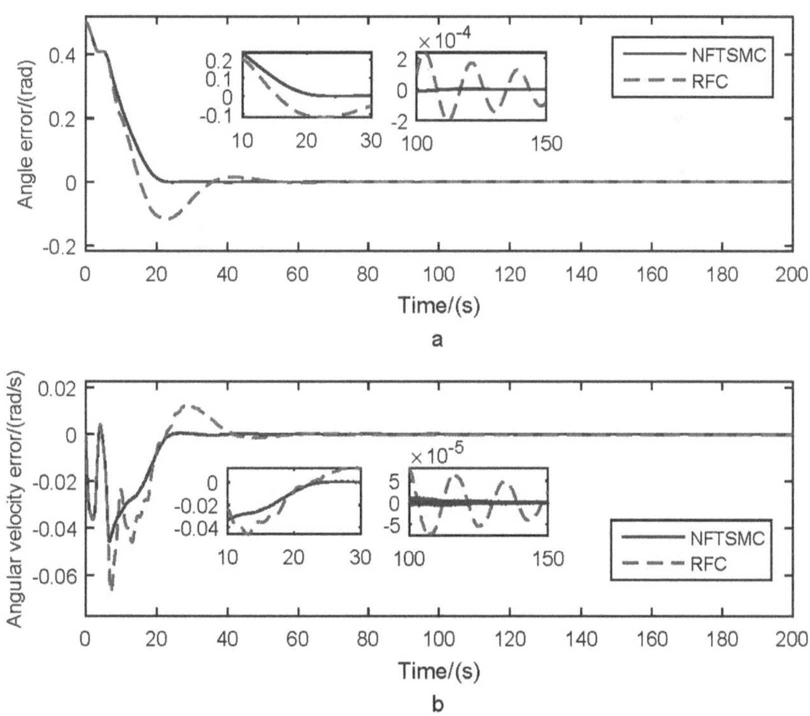

FIG. 8.2 Control effect of the proposed controller (8.17), part I: (A) the angle error; (B) the angular velocity error.

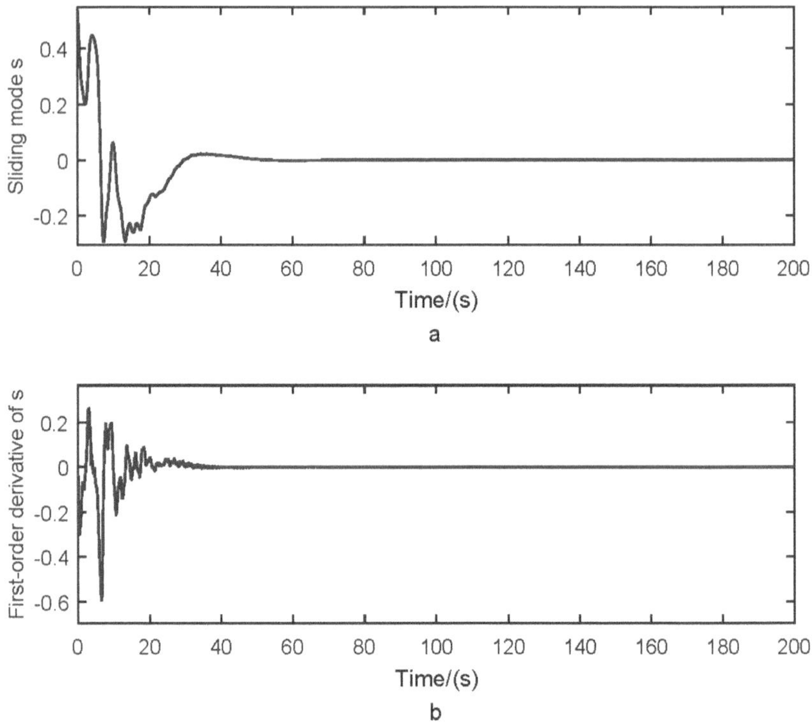

FIG. 8.3 Control effect of the proposed controller (8.31), part II: (A) the time response of the control torque; (B) the tip deflection of the beam.

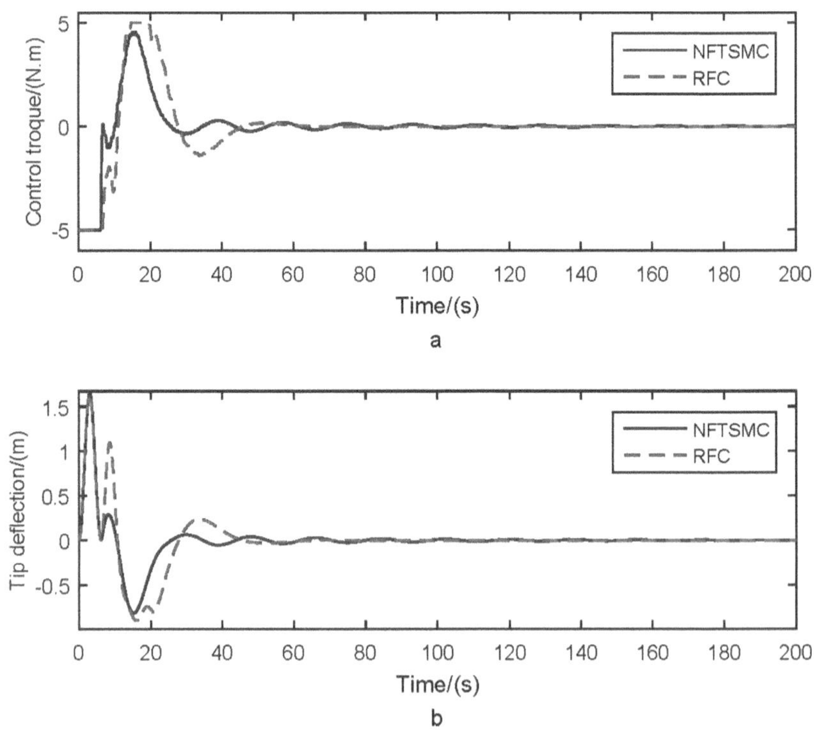

FIG. 8.4 Control effect of the proposed controller (8.17), part III: (A) the time response of sliding mode; (B) the time response of the first-order derivative of *s*.

are effective in attenuating the chattering. From Fig. 8.4B, it can be seen that the vibration of the appendage is obviously decreased and gradually attenuated at last, which means the inhibition of the chattering indeed decreases the vibration of the flexible appendage. It is worthy to be mentioned that when the chattering and the vibration are attenuated, the control accuracy of the attitude angle error and angular velocity error is slightly decreased, and the convergence speed becomes slower, which are resulted from the decrease of the control gain. To meet the requirement of the on-orbit missions, we should adjust the control parameters to achieve the best trade-off between the control accuracy, convergence time and the chattering, and vibration of the appendage.

A comparison simulation is then presented to test the advantage of the proposed controller. The first-order rigid–flexible coupling model is controlled by the robust output feedback controller (RFC) in Ref. [2]. The simulation result is presented as dashed lines in Figs. 8.2 and 8.4. It is easy to see that the nonsingular fast terminal sliding mode controller (NFTSMC) has obvious advantage in convergence time over the RFC. Although the attitude point accuracy and attitude stability of these two controllers are on the same level, the proposed controller still has better performance. As to the control torque, the proposed controller has less saturation interval. For the tip deflection, the maximum value of these two controllers is both 1.5 m, but NFTSMC can attenuate more quickly than the RFC.

In order to further show the rejection of the disturbance and robustness of the model uncertainty of the proposed finite time controller, we give simulations by increasing both the model uncertainty and external disturbance. The simulation results are shown in Table 8.1, where the first part in parentheses is external disturbance, and $d = 0.4\dot{\theta}\sin(0.1t)\mathrm{N}\cdot\mathrm{m}$. The other one is model uncertainty, $\Delta J_{mo}/J_{mo}$. We define T_c, P_a, P_{av} to indicate the performance of the controller in different situations, where T_c is the convergence time when both angle error and angular velocity error converge to 1% of the initial values, P_a and P_{av} are the average point accuracy and attitude stability between 120 and 160 s. From Table 8.1, as the external disturbance and model uncertainty are increasing, the indicators of the proposed observer based controller varies accordingly, but the variation is limited in a small range, and the control performance can still maintain in the same level. Thus, the rejection of the disturbance and robustness of the model uncertainty of the proposed controller are proved in the simulation.

8.3.2 Experimental simulation

In this section, considering the coupling effect of the rigid–flexible coupling satellite, based on the proceeding simulations, a ground experiment is established to further verify the effectiveness of the proposed control scheme. The ground experimental system for the rigid–flexible coupling satellite is set in Fig. 8.5, and it is mainly based on a single-axis air-bearing table providing a

TABLE 8.1 Control performance on different external disturbance and model uncertainty.

Parameters	(0, 0)	(d, 0)	(5d, 0)	(0, 50%)	(0, 100%)	(d, 50%)	(d, 100%)	(5d, 50%)	(5d, 100%)
T(s)	15	17	17	17	19	18	21	21	23
P_a($\times 10^{-4}$rad)	0.8	0.9	1.1	1.0	1.2	1.3	1.3	1.3	1.3
P_{av}($\times 10^{-5}$rad/s)	4.7	5.0	5.3	5.1	5.2	5.4	5.4	5.6	5.7

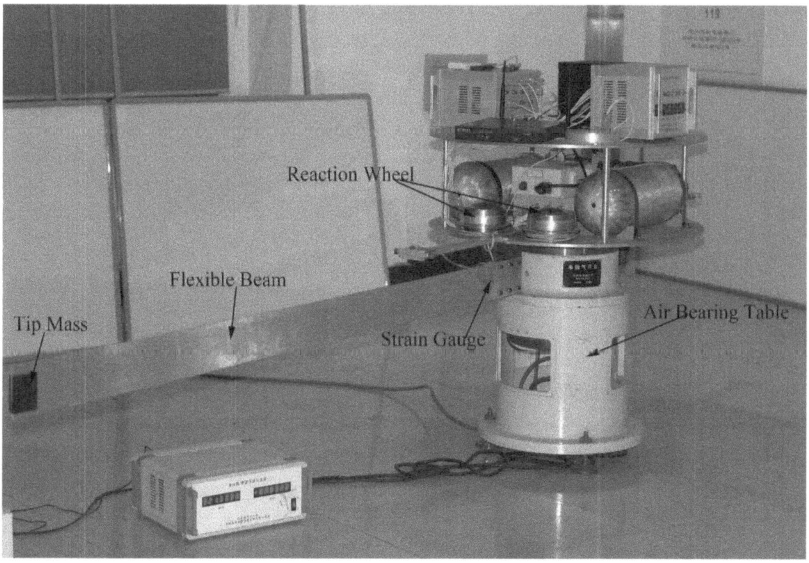

FIG. 8.5 Ground experimental system of a single-axis air-bearing table.

frictionless motion over the flat surface, which can simulate the microgravity environment on-orbit. The rigid hub acts as the central rigid body of the satellite, and the flexible aluminum beam with tip mass attached to the end acts as the flexible appendage like solar panels or flexible antennas. The actuator system consists of two reaction wheels, which are widely used in many large long-life satellites. The rotatory angle is measured by the combination of an inductosyn with an angle digital display meter, while the rotatory angular velocity is measured by optical fiber rate gyro. A laptop based on xPC is acted as the central process unit and it communicates with sensors and actuators by giving control orders and accepting experimental results. The physical parameters of the ground system are listed in Table 8.2. The vibration of the flexible appendage is measured by the strain gauge attached to the root of the beam as in Fig. 8.6. Comparing with the accelerometer attached to the tip of the aluminum beam, the strain gauge has several advantages: the strain measurement is robust to disturbances and has high signal noise ratio (SNR) than the accelerometer signal, which can be easily immersed by random noises when the vibration is small. The strain gauge is only sensitive to the beam deformation, while the accelerometer signal is mixed with rotational acceleration signal of the rigid hub. Therefore, the strain gauge can better reflect the vibration condition of the flexible beam.

It is worth mentioning that engineering application is fully considered during the design of this experimental system. Several key components like reaction wheel and rate gyro are the same as the ones used in satellites. Part of the equipment is chosen as off-the-shelve products to reduce the cost while not

TABLE 8.2 Parameters of the ground experimental system.

Parameters	Value
The moment of inertial of the air bearing table	$11\,kg\,m^2$
The radius of the air bearing table	450 mm
The size of the beam	1850 mm × 100 mm × 2.85 mm
The weight of the tip mass	0.3 kg
The maximum output of the reaction wheel	0.05 N·m
The control cycle of xPC	100 ms

failing to meet the accuracy demand of the experiment. Hence, the entire system can successfully simulate the kinemics and dynamics of an on-orbit rigid–flexible coupling satellite.

In order to test the effectiveness of the engineering application of the proposed continues finite-time control scheme, a ground experiment is set to perform an attitude regulation control for a rigid–flexible coupling satellite in large-angle maneuver. The initial attitude of the air-bearing table is $\theta(t_0)=0$ rad, and the desired attitude is $\theta_d = -0.5$ rad. The table starts rotating from a static state.

The angle error and angular velocity error are shown in Fig. 8.7. From Fig. 8.7A, it can be seen that the large-angle maneuver is achieved within 7 s, and the attitude point accuracy is 5×10^{-4} *rad*. As to the angular velocity in Fig. 8.7B, the attitude stability is 3.9×10^{-4} *rad/s*. The vibration result of the beam is shown in Fig. 8.8B, from which we can see that the maximum vibration of the first-order mode is $0.8\mu\varepsilon$, and then converges rapidly. Fig. 8.8A shows the experimental results of the control torque, which has output saturation about 5 s in the initial period.

The fast finite-time stability is verified in the experiment, and the result is consistent with the one in the numerical simulation. The vibration of the beam is relatively large, but will damp rapidly. In addition, the residual vibration is $0.1\,\mu\varepsilon$, which is small considering the measurement noise. It also should be noticed that although the attitude point accuracy of the experiment is not as good as the simulation results, nearly one order lower, the point accuracy is very high for a real satellite considering all errors of sensors as well as actuators. Sensors and actuators like gyros, angle digital display meter and reaction wheels all have errors that will deteriorate the control accuracy of the proposed controller. Taking all the engineering factors into consideration, this experiment successfully shows the fast finite-time stability, high-attitude control accuracy, vibration reduction, and robustness to external disturbance and model uncertainties of the proposed control scheme.

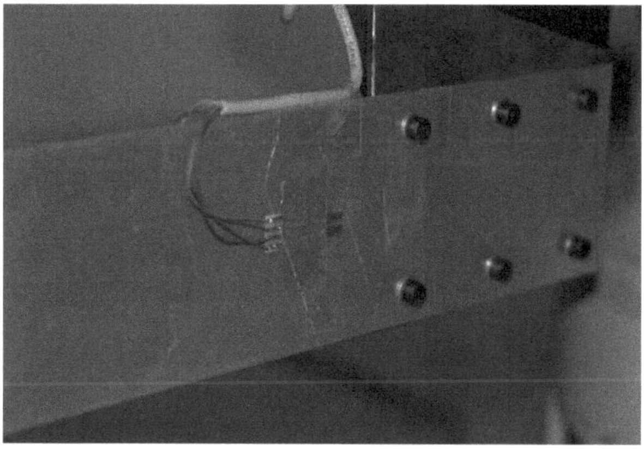

FIG. 8.6 Strain gauge.

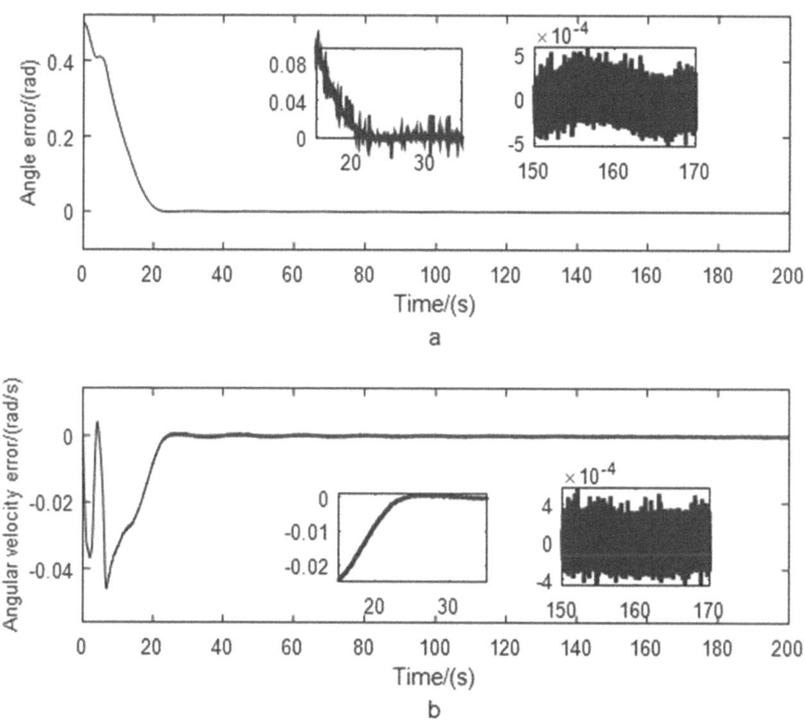

FIG. 8.7 Experiment control effect of the proposed controller (8.17), part I: (A) experimental result of the angle error; (B) experimental result of the angular velocity error.

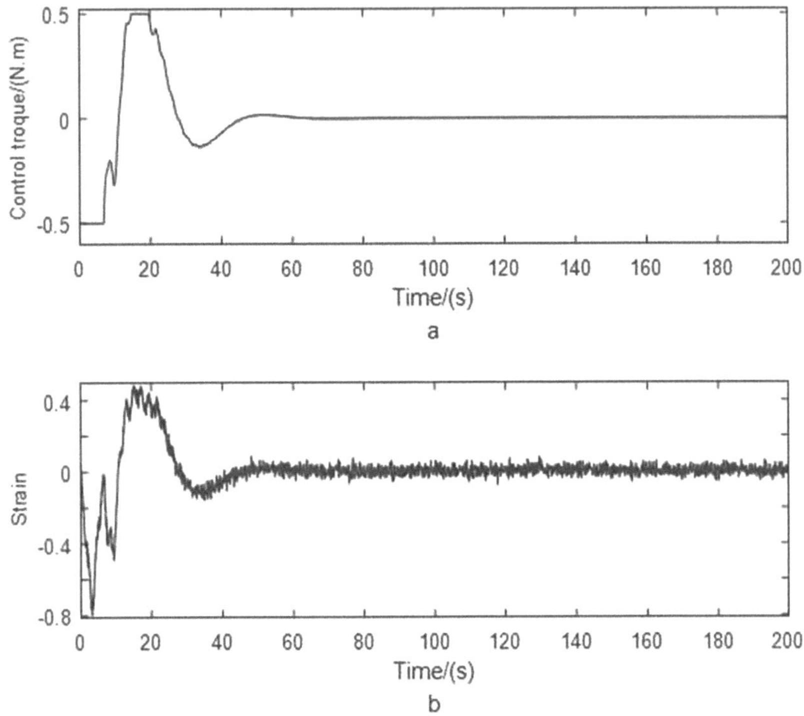

FIG. 8.8 Experiment control effect of the proposed controller (8.31), part II: (A) experimental result of control torque; (B) experimental response of strain at root of the beam.

Remark 8.5 From the above experiment results, it can be seen that the proposed continuous finite time controller is highly applicable considering the small convergence time and high accuracy control performance. This is not only because the observer we designed can estimate the system perturbation without using too much computing resource but also for the chattering attenuation effect of the proposed controller.

8.4 Conclusion

In light of the technical requirements for rigid–flexible coupling satellites in large-angle rapid maneuver mode, the continuous fast finite-time control law with high precision is studied. The proposed rigid–flexible coupling satellite dynamics and control have several advantages: (1) the first-order approximation rigid–flexible coupling dynamic model is more accurate in rapid maneuver, since it can fully reflect the coupling effect. (2) By the application of the ESO, the total perturbation including flexible coupling terms, external disturbances, and model uncertainties can be compensated without knowing the bound in prior. (3) Then based on the second-order sliding mode character of the fast power reaching law, the proposed continuous NFTSM controller can guarantee the fast finite-time stability of the system. (4) The compensation effect of the ESO, which makes the control gain decrease, and the continuity contributed by the fast power reaching law together attenuate the chattering that exists in the sliding mode control. (5) The fast finite-time stability of the attitude with high precision, the attenuation of the chattering, and damping of the vibration of the flexible appendage are thoroughly verified by both the numerical simulation and the ground experiment.

The further research will focus on two aspects: (1) consider the use of the proposed controller in the case when the angular velocity is not available and (2) fault-tolerant control of the rigid–flexible coupling satellite.

References

[1] H. Bang, C.K. Ha, J.H. Kim, Flexible spacecraft attitude maneuver by application of sliding mode control, Acta Astronaut. 57 (11) (2005) 841–850.

[2] Y. Xiao, D. Ye, Z.X. Yang, et al., Research on attitude adjustment control for large angle maneuver of rigid flexible coupling spacecraft, in: Proceedings of the International Conference on Intelligent Robotics and Applications, Springer, Portsmouth, UK, August 2015, pp. 24–33.

[3] Q.L. Hu, Robust adaptive sliding mode attitude maneuvering and vibration damping of three-axis-stabilized flexible spacecraft with actuator saturation limits, Nonlinear Dyn. 55 (4) (2009) 301–321.

[4] C.X. Zhong, G. Yu, Y. Zhen, et al., Vibration suppression and sliding mode attitude control of flexible spacecraft with unknown disturbance and uncertainty, in: Proceedings of the 2014 33rd Chinese control Conference (CCC), IEEE, Nanjing, China, July 2014, pp. 3771–3776.

[5] C.X. Zhong, A.F. Lai, G. Yu, et al., On attitude maneuver control of flexible spacecraft without angular velocity sensors, in: Proceedings of the 2013 IEEE/SICE International Symposium on System Integration (SII), IEEE, Kobe, Japan, December 2013, pp. 318–323.

[6] Y. Feng, X.H. Yu, Z.H. Man, Non-singular terminal sliding mode control of rigid manipulators, Automatica 38 (12) (2002) 2159–2167.

[7] L. Yang, J.Y. Yang, Nonsingular fast terminal sliding-mode control for nonlinear dynamical systems, Int. J. Robust Nonlinear Control 21 (16) (2011) 1865–1879.

[8] S.N. Wu, G.Q. Wu, S.J. Tan, et al., Quaternion-based adaptive terminal sliding mode control for spacecraft attitude tracking, in: Proceedings of the 2013 10th IEEE International Conference on Control and Automation (ICCA), IEEE, Hangzhou, China, June 2013, pp. 913–917.

[9] K.F. Lu, Y.Q. Xia, Adaptive attitude tracking control for rigid spacecraft with finite-time convergence, Automatica 49 (12) (2013) 3591–3599.

[10] C. Pukdeboon, P. Siricharuanun, Nonsingular terminal sliding mode based finite-time control for spacecraft attitude tracking, Int. J. Control. Autom. Syst. 12 (3) (2014) 530–540.

[11] P.M. Tiwari, S. Janardhanan, M.U. Nabi, Rigid spacecraft attitude control using adaptive integral second order sliding mode, Aerosp. Sci. Technol. 42 (2015) 50–57.

[12] B. Xiao, S. Yin, O. Kaynak, Tracking control of robotic manipulators with uncertain kinematics and dynamics, IEEE Trans. Ind. Electron. 63 (10) (2016) 6439–6449.

[13] C.X. Zhong, Y. Guo, Z. Yu, et al., Finite-time attitude control for flexible spacecraft with unknown bounded disturbance, Trans. Inst. Meas. Control. 38 (2) (2016) 240–249.

[14] P. Ben-Tzvi, S.J. Bai, Q.K. Zhou, et al., Fuzzy sliding mode control of rigid-flexible multibody systems with bounded inputs, J. Dyn. Syst. Meas. Control. 133 (6) (2011), 061012.

[15] Z. Zhu, Y.Q. Xia, M.Y. Fu, Adaptive sliding mode control for attitude stabilization with actuator saturation, IEEE Trans. Ind. Electron. 58 (10) (2011) 4898–4907.
[16] Y.Q. Xia, Z. Zhu, M.Y. Fu, et al., Attitude tracking of rigid spacecraft with bounded disturbances, IEEE Trans. Ind. Electron. 58 (2) (2011) 647–659.
[17] J.Q. Han, Active Disturbance Rejection Control Technique—The Technique for Estimating and Compensating the Uncertainties, National Defense Industry Press, Beijing, 2008, pp. 197–270.
[18] Y. Huang, A new synthesis method for uncertain systems the selfstable region approach, Int. J. Syst. Sci. 30 (1) (1999) 33–38.
[19] S.H. Yu, X.H. Yu, B. Shirinzadeh, et al., Continuous finite-time control for robotic manipulators with terminal sliding mode, Automatica 41 (11) (2005) 1957–1964.
[20] W.B. Gao, J.C. Hung, Variable structure control of nonlinear systems: a new approach, IEEE Trans. Ind. Electron. 40 (1) (1993) 45–55.
[21] Y.J. Shen, X.H. Xia, Semi-global finite-time observers for nonlinear systems, Automatica 44 (12) (2008) 3152–3156.
[22] Y.B. Shtessel, I.A. Shkolnikov, A. Levant, Smooth second-order sliding modes: missile guidance application, Automatica 43 (8) (2007) 1470–1476.
[23] S.S. Xu, C. Chen, Z. Wu, Study of nonsingular fast terminal sliding-mode fault-tolerant control, IEEE Trans. Ind. Electron. 62 (6) (2015) 3906–3913.
[24] A.M. Zou, K.D. Kumar, Z.G. Hou, et al., Finite-time attitude tracking control for spacecraft using terminal sliding mode and Chebyshev neural network, IEEE Trans. Syst. Man Cybern. B Cybern. 41 (4) (2011) 950–963.
[25] S.P. Bhat, D.S. Bernstein, Continuous finite-time stabilization of the translational and rotational double integrators, IEEE Trans. Autom. Control 43 (5) (1998) 678–682.
[26] Z.Y. Liu, J.Z. Hong, J.Y. Liu, Finite element formulation for dynamics of planar flexible multibeam system, Multibody Syst. Dyn. 22 (1) (2009) 1–26.
[27] D. Ye, Y. Xiao, Robust output feedback attitude tracking control for rigid-flexible coupling spacecraft, J. Frankl. Inst. 355 (18) (2018) 9209–9223.

Chapter 9

Neural network-based finite-time fault-tolerant attitude control

9.1 Introduction

In the endeavor to improve the attitude control of the rigid–flexible coupling satellite, some literatures proposed different methods [1–9]. However, most of literatures do not consider satellite faults when proposed to handle the attitude control of the rigid–flexible coupling satellite. Since the environment of the out space that satellites work in is very harsh and the structure of the rigid–flexible coupling satellite is much more complex compared to the rigid one, it is nontrivial that there will be some emergences and component aging, which will result in the fault of satellites [10–12]. Among these various satellite faults, the most typical one is the actuator fault, like thruster fault and reaction wheel fault [13]. Actuator is the most important component of the satellite control system, and if there is not a proper fault response mechanism when actuator fault happens, the actuator fault will decrease the attitude control accuracy or even lead to attitude instability, then terminate on-orbit missions [14–16]. Hence, the study of the fault-tolerant control (FTC) for rigid–flexible coupling satellites in actuator faults is both challenging and meaningful.

The FTC can be divided into two categories [17]. The first category is the passive fault-tolerant control (PFTC), and it is applied when the detailed knowledge of the fault is not available. Under the PFTC, the actuator fault is treated as a kind of disturbance, and what the PFTC does is to design a controller to compensate the fault. Since the explicit knowledge of the actuator fault is unknown, the compensate part of the PFTC is always much bigger than the real fault that need to be counteracted, so the control gain is relatively large which makes the PFTC conservative and part of the limited control energy is squandered [18]. Compared to the PFTC, the active fault-tolerant control (AFTC) can estimate the actuator fault online with a fault observer, which makes the AFTC more accurate and efficient in practical applications [19]. Given the control problem for satellites under the condition of actuator faults, some researchers propose

Fast Satellite Attitude Maneuver and Control. https://doi.org/10.1016/B978-0-323-95455-6.00012-3

related strategies. An adaptive backstepping FTC method is proposed in [20], when external disturbances and model uncertainties are considered. However, there are two disadvantages for the adaptive control scheme, one is that the nonlinear term of the model uncertainties should satisfy the need of linear parameterization or the regression should be known a priori; the other one is that the adaptive control lacks robustness for external disturbances and unmodeled dynamics. Xiao et al. [21] proposes an attitude fault-tolerant controller when loss of control effectiveness and additive fault occur in actuators simultaneously. Then it introduces a first-order differentiation filter to account for the unmeasured magnitude of the angular velocity and thus resolves the fault-tolerant problem when the angular velocity is partly unmeasurable. It should be mentioned that disturbances and model uncertainties in the existing fault-tolerant literatures are always lower frequency excitations close to constant signals, so the bound of these signals should be known during the FTC process [20–22]. With practical consideration, however, the upper bound of disturbance signal is difficult to obtain during the real satellite control process. Thus, in order to counteract disturbances and model uncertainties, larger control gain will be chosen, which makes the control scheme conservative. What is more, large control gain not only leads to waste of the limited energy but also aggravates the chattering of the control torque.

Motivated from the foregoing problems that the upper bond of the total disturbance exist in the model is difficult to obtain and the control gain is too large, some literatures present improved methods [23–26]. Fuzzy mechanism is introduced in [24] to approximate the discontinuous term in the controller. It is important to note, however, the fuzzy mechanism is a method overwhelmingly dependent on experience, and this means the approximation effect of the fuzzy mechanism is largely dependent upon the choice of the fuzzy set and also sensitive to the adjustment of parameters. Some researchers propose neural network (NN), which is more intelligent and has stronger self-learning ability, to estimate the total disturbance including coupling terms, external disturbances, unmodeled dynamics, and model uncertainties. To control the attitude of the three-axis-stabilized flexible satellite, a sliding mode controller is designed in [25], where a fuzzy-NN based estimator is introduced to estimate the model nonlinear terms, thus the elastic vibration is suppressed. It is worthy to be mentioned that, when the traditional NN is utilized to estimate the model nonlinear terms, the computation is large and complex for its multi-layer structure. Moreover, plenty of iterations are needed to finish the training work of the self-learning ability of the NN, thus makes the estimation process inefficient and time consuming [27,28]. Compared to the multilayer traditional NN, the functional link NN also has a good performance in function approximation while has less computational cost and faster convergence speed [29]. So in this chapter, we will propose a functional link NN called Legendre polynomial-based neural network (LeNN) to approximate the unknown part in the control system [30]. To the best knowledge of the author, this is the first time that LeNN is applied in the rigid–flexible coupling satellite attitude control system.

Thus, for rigid–flexible spacecraft with actuator faults and unknown system total disturbance, this chapter designs a LeNN based finite time terminal sliding mode controller. The main contributions of this chapter can be summarized as follows: (1) by employing the NFTSMC, the attitude stability of the rigid–flexible coupling satellites in the condition of actuator faults can be achieved in finite time with high accuracy and fast convergence speed; (2) compared to the traditional NN algorithms and other estimate methods, the LeNN not only is simple for application with its single layer structure but also possesses the advantage of NN that it can estimate the system nonlinear term in arbitrary accuracy. (3) A switch mechanism is designed to switch between the LeNN-based terminal sliding mode controller and the robust controller, thus efficiently avoids the overestimate of the total disturbance at the initial learning stage of the LeNN. Hence, compared to the existing literatures, the control scheme proposed in this chapter can guarantee the attitude of the rigid–flexible coupling satellite converge to the equilibrium point in finite time, while is more computationally favorable and practical for application.

9.2 Problem formulation

Legendre polynomial-based neural network (LeNN).

The NN used in this chapter is a single layer NN, and it is also a functional link artificial NN which based on a set of Legendre orthogonal functions [25]. The LeNN is made up of two parts. The first part is the expansion part, where part or all of the inputs of the NN is enhanced by a set of Legendre orthogonal functions. Compared to a multilayer perception (MLP), the LeNN is with much less computational cost due to the elimination of the hidden layers, which is resulted from transforming the input space to a higher dimensional space. The second part is the learning part, where the weighs of the LeNN are updated online by proper algorithms. Thus, the output of the LeNN can approach the target function step by step [30].

The Legendre orthogonal functions are designed as a set of polynomials, and let $L_n(x)$ be the nth order Legendre polynomial, then the Legendre polynomials can be obtained by solving the differential equation

$$\frac{\mathrm{d}}{\mathrm{d}x}\left[\left(1-x^2\right)\frac{\mathrm{d}y}{\mathrm{d}x}\right]+n(n+1)y=0$$

When we use the LeNN, we always set the zeroth and the first-order Legendre polynomials as $L_0(x)=1$, $L_1(x)=x$, respectively. Then we use a recursive formula below to find the high order Legendre polynomials [30] (Fig. 9.1).

$$L_{n+1}(x)=\frac{1}{n+1}[(2n+1)xL_n(x)-nL_{n-1}(x)] \tag{9.1}$$

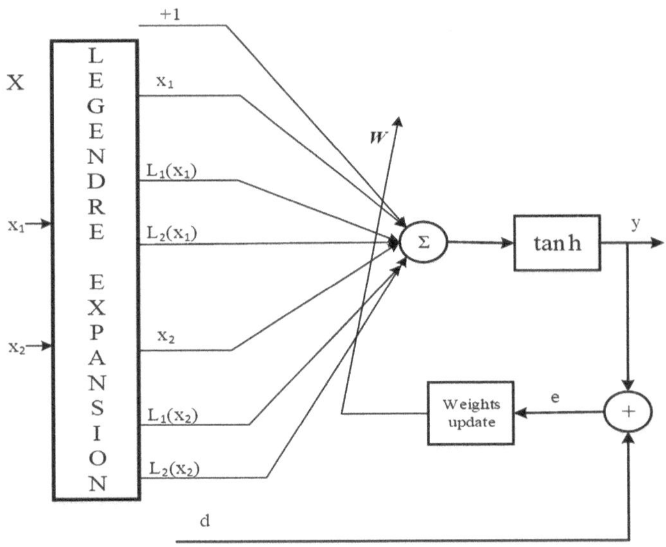

FIG. 9.1 Structure of the LeNN.

Based on Eq. (9.1), the high-order polynomials are as follows:

$$L_2(x) = \frac{1}{2}\left(3x^2 - 1\right)$$

$$L_3(x) = \frac{1}{2}\left(5x^3 - 3x\right)$$

$$L_4(x) = \frac{1}{8}\left(35x^4 - 30x^2 + 3\right)$$

$$\vdots$$

If the input is $X = [x_1, x_2, \cdots, x_m]^T$, then by the expansion effect of the nth Legendre polynomials, the m-dimensional X can be enhanced to

$$X^e = \left[1, L_1(x_1), L_2(x_1), \cdots, L_n(x_1), \cdots L_1(x_m), L_2(x_m), \cdots, L_n(x_m)\right]^T$$

where X^e is called the Legendre polynomial basis function, and n is the order of Legendre polynomials.

Let $f(x)$ be a continuous nonlinear function, then based on the approximation of the LeNN, $f(x)$ can be estimated as

$$f(x) = W^* X^e(x) + \xi \tag{9.2}$$

where W^* is the optimal weigh matrix of the LeNN, and ξ represents the up bound of the approximation error of the LeNN.

We assume that the fault of actuators of the satellite is effectively treated by designing proper fault detection, isolation, and estimation system. The dynamic Eq. (2.40) can be rewritten as

$$\begin{cases} \dot{x}_1 = x_2 \\ \dot{x}_2 = f + J_{\text{mo}}^{-1} T_{\text{c}} + d_1 \end{cases} \tag{9.3}$$

where $x_1 = \theta_e, f = -J_{\text{mo}}^{-1} F \dot{\theta}_e$ is the nominal part of the system dynamics, and $d_1 = g + \Delta M - J_{mo}^{-1} H$ is the total disturbance including external disturbance, model uncertainty, and flexible coupling terms.

By the processing effect of the fault detection, isolation, we divide the control torque of the flexible satellite into two parts, namely, the nominal part T_{N}, and the fault part T_{F}. Then the fault mode dynamic equation can be written as

$$\begin{cases} \dot{x}_1 = x_2 \\ \dot{x}_2 = f + P_{\text{N}}(x) T_{\text{N}} + P_{\text{F}}(x) T_{\text{F}} + d_1 \end{cases} \tag{9.4}$$

where we assume the flexible satellite has m actuators, of which k actuators function well, and the rest m-k have faults. ${P_{\text{N}}}^{\text{T}} \in \mathfrak{R}^k$ and ${P_{\text{F}}}^{\text{T}} \in \mathfrak{R}^{m-k}$ are all continuous functions. $T_{\text{N}} \in \mathfrak{R}^k$ and $T_{\text{F}} \in \mathfrak{R}^{m-k}$ are nominal control torque and fault control torque, respectively. Torque distribution is not considered in detail here. Since we study the FTC of a single-axis rigid–flexible coupling satellite, so $1 < k < m$. When the actuator faults happen, actuators that work normally can sufficiently match the need of attitude fault-tolerant controller design.

Then with the fault estimation, we further divide the fault control torque into two parts as

$$T_{\text{F}} = \widehat{T}_{\text{F}} + \Delta T_{\text{F}} \tag{9.5}$$

where $\widehat{T}_{\text{F}}$ is the estimation value of the fault control torque that is available through the fault estimation system, and ΔT_{F} is the estimation error of the fault control torque. Then the fault mode dynamic Eq. (9.4) can be rewritten as

$$\begin{cases} \dot{x}_1 = x_2 \\ \dot{x}_2 = f + P_{\text{N}}(x) T_{\text{N}} + P_{\text{F}}(x) \left(\widehat{T}_{\text{F}} + \Delta T_{\text{F}} \right) + d_1 \end{cases} \tag{9.6}$$

With the fault problem formulated above, our goal is to design a fault-tolerant controller, such that the finite time attitude convergence of the rigid–flexible coupling satellite can be guaranteed when there are actuator faults.

9.3 LeNN-based finite time fault-tolerant controller design

In this section, we will first design a finite-time controller based on the nonsingular fast terminal sliding mode theory. And then we incorporate the LeNN to approximate the total disturbance of the fault dynamics system including the fault control torque estimation error. With the fault-tolerant controller combining the NFTSM with the LeNN, the attitude stability of the flexible satellites in actuator fault mode can be achieved in finite time.

Based on the analysis of the fault mode dynamic Eq. (9.6), the estimation value of the fault control torque $\widehat{T}_{\mathrm{F}}$ can be attained by the fault detection, isolation, and estimation system, so the total disturbance of the fault system can be presented as

$$d_2 = d_1 + P_{\mathrm{F}}(x)\Delta T_{\mathrm{F}} \tag{9.7}$$

when the attitude stability of the satellite is reached, namely $\theta_e = 0$, and $\dot{\theta}_e = 0$, d_2 will converge to a continuous nonlinear function $\overline{d}_2$ which is based on the desired attitude. Using the LeNN, the estimation value of $\overline{d}_2$ can be approximated as

$$\widehat{\overline{d}}_2 = W\boldsymbol{X}^e(\theta_d) \tag{9.8}$$

where W is the estimation of W^*. Let the LeNN weigh approximation error matrix be $\widetilde{W} = W^* - W$, and the auxiliary variable be $\phi = d_2 - \overline{d}_2$. It should be mentioned that, since the external disturbance, the uncertain part of the rotational inertia, the elastic deformation are all bounded, and the estimation error of the fault control torque is also bounded for the normal operation of the fault detection, isolation and estimation system, so the total disturbance of the fault system is bounded, which satisfies $|\overline{d}_2| \le d_{\mathrm{max}}$, $|\phi| < \phi_{\mathrm{max}}$, where d_{max} and ϕ_{max} are positive constants.

Using the NFTSM defined by Eq. (7.6), the LeNN-based finite time fault-tolerant controller is given as

$$T_N = T_{eq} + T_{LeNN} \tag{9.9}$$

$$T_{eq} = -P_{\mathrm{N}}^{+}(x)\left[f + P_{\mathrm{F}}(x)\widehat{T}_{\mathrm{F}} + J_{\mathrm{mo}}(b\gamma_2)^{-1}\left(1 + a\gamma_1|\theta_{\mathrm{e}}|^{\gamma_1 - 1}\right)sig\left(\dot{\theta}_{\mathrm{e}}\right)^{2-\gamma_2}\right] \tag{9.10}$$

$$T_{LeNN} = -P_N^{+}(x)[l_1 s + l_2 s^{\mu} + m(t)(W\boldsymbol{X}^e + \psi_1) + (1 - m(t))\psi_2] \tag{9.11}$$

where $P_{\mathrm{N}}^{+}(x) = P_{\mathrm{N}}^{\mathrm{T}}(x)[P_{\mathrm{N}}(x)P_{\mathrm{N}}^{\mathrm{T}}(x)]^{-1}$, ψ_1 and ψ_2 are both robust controllers, l_1 and l_2 are positive constants, $\mu = p/q$, odd integers p and q are chosen such that $1 < q/p < 2$. $m(t)$ is the switching function, which is defined as

$$m(t) = \begin{cases} 0, & |W\boldsymbol{X}^e| > d_{\mathrm{max}} \\ 1, & |W\boldsymbol{X}^e| \le d_{\mathrm{max}} \end{cases} \tag{9.12}$$

With the switching operation of $m(t)$, the fault-tolerant controller can switch between the LeNN-based controller and robust controller ψ_2. The adaptive law for W is given as

$$\dot{W} = m(t)\left(\upsilon_1 s\xi^T - \upsilon_1\upsilon_2 W\right) \tag{9.13}$$

where υ_1 and υ_2 are positive constants.

To facilitate the stability proof of the closed-loop system, several assumptions are introduced below.

Assumption 9.1 The LeNN optimal weight matrix is bounded and satisfies

$$\mathrm{tr}\{W^{*\mathrm{T}}W^{*}\} \le W_{\max} \tag{9.14}$$

where $W_{\max}$ is a positive constant.

Assumption 9.2 The LeNN approximation error is bounded with $|\xi| \le \xi_{\max}$, where $\xi_{\max}$ is a positive constant.

The robust term ψ_1 in the LeNN controller is introduced to compensate the LeNN estimation error and is designed as

$$\psi_1 = \sigma_1 \tanh\left(\frac{3k_u\sigma_1 s}{\varepsilon}\right) \qquad k_u = 0.2785 \tag{9.15}$$

where σ_1 is a positive constant with $\sigma_1 \ge \xi_{\max}$, and ε is also a positive constant.

The robust controller ψ_2 comes in when the approximation value of the LeNN is out of the bound of the total disturbance, and it is designed as

$$\psi_2 = \sigma_2 \tanh\left(\frac{3k_u\sigma_2 s}{\varepsilon}\right) \qquad k_u = 0.2785 \tag{9.16}$$

where σ_2 is a positive constant with $\sigma_2 \ge d_{\max}$. According to literature [31], it is known that both ψ_1 and ψ_2 satisfy following conditions:

$$\begin{aligned} &(1)\ \ s\psi_i \ge 0; \\ &(2)\ \ \sigma_i|s| - s\psi_i \le \varepsilon/3; \qquad i = 1, 2. \end{aligned} \tag{9.17}$$

Theorem 9.1 *Consider the first-order approximation rigid–flexible coupling dynamic model (7.1), (7.2) undergoes actuator faults. Under assumptions 9.1 and 9.2, and with controller (9.9)–(9.11), and the adaptive law for* W*(9.13), if the initial condition of the system satisfies*

$$s^2(0) + \mathrm{tr}\left\{\widetilde{W}^{\mathrm{T}}(0)\widetilde{W}(0)\right\} \le V_{\max} \tag{9.18}$$

then s and $\widetilde{\mathrm{W}}$ *are uniformly ultimately bounded (UUB).*

Proof: Design the Lyapunov candidate as

$$V = \frac{1}{2}s^2 + \frac{1}{2\upsilon_1}\mathrm{tr}\left\{\widetilde{W}^{\mathrm{T}}\widetilde{W}\right\} \tag{9.19}$$

Calculating the time derivative of the Lyapunov function candidate (9.19) and substituting the controller (9.9)–(9.11) and the adaptive law (9.13) into it, we obtain

$$\begin{aligned}
\dot{V} &= s\left[\left(1+a\gamma_1|\theta_e|^{\gamma_1-1}\right)\dot{\theta}_e + b\gamma_2\left|\dot{\theta}_e\right|^{\gamma_2-1}\left(f+P_N(x)T_N+P_F(x)\left(\widehat{T}_F+\Delta T_F\right)+d_1\right)\right] \\
&\quad -\frac{1}{\upsilon_1}\mathrm{tr}\left\{\widetilde{W}^{\mathrm{T}}\dot{W}\right\} \\
&= sb\gamma_2\left|\dot{\theta}_e\right|^{\gamma_2-1}\left[\left(d_2-\overline{d}_2\right)+m(t)\left(\widetilde{W}\boldsymbol{X}^e+\xi-\psi_1\right)+(1-m(t))\left(\overline{d}_2-\psi_2\right)\right] \\
&\quad -sb\gamma_2\left|\dot{\theta}_e\right|^{\gamma_2-1}\left(l_1 s+l_2 s^{\mu}\right)-\frac{1}{\upsilon_1}\mathrm{tr}\left\{\widetilde{W}^{\mathrm{T}}\dot{W}\right\} \\
&\le \varphi\left[s\phi+\frac{\varepsilon}{3}-l_1 s^2-l_2 s^{\mu+1}\right]+\upsilon_2 m(t)\mathrm{tr}\left\{\widetilde{W}^{\mathrm{T}}W\right\} \\
&\le \varphi\left[s^2+\frac{\phi_{\max}^2}{4}+\frac{\varepsilon}{3}-l_1 s^2-l_2 s^{\mu+1}\right]+\frac{\upsilon_2}{2}W_{\max} \\
&< -\varphi(l_1-1)s^2+c_1
\end{aligned} \tag{9.20}$$

For the deduction above, the following inequalities are used

$$s\xi - s\psi_1 \le \frac{\varepsilon}{3} \tag{9.21}$$

$$s\overline{d}_2 - s\psi_2 \le \frac{\varepsilon}{3} \tag{9.22}$$

$$s\phi \le s^2 + \frac{\phi_{\max}^2}{4} \tag{9.23}$$

$$\begin{aligned}
\upsilon_2 m(t)\mathrm{tr}\left\{\widetilde{W}^{\mathrm{T}}W\right\} &\le -\frac{\upsilon_2 m(t)}{2}\mathrm{tr}\left\{\widetilde{W}^{\mathrm{T}}\widetilde{W}\right\}+\frac{\upsilon_2 m(t)}{2}W_{\max} \\
&\le \frac{\upsilon_2}{2}W_{\max}
\end{aligned} \tag{9.24}$$

where $\varphi = \inf\left\{b\gamma_2\left|\dot{\theta}_e\right|^{\gamma_2-1}\right\}$, $c_1 = \varphi(\phi_{\max}^2/4+\varepsilon/3)+\upsilon_2 W_{\max}/2$. It is proved in [32] that $\dot{\theta}_e = 0$ is not an attractor, so $\dot{V}$ is negative outside the set $\Omega_s = \left\{s\middle||s|\le\sqrt{c_1/\varphi(l_1-1)}\right\}$, then s and $\widetilde{W}$ are UUB according to the Lyapunov stability theory.

Remark 9.1 However, Theorem 9.1 can only prove the stability of the closed-loop system, while the finite time stability of the system is still not determined. Since $\boldsymbol{X}^e$ is bounded for the bounded θ_d, then it is easy to get that $\left|\widetilde{W}\boldsymbol{X}^e\right|$ is also bounded. Based on the Theorem 9.1, the following Theorem 9.2 is presented to prove the finite time stability of the closed-loop system.

Theorem 9.2 *Consider the first-order approximation rigid–flexible coupling dynamic model (7.1), (7.2) undergoes actuator faults. Under assumptions 9.1 and 9.2, and with controller (9.9)–(9.11), and the adaptive law for* W *(9.13), if the initial condition of the system satisfies*

$$s^2(0)+\mathrm{tr}\left\{\widetilde{W}^{\mathrm{T}}(0)\widetilde{W}(0)\right\}\le V_{\max} \tag{9.25}$$

and σ_1 is selected as $\sigma_1 \geq \xi_{\max} + \left|\widetilde{W}X^e\right|$, *then the sliding surface* s *will converge to a certain region*

$$|s| \leq \Delta \tag{9.26}$$

$$\Delta = \min\left\{\sqrt{\frac{c_2}{\varphi(l_1 - 1)}}, \left(\frac{c_2}{\varphi l_2}\right)^{1/\mu+1}\right\} \tag{9.27}$$

in finite time, where $c_2 = \varphi\varepsilon/3 + \varphi\phi^2/4 > 0$. *In addition, the error angle and the error angular velocity will also converge to*

$$|\theta_e| \leq \Delta_1 = \min\left\{\Delta, \left(\frac{\Delta}{a}\right)^{1/\gamma_1}\right\} \tag{9.28}$$

$$\left|\dot{\theta}_e\right| \leq \Delta_2 = \left(\frac{\Delta_1 + a\Delta_1^{\gamma_1}}{b}\right)^{1/\gamma_2} \tag{9.29}$$

Proof: Choose the Lyapunov candidate as

$$V = \frac{1}{2}s^2 \tag{9.30}$$

Differentiating Eq. (9.30) with time, then inserting controller (9.9)–(9.11) and the adaptive law for W (9.13) result in

$$\begin{aligned}
\dot{V} &= s\left[\left(1 + a\gamma_1|\theta_e|^{\gamma_1-1}\right)\dot{\theta}_e + b\gamma_2\left|\dot{\theta}_e\right|^{\gamma_2-1}\left(f + P_N(x)T_N + P_F(x)\left(\widehat{T}_F + \Delta T_F\right) + d_1\right)\right] \\
&= sb\gamma_2\left|\dot{\theta}_e\right|^{\gamma_2-1}\left[\phi + m(t)\left(\widetilde{W}X^e + \xi - \psi_1\right) + (1 - m(t))\left(\overline{d}_2 - \psi_2\right)\right] \\
&\quad -sb\gamma_2\left|\dot{\theta}_e\right|^{\gamma_2-1}\left(l_1 s + l_2 s^\mu\right) \\
&\leq \varphi s^2 + \varphi\frac{\phi_{\max}^2}{4} + \varphi\frac{\varepsilon}{3} - \varphi l_1 s^2 - \varphi l_2 s^{\mu+1} \\
&= -(l_1 - 1)\varphi s^2 - \varphi l_2 s^{1+\mu} + c_2 \\
&= -2(l_1 - 1)\varphi\frac{s^2}{2} - 2^{(1+\mu)/2}\varphi l_2\left(\frac{s^2}{2}\right)^{(1+\mu)/2} \\
&= -b_1 V - b_2 V^{(1+\mu)/2} + c_2
\end{aligned} \tag{9.31}$$

where $\varphi = \inf\left\{b\gamma_2\left|\dot{\theta}_e\right|^{\gamma_2-1}\right\}$ is the same as one defined in the proof of Theorem 9.1. $b_1 = 2(l_1 - 1)\varphi$, $b_2 = 2^{(1+\mu)/2}\varphi l_2$. Rewrite Eq. (9.31) into the following two forms

$$\dot{V} \leq -\left(b_1 - \frac{c_2}{V}\right)V - b_2 V^{\frac{1+\mu}{2}} \tag{9.32}$$

$$\dot{V} \leq -b_1 V - \left(b_2 - \frac{c_2}{V^{(1+\mu)/2}} \right) V^{(1+\mu)/2} \tag{9.33}$$

When $b_1 - c_2/V > 0$, according to Lemma 7.2 Eq. (9.32) satisfies the form of fast finite time stability, thus the sliding manifold s will converge to a small neighborhood of the sliding face $s = 0$

$$|s| \leq \sqrt{\frac{c_2}{\varphi(l_1 - 1)}} \tag{9.34}$$

in finite time. Similarly, when $b_2 - c_2/V^{(1+\mu)/2} > 0$, Eq. (9.33) accords with the form of fast finite time stability based on Lemma 7.2, so s will converge to a small neighborhood of the sliding face $s = 0$

$$|s| \leq \left(\frac{c_2}{\varphi l_2} \right)^{1/(\mu+1)} \tag{9.35}$$

in finite time. Then based on the definition of the NFTSM in Eq. (8.14), there exists constant ϑ which can be chosen as $|\vartheta| \leq \Delta$, such that we have

$$\theta_e + a sig(\theta_e)^{\gamma_1} + b sig(\dot{\theta}_e)^{\gamma_2} = \vartheta \tag{9.36}$$

The foregoing equation can be evaluated into two forms as

$$b|\dot{\theta}_e|^{\gamma_2} = \left(1 - \frac{|\vartheta|}{|\theta_e|} \right) |\theta_e| + a|\theta_e|^{\gamma_1} \tag{9.37}$$

$$b|\dot{\theta}_e|^{\gamma_2} = |\theta_e| + \left(a - \frac{|\vartheta|}{|\theta_e|^{\gamma_1}} \right) |\theta_e|^{\gamma_1} \tag{9.38}$$

When $1 - |\vartheta|/|\theta_e| > 0$, Eq. (9.37) is in the form of fast finite time stability by Lemma 7.2, such that θ_e can converge to the region

$$|\theta_e| \leq |\vartheta| \leq \Delta \tag{9.39}$$

in finite time. When $a - |\vartheta|/|\theta_e|^{\gamma_1} > 0$, the other Eq. (9.38) accords with the form of the fast finite time stability based on Lemma 7.2, and thus θ_e will converge to the region

$$|\theta_e| \leq \left(\frac{|\vartheta|}{a} \right)^{1/\gamma_1} \leq \left(\frac{\Delta}{a} \right)^{1/\gamma_1} \tag{9.40}$$

in finite time. Hence, θ_e can converge to the region $|\theta_e| \leq \Delta_1$ in finite time. Then based on Eq. (9.36), $\dot{\theta}_e$ will converge to the region

$$|\dot{\theta}_e| \leq \left(\frac{\Delta_1 + a\Delta_1^{\gamma_1}}{b} \right)^{1/\gamma_2} = \Delta_2 \tag{9.41}$$

in finite time. Thereby, the proof is completed.

Remark 9.2 Since the sliding mode we use is NFTSM, so the NFTSM-based controller present in this chapter has fast convergence speed and no singularity, if we choose parameters to satisfy certain condition as $a, b > 0, 2 > \gamma_2 > 1$, and $\gamma_1 > \gamma_2$. Proper control parameters can be found by trial and error, and we can also tune the control parameters by some adaptive methods [33].

Remark 9.3 During the proof of finite time stability, the PFS theory is introduced. It means that under the control effect of the proposed controller, the sliding manifold s, the angle error θ_e, and the angular velocity error $\dot{\theta}_e$ can both converge to different vicinities of zero in finite time, respectively. All these small regions depend on control parameters like a, b, γ_1, γ_2, l_1, and l_2 which means control accuracy can be made as small as possible by tuning these parameters. And this in a way increases the application availability of the control scheme.

Remark 9.4 Generally, the controller presented in this chapter is an output feedback controller, which uses attitude and angular velocity only. Compared to lots of existing literatures treating flexible measurements as states in the controller design, the controller (9.9)–(9.11) is easier for implementation, since additional instruments to get flexible measurements are not needed. It also should be mentioned that the single- layer LeNN we proposed not only has elegant structure but also possesses high approximation ability as the traditional multilayer NN, and this means we can get approximation of the unknown perturbation with high accuracy while does not involve demand expensive online computations.

Remark 9.5 It should be noted that in the LeNN-based finite-time fault-tolerant controller (9.9)–(9.11), when the LeNN-based estimator fall to work, namely $m(t) = 0$, we have the robust finite-time fault-tolerant controller as

$$T_r = -P_N^{+}(x)\Big[f + P_F(x)\widehat{T}_F + J_{mo}(b\gamma_2)^{-1}\left(1 + a\gamma_1|\theta_e|^{\gamma_1-1}\right)sig\left(\dot{\theta}_e\right)^{2-\gamma_2} + l_1 s + l_2 s^{\mu} + \psi_2\Big] \tag{9.42}$$

From the previous proof, it is easy to know that under the robust finite-time controller, the trajectory of the system (7.1), (7.2) can converge to a vicinity of the equilibrium. Nevertheless, the convergence speed and accuracy will be degraded due to the lack of the compensation effect from the LeNN, and this will be verified by the following simulation.

9.4 Simulation results

Simulations will be set in this section to verify the effectiveness of the LeNN-based finite time fault-tolerant controller. We will first present the results of fault detection, isolation, and estimation system, then make comparison between the LeNN-based fault-tolerant controller (9.9)–(9.11) and the robust fault-tolerant controller (9.42) in the presence of the actuator faults. The fault detection, isolation, and estimation system used in the simulation is based on [34]. Since this chapter investigates the attitude FTC of a single axis rigid–flexible coupling satellite, there are two actuators in this simulation, and the output control torque are u_1 and u_2, respectively. During the attitude control, assume that the second actuator is out of order. The fault observer and the residual signal r can be designed based on [34] as

$$\dot{\xi} = f + P_{\mathrm{N}}u_1 + P_{\mathrm{F}}u_2 + \tau(x_2 - \xi) \tag{9.43}$$

$$r = x_2 - \xi \tag{9.44}$$

where $k_2 > 0$. According to the result in [34], the fault detection, isolation, and estimation system (9.43) and (9.44) can successfully detect, isolate the actuator fault, and then give the estimation of the faulty actuator control torque as $u_2 + \tau r$. When the magnitude of the residual signal r is in excess of a preset threshold, the fault-tolerant mode is triggered. In this simulation, we assume that the second actuator u_2 goes out of action when $t = 4s$. During the attitude adjustment, the initial attitude information of the rigid–flexible coupling satellite are $\theta(t_0) = 0$ rad, $\dot{\theta}(t_0) = 0$ rad/s . The desired attitude angle is $\theta_{\mathrm{d}} = -0.5$ rad. Without the loss of generality, external disturbance is assigned as $T_{\mathrm{d}} = 0.4\dot{\theta}\sin(0.1t)\mathrm{N}\cdot\mathrm{m}$, and uncertainty of the moment of inertia is given as $\Delta J_{mo}/J_{mo} = 20\%$. Parameters of the controller and the fault observer are chosen as $a = 0.2$, $b = 28$, $\gamma_1 = 2.3$, $\gamma_2 = 7/5$, $l_1 = 30$, $l_2 = 15$, $\mu = 7/5$, $P_{\mathrm{N}} = P_{\mathrm{F}} = 0.5$, and $\tau = 10$. Simulation results are listed from Figs. 9.2 to 9.9 and Table 9.1.

From the simulation results in Figs. 9.4 and 9.5, it can be seen that the fault can be detected at about $t = 4.3$s. Then Figs. 9.2 and 9.3 show that the LeNN can approximate the total disturbance of the fault dynamic system including the estimation error of the fault control torque in a short period of time, with the approximation error converging quickly.

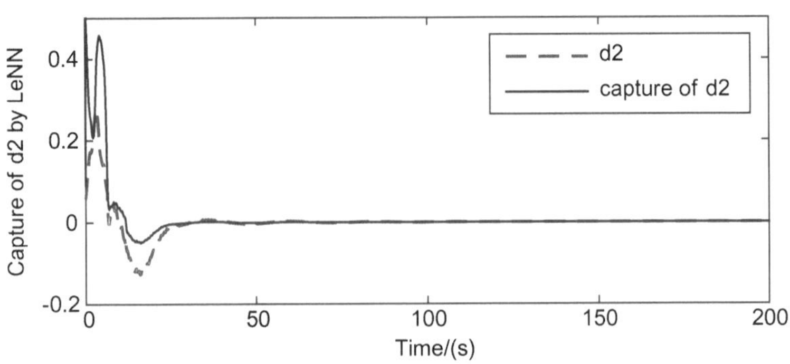

FIG. 9.2 Estimation of the total disturbance.

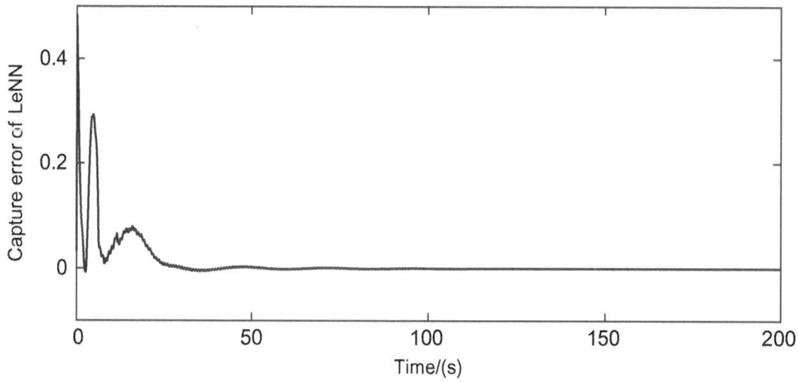

FIG. 9.3 Estimation error of the total disturbance.

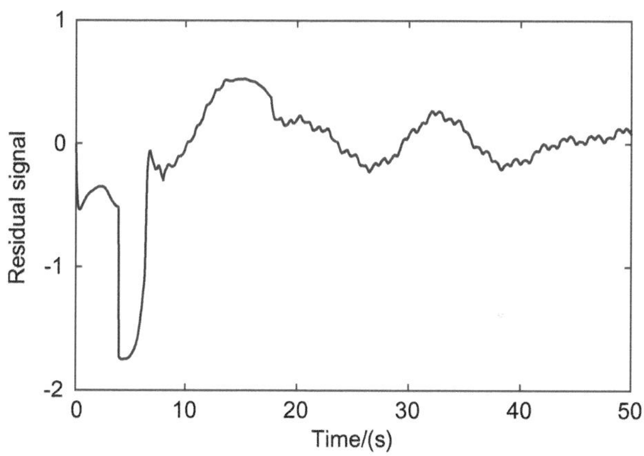

FIG. 9.4 Residual signal of actuator faults.

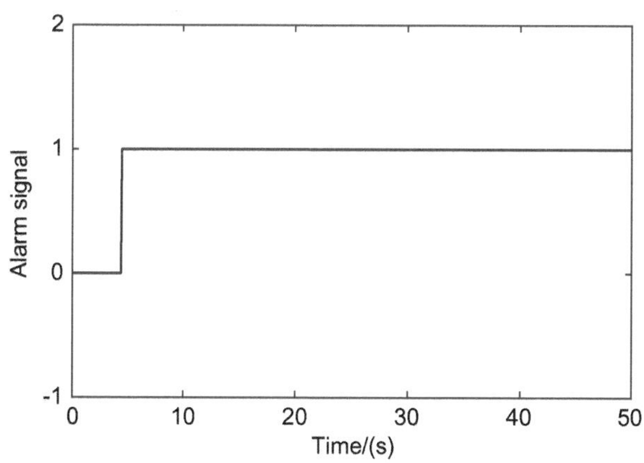

FIG. 9.5 Alarm signal.

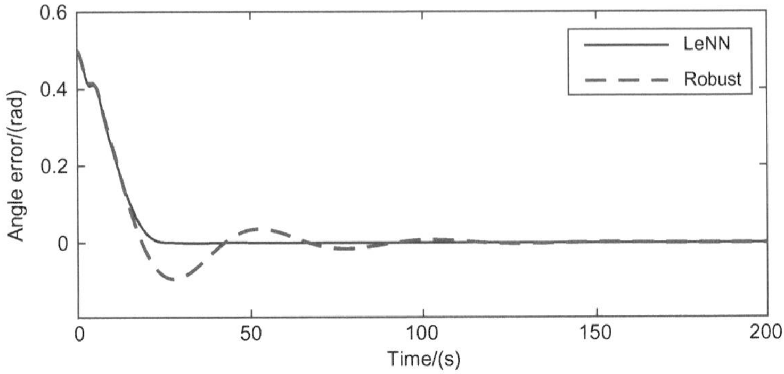

FIG. 9.6 Angle error.

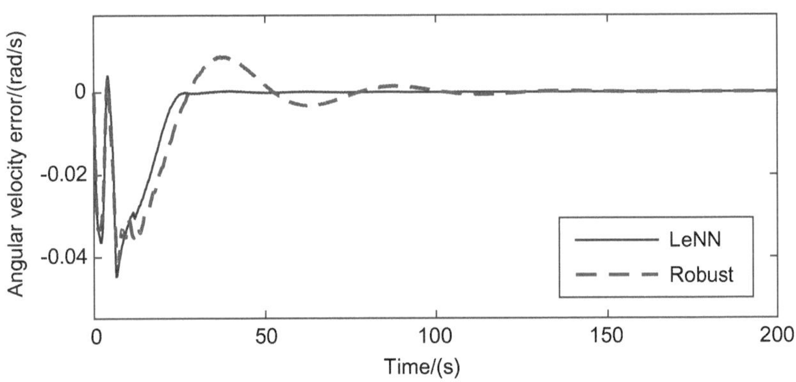

FIG. 9.7 Angular velocity error.

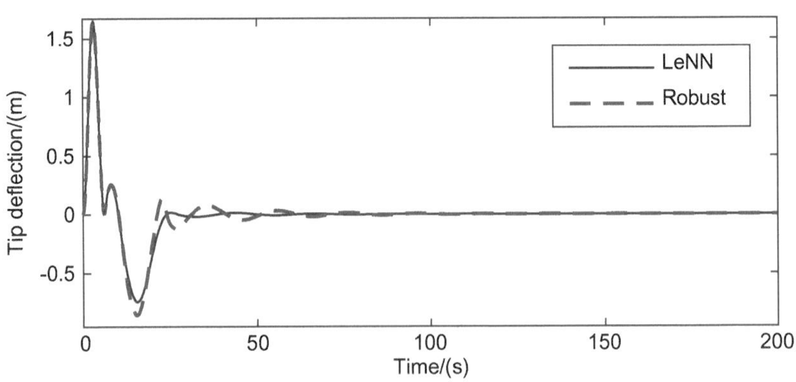

FIG. 9.8 Tip deflection.

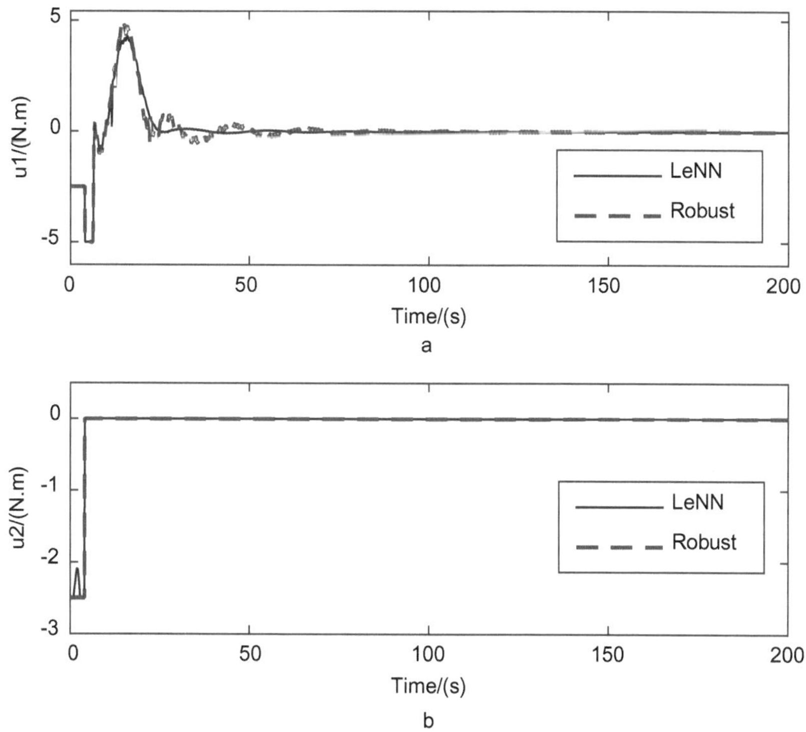

FIG. 9.9 Control torque (A) Control torque u_1; (B) control torque u_2.

TABLE 9.1 The comparisons of convergence speed and control accuracy between LeNN-based FTC and robust FTC.

	Convergence time of angle error	Attitude point accuracy	Convergence time of angular velocity error	Attitude stability
LeNN-based FTC (31)–(33)	20s	1.5×10^{-4}rad	21 s	5×10^{-5}rad/s
Robust FTC (64)	103 s	6×10^{-4}rad	107 s	1×10^{-4}rad/s

After the detection and estimation of the actuator faults, the time history of the angle error and the angular velocity error in actuator fault mode are shown in Figs. 9.6 and 9.7, and the comparisons of the convergence time and control accuracy are listed in Table 9.1. It is obvious that, compared to the robust fault-tolerant controller, with the control effort of the proposed LeNN-based fault-tolerant controller, the stability of the closed loop can be reached in finite time with faster convergence speed and higher accuracy that can meet the need of the control accuracy.

The time histories of control torques are showed in Fig. 9.9, which imply that actuator u_2 undergoes faults and loses control effect at $t = 4\text{s}$ and u_1 becomes the dominant controller since then. It can be noticed that while uncertain rotational inertia and state-related external disturbances are taken into account, for the compensation effect of the LeNN, the chattering of the controller is attenuated. Since the LeNN is an estimation method independent of states, there exists weak vibration of the flexible appendage at the early control stage, but it damps quickly and this can be saw in Fig. 9.8.

Remark 9.6 It is worthy to be mentioned that, in the condition when the LeNN falls to work, the robust fault tolerant controller can be a temporary replacement. Although the control performance is not as good as the LeNN-based controller, it can maintain the control system in a relative normal condition, and thus give the LeNN time to recover.

9.5 Conclusion

This chapter studies the FTC of the rigid–flexible satellite when part of its actuators fall to work normally. First, a rigid–flexible coupling dynamic model of the satellite is set based on the first-order approximate method, which is more accurate in large-angle maneuver. Then a finite-time fault tolerant attitude controller is designed based on the NFTSM in the condition of actuator faults. In this FTC scheme, for the first time, LeNN is incorporated to approximate the total disturbance of the fault mode system including the estimation error of the fault control torque, high-order coupling terms, external disturbances, and uncertain rotational inertia. Furthermore, a switch mechanism that switches between the LeNN based finite time fault tolerant controller and a robust controller is designed to avoid the overestimate of the total disturbance at the early control stage, this means when the estimation value is greater than the up bound of the total disturbance, the control system will be taken over by a robust controller. The finite-time stability of the closed-loop system is rigorously proved by the Lyapunov finite stability theory. Finally, the effectiveness of the proposed control scheme is verified by the numerical simulation, which shows that the proposed finite time FTC scheme can provide the closed loop in actuator fault mode the finite-time stability with fast convergence speed, high control accuracy, disturbance rejection, robustness to model uncertainties, and also has good behavior in chattering attenuation and vibration dapping. It should be mentioned that since the LeNN proposed in this chapter is a single layer and functional based NN, compared to the traditional NN, it is more computationally favorable and practical for application.

References

[1] H.B. Du, S.H. Li, Attitude synchronization for flexible spacecraft with communication delays, IEEE Trans. Autom. Control 61 (11) (2016) 3625–3630.

[2] S.N. Singh, Robust nonlinear attitude control of flexible spacecraft, IEEE Trans. Aerosp. Electron. Syst. AES-23 (3) (1987) 380–387.

[3] S. Di Gennaro, Output attitude tracking for flexible spacecraft, Automatica 38 (10) (2002) 1719–1726.

[4] H.B. Du, S.H. Li, Attitude synchronization control for a group of flexible spacecraft, Automatica 50 (2) (2014) 646–651.

[5] M. Celia, D. Gadelha, Using of H-infinity control method in attitude control system of rigid-flexible satellite, Math. Probl. Eng. 2009 (2009) 266–287.

[6] C.B. Zhou, D. Zhou, B. Huang, Robust adaptive dynamic surface attitude control of flexible spacecraft, in: 54th Annual Conference of the Society of Instrument and Control Engineers of Japan (SICE), Hangzhou, July, 2015, pp. 510–517.

[7] P. Guan, X.J. Liu, J.Z. Liu, Adaptive Fuzzy Sliding Mode Control for Flexible Satellite, 18, Pergamon Press, 2005, pp. 451–459.

[8] S.H. Wen, S.N. Wu, Robust H_∞ output feedback control for attitude stabilization of a flexible spacecraft, Nonlinear Dyn. 84 (2016) 405–412.

[9] T. Ohtani, Y. Hamada, T. Nagashio, et al., Robust attitude control using mu-synthesis for the large flexible satellite ETS-VIII, J. Space Technol. Sci. 25 (2009) 27–40.

[10] S.N. Wu, Z.G. Wu, G. Radice, et al., Adaptive control for spacecraft relative translation with parametric uncertainty, Aerosp. Sci. Technol. 31 (1) (2013) 53–58.

[11] S. Yin, B. Xiao, S.X. Ding, et al., A review on recent development of spacecraft attitude fault tolerant control system, IEEE Trans. Ind. Electron. 63 (5) (2016) 3311–3320.

[12] S.N. Wu, R. Wang, G. Radice, et al., Robust attitude maneuver control of spacecraft with reaction wheel low-speed friction compensation, Aerosp. Sci. Technol. 43 (2015) 213–218.

[13] J.D. BoSkoviC, S.M. Li, R.K. Mehra, Intelligent control of spacecraft in the presence of actuator failures, in: Proceedings of the 38th IEEE Conference on Decision and Control, Phoenix, December, 1999, pp. 4472–4477.

[14] Y.J. Ma, B. Jiang, G. Tao, et al., Actuator failure compensation and attitude control for rigid satellite by adaptive control using quaternion feedback, J. Frankl. Inst. 351 (1) (2014) 296–314.

[15] J. Yang, X.P. Shi, Integrity fault tolerant attitude controller of satellite with disturbance attenuation, J. Syst. Eng. Electron. 35 (1) (2013) 126–131.

[16] J. Jiang, Q. Zhao, Design of reliable control systems possessing actuator redundancies, J. Guid. Control. Dyn. 23 (4) (2000) 709–718.

[17] J. Jiang, X. Yu, Fault-tolerant control systems: a comparative study between active and passive approaches, Annu. Rev. Control. 36 (1) (2012) 60–72.

[18] M. Mirshams, M. Khosrojerdi, M. Hasani, Passive fault-tolerant sliding mode attitude control for flexible spacecraft with faulty thrusters, Proc. Inst. Mech. Eng. G J. Aerospace Eng. 228 (2014) 2343–2357.

[19] J. Zhou, X. Li, R. Liu, et al., Active fault-tolerant satellite attitude control based on fault effect classification, Proc. Inst. Mech. Eng. G J. Aerospace Eng. 231 (2016) 1917–1934.

[20] Y. Jiang, Q.L. Hu, G.F. Ma, Adaptive backstepping fault-tolerant control for flexible spacecraft with unknown bounded disturbances and actuator failures, ISA Trans. 49 (1) (2010) 57–69.

[21] B. Xiao, Q.L. Hu, Y.M. Zhang, Fault-tolerant attitude control for flexible spacecraft without angular velocity magnitude measurement, J. Guid. Control. Dyn. 34 (5) (2012) 1556–1561.

[22] Q. Meng, T. Zhang, J.Y. Song, Modified model-based fault-tolerant time-varying attitude tracking control of uncertain flexible satellites, Proc. Inst. Mech. Eng. G J. Aerospace Eng. 227 (2013) 1827–1841.

[23] B. Xiao, S. Yin, Velocity-free fault-tolerant and uncertainty attenuation control for a class of nonlinear systems, IEEE Trans. Ind. Electron. 63 (7) (2016) 4400–4411.

[24] Y.J. Ma, B. Jiang, G. Tao, et al., Uncertainty decomposition-based fault-tolerant adaptive control of flexible spacecraft, IEEE Trans. Aerosp. Electron. Syst. 51 (2) (2015) 1053–1068.

[25] S.J. Bai, P. Ben-Tzvi, Q.K. Zhou, et al., Fuzzy sliding mode control of a flexible spacecraft with input saturation, in: Proceedings of the ASME 2009 International Mechanical Engineering Congress & Exposition, Lake Buena Vista, November, 2009, pp. 1055–1061.

[26] Y. Chen, C.Y. Dong, A fuzzy-neural network sliding mode control for flexible spacecraft, in: 2nd International Conference on Information Engineering and Computer Science (ICIECS), Wuhan, December, 2010, pp. 1–4.

[27] Y.H. Pao, Adaptive Pattern Recognition and Neural Networks, Addison-Wesley, Boston, MA, 1989.

[28] L. Zhang, Y. Zhu, X.Z. Wei, State estimation of discrete-time switched neural networks with multiple communication channels, IEEE Trans. Cybern. (2016) 1–13.

[29] J.C. Patra, R.N. Pal, B.N. Chatterji, et al., Identification of nonlinear dynamic systems using functional link artificial neural network, IEEE Trans. Syst. Man Cybern. B Cybern. 29 (1999) 254–262.

[30] J.C. Patra, C. Bornand, Nonlinear dynamic system identification using Legendre neural network, in: The 2010 International Joint Conference on Neural Networks (IJCNN), Barcelona, October, 2010, pp. 1–7.

[31] H. Lee, M. Tomizuka, Robust adaptive control using a universal approximator for SISO nonlinear systems, IEEE Trans. Fuzzy Syst. 8 (1) (2000) 95–106.

[32] H.B. Du, G.H. Wen, Y.Y. Cheng, et al., Distributed finite-time cooperative control of multiple high-order nonholonomic mobile robots, IEEE Trans. Neural Netw. Learn. Syst. 28 (12) (2016) 2998–3006.

[33] Y. Liang, J.Y. Yang, Nonsingular fast terminal sliding-mode control for nonlinear dynamical systems, Int. J. Robust Nonlinear Control 21 (16) (2011) 1865–1879.

[34] Y.W. Liang, S.D. Xu, C.L. Tsai, Study of VSC reliable designs with application to spacecraft attitude stabilization, IEEE Trans. Control Syst. Technol. 15 (2007) 332–338.

Part IV

Hybrid actuator-based fast maneuver and stabilization control for rigid satellites

Chapter 10

Hybrid thrusters and reaction wheel-based attitude control

10.1 Introduction

Three-axis, large-angle maneuvers of agile spacecraft have been receiving significant focus for several decades. A crucial issue appearing in the design process of feedback control laws for large-angle maneuvers is control torque saturation. Boskovic proposed a solution to the torque magnitude constraint problem by removing the discontinuous term in the traditional sliding mode control scheme [1,2]. Tandale developed an adaptive control methodology that facilitates correct adaptation in the presence of actuator saturation constraints based on a dynamic inversion approach [3], where saturation is avoided by modifying the reference and the tracking error due to saturation that does not influence the update of the learning parameters. In Ref. [4], Wallsgrove resolved the control torque saturation problem by carefully designing the structure of the feedback control law such that its magnitude cannot exceed the saturation limit.

Recently, a combination of various actuators has been drawing much attention for large-angle rapid reorientation [5,6]. A combined control algorithm using reaction wheels and magnetic torquers as the control actuators was adopted to reduce effectively the reaction wheel speeds during large-angle attitude maneuvers [7]. A similar concept was used for agile satellites utilizing control moment gyros (CMG) in order to compensate for gimbal angle deviations during external disturbances in Ref. [8]. The most common actuator to rotate the spacecraft is the thruster that typically can exert much larger control torque than reaction wheels. But due to the discontinuous nature of thrusters, they cannot be used to achieve high control accuracy [9]. Another disadvantage of thrusters is their fuel consumption. In contrast, although the torque that reaction wheels can offer is one or two order of magnitude lower than that of thrusters and CMG [10], reaction wheels are capable of finer control for high precision control demand. In addition, reaction wheels consume only electrical energy. The idea of incorporating both thrusters and reaction wheels as actuators seeks to take advantage of the strong point of each individual type of actuators, when a rapid maneuver of a rigid spacecraft with high precision is demanded. In that case,

Fast Satellite Attitude Maneuver and Control. https://doi.org/10.1016/B978-0-323-95455-6.00007-X

thrusters would offer the primary reorientation maneuver torque to make the spacecraft to rotate rapidly, while reaction wheels provide the fine control torque to evaluate the control precision.

The chapter is organized as follows. In Section 10.2, we give a detailed statement of the strategy using hybrid thrusters and reaction wheels as actuators. In Section 10.3, we first cover the dynamics and kinematics of the system incorporating both thrusters and reaction wheels as actuators. Quaternions are used for the attitude representation. Then, the equation of the reference motion is introduced. In Section 10.4, we develop a tracking law based on variable structure control and the Lyapunov method to drive the complete system globally asymptotically stable. In Section 10.5, numerical simulations are presented to describe the performance of the proposed theoretical results followed by the concluding statement in Section 10.6.

10.2 Problem formulation

10.2.1 Control strategy

In this section, a strategy integrating thrusters and reaction wheels as actuators is put forward. In the control strategy, first, the reference motion is generated with the information including the current attitude and the angular velocity of the spacecraft, the target attitude and angular velocity, and the dynamic parameters of the spacecraft. Then, the thrusters and the reaction wheels will exert the torque to the spacecraft with the purpose of high precision and large-angle rapid reorientation. A detailed block diagram of the control strategy is shown in Fig. 10.1.

The "reference motion" block including reference attitude, reference angular velocity, and nominal control torque are all calculated offline before the implementation of the control system. There are a number of methods to plan a trajectory to accomplish the maneuver. In particular, in order to achieve the minimum-time maneuver of a three-axis rigid spacecraft, the offline reference trajectory of the optimal attitude control is generated via "General Pseudospectral Optimal Control Software" (short for GPOPS), which is an open-source MATLAB optimal control software for solving optimal control problem [11].

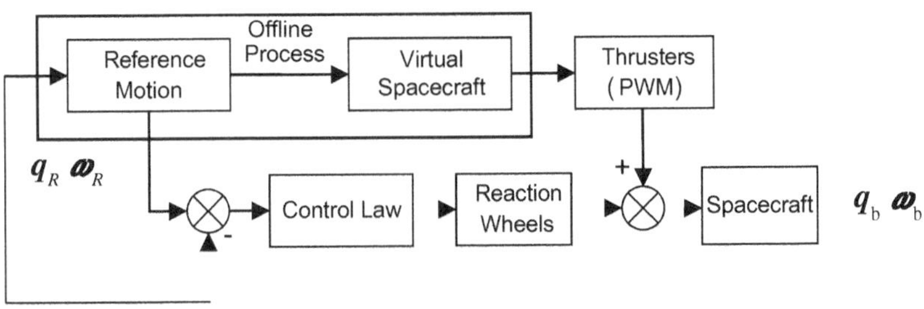

FIG. 10.1 Control organization for hybrid thrusters and reaction wheels strategy.

The "Thrusters" block exerts the torque generated on the spacecraft via off-line process. Without initial condition errors, thrusters would render the spacecraft maneuver along the desired reference motion if perfect knowledge of the system is available. However, in practical implementations, both initial condition errors and system parameter uncertainty will cause the spacecraft to diverge from the reference motion. Furthermore, due to the discontinuous natural defect of thrusters, there exist biases between the torque generated by thrusters and the required torque. These unfavorable factors should be taken into account during the design process of the control system. In this chapter, the thrusters are used in the PWM mode, thus they can be approximated as linear actuators.

The "Control Law" block gives the feedback control in the closed loop to compensate tracking errors between the reference and actual trajectories caused by initial conditions. And the bias between the torque generated by thrusters and the required torque will also be handled by the variable structure control. The "Reaction wheels" block exerts the torque calculated by the "Control Law" block.

10.2.2 Equations of motion

For the convenience of building the dynamics of the system, we use $\boldsymbol{J}_{\mathrm{b}}$ to denote the inertia matrix of spacecraft not including the reaction wheels; $J_{wi}, i=1, 2,\cdots,\mathrm{N}$ represent the axial moments of inertia of the reaction wheels, and the moments of inertia set of the reaction wheels is denoted by $\boldsymbol{J}_{\mathrm{w}}=\mathrm{diag}\{J_{\mathrm{w1}},\cdots,J_{\mathrm{wN}}\}$.

The angular momentum of the reaction wheels can be written as

$$\boldsymbol{h}_{\mathrm{w}} = \boldsymbol{J}_{\mathrm{w}}\left(\mathrm{A}^{\mathrm{T}}\boldsymbol{\omega}_b + \boldsymbol{\omega}_{\mathrm{w}}\right) \tag{10.1}$$

where $\boldsymbol{\omega}_b \in \Re^3$ denotes the angular velocity of the body-fixed frame B with respect to the inertial frame N; $\mathrm{A} \in \Re^{3\times\mathrm{N}}$ is the installation matrix of the reaction wheels with respect to frame B, each column of which represents the axial unit vector of the reaction wheel; and $\boldsymbol{\omega}_{\mathrm{w}}=(\omega_{\mathrm{w1}},\omega_{\mathrm{w2}},\cdots,\omega_{\mathrm{wN}})^{\mathrm{T}} \in \Re^{\mathrm{N}}$ expresses the axial angular velocity set of the reaction wheels with respect to frame B.

The angular momentum of the total spacecraft system with hybrid actuators consisting of thrusters and reaction wheels can be expressed as

$$\begin{aligned}\boldsymbol{h} &= \boldsymbol{h}_{\mathrm{b}} + \mathrm{A}\boldsymbol{h}_{\mathrm{w}} \\ &= \boldsymbol{J}_{\mathrm{b}}\boldsymbol{\omega}_b + \mathrm{A}\boldsymbol{J}_{\mathrm{w}}\left(\mathrm{A}^{\mathrm{T}}\boldsymbol{\omega}_b + \boldsymbol{\omega}_{\mathrm{w}}\right)\end{aligned} \tag{10.2}$$

Defining $\boldsymbol{J}=\boldsymbol{J}_{\mathrm{b}}+\mathrm{A}\boldsymbol{J}_{\mathrm{w}}\mathrm{A}^{\mathrm{T}}$, Eq. (10.2) can be rewritten as

$$\boldsymbol{h} = \boldsymbol{J}\boldsymbol{\omega}_b + \mathrm{A}\boldsymbol{J}_{\mathrm{w}}\boldsymbol{\omega}_{\mathrm{w}} \tag{10.3}$$

From Eq. (10.3), the angular velocity of the spacecraft body can be expressed as

$$\boldsymbol{\omega}_b = \boldsymbol{J}_{\mathrm{b}}^{-1}(\boldsymbol{h} - \mathrm{A}\boldsymbol{h}_{\mathrm{w}}) \tag{10.4}$$

According to the moment of momentum theorem

$$\frac{d\boldsymbol{h}}{dt} = \boldsymbol{u}_e \tag{10.5}$$

where $\boldsymbol{u}_e$ is the external torque applied to the spacecraft. The governing dynamics possessing hybrid actuators can be described by

$$\begin{aligned} \dot{\boldsymbol{h}} &= \boldsymbol{h}^{\times}\boldsymbol{J}_b^{-1}(\boldsymbol{h} - \mathrm{A}\boldsymbol{h}_{\mathrm{w}}) + \boldsymbol{u}_e \\ \dot{\boldsymbol{h}}_{\mathrm{w}} &= \boldsymbol{u}_{\mathrm{w}} \end{aligned} \tag{10.6}$$

where $\boldsymbol{u}_{\mathrm{w}} \in \mathfrak{R}^{\mathrm{N}}$ denotes the control torque toward spacecraft generated by the reaction wheels. The notation $\zeta^{\times}$ denotes the following skew-symmetric matrix defined by [12].

$$\zeta^{\times} = \begin{bmatrix} 0 & -\zeta_3 & \zeta_2 \\ \zeta_3 & 0 & -\zeta_1 \\ -\zeta_2 & \zeta_1 & 0 \end{bmatrix} \tag{10.7}$$

which designates the vector cross product operation such that $\zeta^{\times}\varsigma = \zeta \times \varsigma$ for all three-dimensional vectors ζ and ς.

Due to the nonsingular representation through all possible rotation angles, the quaternion is adopted to represent the attitude of the spacecraft. Each quaternion obviously has four parameters where the first element, the scalar part of the quaternion, which is represented by the subscript "0", and rest of them make up the vector part of the quaternion, which is represented by the subscript "$\boldsymbol{v}$" with bold face character, i.e., $\boldsymbol{q} = [q_0 \ \boldsymbol{q}_v]$ [13] and

$$q_0 = \cos(\varphi/2), \quad \boldsymbol{q}_v = \begin{bmatrix} q_1 \\ q_2 \\ q_3 \end{bmatrix} = \upsilon \sin(\varphi/2) \tag{10.8}$$

which denote the result of a virtual rotation by eigenaxis rotation angle φ about a virtual unit axis vector υ (known as eigenaxis) and subjected to the constraint

$$\boldsymbol{q}_v^{\mathrm{T}}\boldsymbol{q}_v + q_0^2 = 1 \tag{10.9}$$

And the inverse quaternion is defined by $\boldsymbol{q}^{-1} = [q_0 \ -\boldsymbol{q}_v]$.

The kinematic evolution equation for quaternion is expressed as

$$\dot{\boldsymbol{q}} = \frac{1}{2}\boldsymbol{E}(\boldsymbol{q})\boldsymbol{\omega}_b \tag{10.10}$$

where the quaternion operator $\boldsymbol{E}(\cdot): \mathfrak{R}^4 \to \mathfrak{R}^{4\times 3}$ is defined by

$$\boldsymbol{E}(\boldsymbol{q}) = \begin{bmatrix} -\boldsymbol{q}_v^T \\ \boldsymbol{q}_v^{\times} + q_0\boldsymbol{I}_{3\times 3} \end{bmatrix} \tag{10.11}$$

The quaternion $\boldsymbol{q}$ relates to the direction cosine matrix $\boldsymbol{C}(\boldsymbol{q})$ through the following formula

$$\boldsymbol{C}(\boldsymbol{q}) = \left(2q_0^2 - 1\right)\boldsymbol{I}_{3\times 3} + 2\boldsymbol{q}_v\boldsymbol{q}_v^{\mathrm{T}} - 2q_0\boldsymbol{q}_v^{\times} \tag{10.12}$$

In this chapter, the quaternion $\boldsymbol{q}_b$ is used to represent the attitude of the spacecraft with respect to frame N, and the reference trajectory is defined by a reference quaternion $\boldsymbol{q}_R$ expressed in the reference coordinate frame R with respect to frame N.

In the process of designing the reference trajectory, we assume that only thrusters exert the control torque to the spacecraft and that reaction wheels are fixed with respect to the platform (i.e. $\omega_{wi}=0$, $i=1, 2, \cdots, N$). In this case, from Eq. (10.3), it follows that $\boldsymbol{h}=\boldsymbol{J}\boldsymbol{\omega}_b$. In frame R, the governing equation of reference motion is defined as [6].

$$\dot{\boldsymbol{h}}_R = \boldsymbol{h}_R^{\times}\boldsymbol{J}^{-1}\boldsymbol{h}_R + \boldsymbol{u}_R \tag{10.13}$$

where $\boldsymbol{h}_R=\boldsymbol{J}\boldsymbol{\omega}_R$, $\boldsymbol{\omega}_R$ represents the reference angular velocity with respect to frame N expressed in frame R. The desired trajectory to be tracked can be derived from the motion of the reference spacecraft. In Eq. (10.13), $\boldsymbol{u}_R$ is the desired nominal control torque. Hence, if $\boldsymbol{u}_R$ is applied to the spacecraft with the same conditions (including the initial orientation, angular velocity, and the moment of inertia) and non-rotating wheels, the desired reference trajectory will be generated.

From Eq. (10.13), we have

$$\dot{\boldsymbol{\omega}}_R = \boldsymbol{J}^{-1}\boldsymbol{h}_R^{\times}\boldsymbol{J}^{-1}\boldsymbol{h}_R + \boldsymbol{J}^{-1}\boldsymbol{u}_R \tag{10.14}$$

The kinematic equation of motion for the reference quaternion $\boldsymbol{q}_R$ is given by

$$\dot{\boldsymbol{q}}_R = \frac{1}{2}\boldsymbol{E}(\boldsymbol{q}_R)\boldsymbol{\omega}_R \tag{10.15}$$

10.3 Hybrid tracking controller design

The tracking error is the difference between the actual spacecraft attitude motion and the reference trajectory defined by the error quaternion $\widetilde{\boldsymbol{q}}$ denoting the orientation of frame B with respect to frame R. That is to say $\widetilde{\boldsymbol{q}} = \boldsymbol{q}_R^{-1} \otimes \boldsymbol{q}_b$. The operator $\otimes$ is the quaternion multiplication operator of two quaternions. The tracking error of angular velocity in frame B is defined through

$$\widetilde{\boldsymbol{\omega}} = \boldsymbol{\omega}_b - \boldsymbol{C}_R^B(\widetilde{\boldsymbol{q}})\boldsymbol{\omega}_R \tag{10.16}$$

The kinematics equation about tracking error can be described by

$$\dot{\widetilde{\boldsymbol{q}}} = \frac{1}{2}\boldsymbol{E}(\widetilde{\boldsymbol{q}})\widetilde{\boldsymbol{\omega}} \tag{10.17}$$

We define the tracking error of the angular momentum expressed in frame B as [6].

$$
\begin{aligned}
\widetilde{\boldsymbol{h}} &= \boldsymbol{h} - \boldsymbol{J}_b \boldsymbol{C}_R^B(\widetilde{\boldsymbol{q}})\boldsymbol{\omega}_R \\
&= \boldsymbol{J}_b\boldsymbol{\omega}_b + \mathrm{A}\boldsymbol{J}_{\mathrm{w}}\left(\mathrm{A}^{\mathrm{T}}\boldsymbol{\omega}_b + \boldsymbol{\omega}_{\mathrm{w}}\right) - \boldsymbol{J}_b \boldsymbol{C}_R^B(\widetilde{\boldsymbol{q}})\boldsymbol{\omega}_R \\
&= \boldsymbol{J}_b\left(\boldsymbol{\omega}_b - \boldsymbol{C}_R^B(\widetilde{\boldsymbol{q}})\boldsymbol{\omega}_R\right) + \mathrm{A}\boldsymbol{J}_{\mathrm{w}}\left(\mathrm{A}^{\mathrm{T}}\boldsymbol{\omega}_b + \boldsymbol{\omega}_{\mathrm{w}}\right)
\end{aligned}
\tag{10.18}
$$

Combining with Eq. (10.1) and Eq. (10.16), Eq. (10.18) can be further rewritten as

$$
\widetilde{\boldsymbol{h}} = \boldsymbol{J}_b\widetilde{\boldsymbol{\omega}} + \mathrm{A}\boldsymbol{h}_w \tag{10.19}
$$

Taking the derivative of Eq. (10.18) in conjunction with Eq. (10.14), we can get

$$
\begin{aligned}
\dot{\widetilde{\boldsymbol{h}}} &= \dot{\boldsymbol{h}} - \boldsymbol{J}_b\frac{d\boldsymbol{C}_R^B(\widetilde{\boldsymbol{q}})}{dt}\boldsymbol{\omega}_R - \boldsymbol{J}_b\boldsymbol{C}_R^B(\widetilde{\boldsymbol{q}})\dot{\boldsymbol{\omega}}_R \\
&= \dot{\boldsymbol{h}} - \boldsymbol{J}_b\boldsymbol{\omega}_b^{\times}\widetilde{\boldsymbol{\omega}} - \boldsymbol{J}_b\boldsymbol{C}_R^B(\widetilde{\boldsymbol{q}})\dot{\boldsymbol{\omega}}_R \\
&= \boldsymbol{h}^{\times}\boldsymbol{J}_b^{-1}(\boldsymbol{h} - \mathrm{A}\boldsymbol{h}_{\mathrm{w}}) + \boldsymbol{u}_e - \boldsymbol{J}_b\boldsymbol{\omega}_b^{\times}\widetilde{\boldsymbol{\omega}} - \boldsymbol{J}_b\boldsymbol{C}_R^B(\widetilde{\boldsymbol{q}})\left(\boldsymbol{J}^{-1}\boldsymbol{h}_R^{\times}\boldsymbol{J}^{-1}\boldsymbol{h}_R + \boldsymbol{J}^{-1}\boldsymbol{u}_R\right) \\
&= \boldsymbol{h}^{\times}\boldsymbol{J}_b^{-1}(\boldsymbol{h} - \mathrm{A}\boldsymbol{h}_{\mathrm{w}}) + \boldsymbol{u}_e - \boldsymbol{J}_b\boldsymbol{\omega}_b^{\times}\widetilde{\boldsymbol{\omega}} - \boldsymbol{J}_b\boldsymbol{C}_R^B(\widetilde{\boldsymbol{q}})\boldsymbol{J}^{-1}\boldsymbol{h}_R^{\times}\boldsymbol{J}^{-1}\boldsymbol{h}_R - \boldsymbol{J}_b\boldsymbol{C}_R^B(\widetilde{\boldsymbol{q}})\boldsymbol{J}^{-1}\boldsymbol{u}_R
\end{aligned}
\tag{10.20}
$$

In order to produce a maneuver to the reference attitude trajectory in the shortest distance, the switching surface for the variable structure control is chosen as [14,15]

$$
\begin{aligned}
\boldsymbol{s} &= \widetilde{\boldsymbol{\omega}} + k\,\mathrm{sgn}\,(\widetilde{q}_0)\widetilde{\boldsymbol{q}}_v \\
&= \boldsymbol{J}_b^{-1}\left(\widetilde{\boldsymbol{h}} - \mathrm{A}\boldsymbol{h}_w\right) + k\,\mathrm{sgn}\,(\widetilde{q}_0)\widetilde{\boldsymbol{q}}_v
\end{aligned}
\tag{10.21}
$$

where k is a positive scalar constant.

We present the following scalar Lyapunov function

$$
V(s) = \frac{1}{2}\boldsymbol{s}^{\mathrm{T}}\boldsymbol{J}_b\boldsymbol{s} \tag{10.22}
$$

The function is globally positive definite except for $\boldsymbol{s} \neq 0_{3\times 1}$ owing to positive definite of inertia matrix $\boldsymbol{J}_b$. Calculating the time derivative of V gives

$$
\dot{V}(s) = \boldsymbol{s}^{\mathrm{T}}\boldsymbol{J}_b\dot{\boldsymbol{s}} \tag{10.23}
$$

where

$$
\begin{aligned}
\boldsymbol{J}_b\dot{\boldsymbol{s}} &= \boldsymbol{J}_b\left[\boldsymbol{J}_b^{-1}\left(\dot{\widetilde{\boldsymbol{h}}} - \mathrm{A}\dot{\boldsymbol{h}}_w\right) + k\,\mathrm{sgn}\,(\widetilde{q}_0)\dot{\widetilde{\boldsymbol{q}}}_v\right] \\
&= \dot{\widetilde{\boldsymbol{h}}} - \mathrm{A}\dot{\boldsymbol{h}}_w + k\boldsymbol{J}_b\,\mathrm{sgn}\,(\widetilde{q}_0)\dot{\widetilde{\boldsymbol{q}}}_v
\end{aligned}
\tag{10.24}
$$

Substituting Eq. (10.24) into Eq. (10.23) gives

$$\begin{aligned}\dot{V}(s) &= \boldsymbol{s}^{\mathrm{T}}\boldsymbol{J}_b\dot{\boldsymbol{s}}\\ &= \boldsymbol{s}^{\mathrm{T}}\left(\dot{\widetilde{\boldsymbol{h}}} - \mathrm{A}\dot{\boldsymbol{h}}_w + k\,\mathrm{sgn}\,(\widetilde{q}_0)\boldsymbol{J}_b\dot{\widetilde{\boldsymbol{q}}}_v\right)\\ &= \boldsymbol{s}^{\mathrm{T}}\left(\dot{\widetilde{\boldsymbol{h}}} - \mathrm{A}\boldsymbol{u}_w + k\,\mathrm{sgn}\,(\widetilde{q}_0)\boldsymbol{J}_b\dot{\widetilde{\boldsymbol{q}}}_v\right)\\ &= \boldsymbol{s}^{\mathrm{T}}\left(\boldsymbol{h}^{\times}\boldsymbol{J}_b^{-1}(\boldsymbol{h} - \mathrm{A}\boldsymbol{h}_{\mathrm{w}}) + \boldsymbol{u}_e - \boldsymbol{J}_b\boldsymbol{\omega}_b^{\times}\widetilde{\boldsymbol{\omega}}\right.\\ &\quad \left.-\boldsymbol{J}_b\boldsymbol{C}_R^B(\widetilde{\boldsymbol{q}})\boldsymbol{J}^{-1}\boldsymbol{h}_R^{\times}\boldsymbol{J}^{-1}\boldsymbol{h}_R - \boldsymbol{J}_b\boldsymbol{C}_R^B(\widetilde{\boldsymbol{q}})\boldsymbol{J}^{-1}\boldsymbol{u}_R - \mathrm{A}\boldsymbol{u}_w + k\,\mathrm{sgn}\,(\widetilde{q}_0)\boldsymbol{J}_b\dot{\widetilde{\boldsymbol{q}}}_v\right)\end{aligned} \tag{10.25}$$

We choose the thrusters to perform the desired nominal control torque as

$$\boldsymbol{u}_e = \boldsymbol{u}_R \tag{10.26}$$

However, thrusters do not typically have fine resolution. That is to say, in the actual implementation, there will be bias between the actual torque and the nominal torque (10.26). Usually, the bias can be bounded as

$$|d_i| \le \lambda_i, \quad i = 1, 2, 3 \tag{10.27}$$

where λ_i, $i=1, 2, 3$ are the known constants.

Then, a feedback control structure is applied to generate additional feedback control torque commands for the purpose of compensation for deviations from the nominal open-loop trajectory, which comes from the initial condition errors and the above-mentioned torque bias of the thrusters.

Let the feedback control law for the reaction wheels satisfy

$$\begin{aligned}\mathrm{A}\boldsymbol{u}_w &= \boldsymbol{h}^{\times}\boldsymbol{J}_b^{-1}(\boldsymbol{h} - \mathrm{A}\boldsymbol{h}_{\mathrm{w}}) - \boldsymbol{J}_b\boldsymbol{\omega}_b^{\times}\widetilde{\boldsymbol{\omega}} - \boldsymbol{J}_b\boldsymbol{C}_R^B(\widetilde{\boldsymbol{q}})\boldsymbol{J}^{-1}\boldsymbol{h}_R^{\times}\boldsymbol{J}^{-1}\boldsymbol{h}_R\\ &\quad -\boldsymbol{J}_b\boldsymbol{C}_R^B(\widetilde{\boldsymbol{q}})\boldsymbol{J}^{-1}\boldsymbol{u}_R + \boldsymbol{u}_R + k\,\mathrm{sgn}\,(\widetilde{q}_0)\boldsymbol{J}_b\dot{\widetilde{\boldsymbol{q}}}_v + \mathrm{K}_1\boldsymbol{s} + \boldsymbol{\lambda}_{\max}\,\mathrm{sgn}\,(\boldsymbol{s})\end{aligned} \tag{10.28}$$

where $\mathrm{K}_1 \in \mathfrak{R}^{3\times 3}$ is a positive definite diagonal matrix ; $\boldsymbol{\lambda}_{\max} = \mathrm{diag}(\lambda_i)$, $i=1$, 2, 3 is a diagonal matrix of the torque bias magnitude; the vector signum function $\mathrm{sgn}\,(\mathbf{s}) = [\mathrm{sgn}\,(s_1)\,\mathrm{sgn}\,(s_2)\,\mathrm{sgn}\,(s_3)]^{\mathrm{T}}$ has individual components defined by

$$\mathrm{sgn}\,(s_i) = \begin{cases} +1 & s_i > 0 \\ 0 & s_i = 0 \\ -1 & s_i < 0 \end{cases} \qquad i = 1, 2, 3 \tag{10.29}$$

According to the aforementioned definition, the time derivative of V leads to

$$\dot{V}(s) = -\boldsymbol{s}^{\mathrm{T}}\mathrm{K}_1\boldsymbol{s} + \boldsymbol{s}^{\mathrm{T}}(\boldsymbol{d} - \boldsymbol{\lambda}_{\max}\,\mathrm{sgn}\,(\boldsymbol{s})) \le 0 \tag{10.30}$$

Hence, when $t \to \infty$, we can get $\boldsymbol{s}(t) \to 0_{3\times 1}$. That is to say $\lim_{t\to\infty}\widetilde{\boldsymbol{\omega}} = 0_{3\times 1}$, $\lim_{t\to\infty}\widetilde{\boldsymbol{q}}_v = 0_{3\times 1}$, and $\lim_{t\to\infty}\widetilde{q}_0 = 1$ or -1 depending on which equilibrium point is the minimum angular path [16].

Remark 10.1 Due to the imperfect implementation of the signum function in the controller (10.28), the chattering phenomenon may arise. To alleviate it, the discontinuous term sgn($\boldsymbol{s}$) in controller (10.28) can be replaced by the following saturation function

$$\mathrm{sat}(s_i) \equiv \begin{cases} +1 & s_i > \varepsilon_i \\ s_i/\varepsilon_i & |s_i| \le \varepsilon_i \\ -1 & s_i < -\varepsilon_i \end{cases} \qquad i = 1,\ 2,\ 3 \tag{10.31}$$

where ε_i is a small positive quantity called the saturation boundary thickness.

The reaction wheels may be stuck in saturation during rapid maneuver. In that case, the control strategy will lose the ability to elevate the control precision. In order to avoid this problem, when the angular momenta of the reaction wheels exceed some predetermined limiting value (such as 80% of the maximum angular momenta), the reaction wheels will exert an unloading torque. At the same time, the thrusters will also exert the equal torque but opposite in direction to the spacecraft. And from the spacecraft's point view, the attitude was not affected at all by the two control hardware items' simultaneous operation. Thus, in this case, the excess momentum of the reaction wheel will be unloaded, and it does not influence the implementation of the control strategy.

10.4 Simulation results

A time-optimal rest-to-rest maneuver is applied to demonstrate the effectiveness of the control strategy. The inertia matrix of the spacecraft is chosen as $\boldsymbol{J}_b = \mathrm{diag}\{150\ \ 120\ \ 200\}\ \mathrm{kg}\cdot\mathrm{m}^2$. Three reaction wheels aligned with the principal axis are used to exert the feedback control torque in the closed loop, and the axial moment of inertia of the three reaction wheels is all $0.04\ \mathrm{kg}\cdot\mathrm{m}^2$. The objective of the control law is to drive the spacecraft from the initial attitude described by quaternion $[0.0973\ \ 0.9004\ \ 0.4241\ \ -0.0026]$, which can be represented in Euler angles as $[50\ \ 5\ \ 170]^\circ$, to the finial position represented by quaternion $[-0.136\ \ 0.914\ \ 0.319\ \ 0.210]$, where the corresponding Euler angles are $[36.6\ \ -28.1\ \ -172.6]^\circ$. We assume the initial actual attitude of the spacecraft in the simulation is $[-0.0952\ \ -0.9026\ \ -0.4198\ \ 0.0060]$, which has deviation of $[0.5\ \ -0.3\ \ 0.4]^\circ$ in Euler angles from the nominal initial attitude.

The disturbances of the inaccurate thruster on the spacecraft are the uniform distribution on the interval $[-0.05\ \ 0.05]$ generated by $0.1 * \mathrm{rand}(3,1) - 0.05$ at the beginning of each simulation, where $\mathrm{rand}(3,1)$ returns a 3-by-1 matrix containing pseudorandom values drawn from the standard uniform distribution on the open interval $[0,1]$.

In the process of generating the time optimal trajectory by GPOPS, the following parameters are chosen: the torque limit for thruster is 1 Nm, and the maximum angular velocity is chosen as 3 deg./s for security consideration.

The control gains for control law (10.28) are set to $K_1 = \text{diag}\{45\ \ 45\ \ 45\}$, $k=0.8$, $\boldsymbol{\lambda}_{\max} = \text{diag}\{0.05\ \ 0.05\ \ 0.05\}$, and the saturation boundary thicknesses ε_i in Eq. (10.31) are all set to 0.01.

From Figs. 10.2–10.8, we can see that the control law (10.26, 10.28) can accomplish the large-angle attitude fast maneuver well. It is clear from Figs. 10.4 and 10.5 that the attitude and the angular velocity approach the desired trajectory asymptotically.

Next, we simulate the case where the signum function was not incorporated in the switching surface (10.21), then sgn $(\widetilde{q}_0)$ disappears in third term on the right-hand side of the control law (10.28), and the simulation results are shown in Figs. 10.9 and 10.10. From the curve of $\widetilde{\boldsymbol{q}}_v$ in Fig. 10.9, we can see that the spacecraft fails to track the desired trajectory between 5 s and 30s. Of course, this is quite reasonable and is expected to happen because the controller (10.28) without signum function drives the first element of the quaternion $\widetilde{q}_0$ from -1 to 1. It equals to rotate the spacecraft 360° uselessly that can be easily seen from the corresponding Euler angles in Fig. 10.10. Thus, compared to the

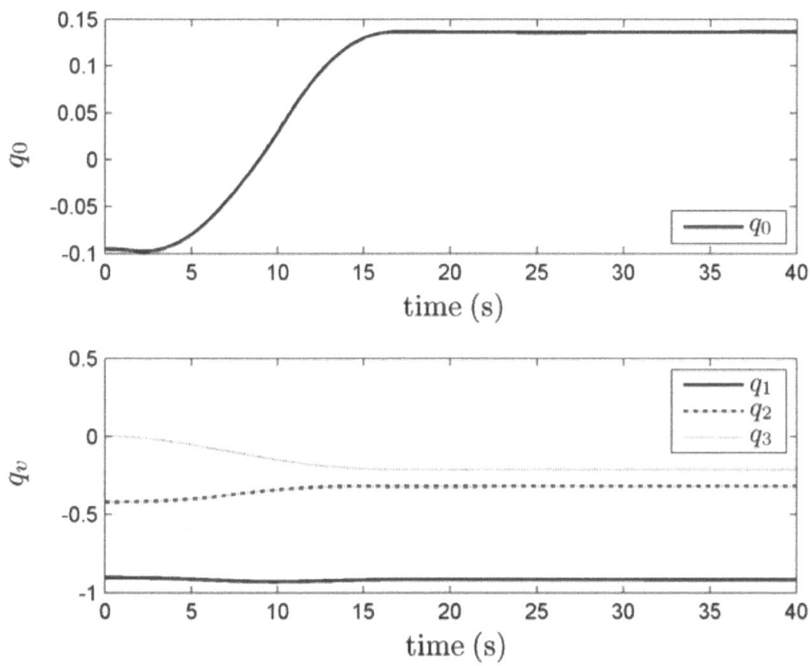

FIG. 10.2 Attitude of the spacecraft.

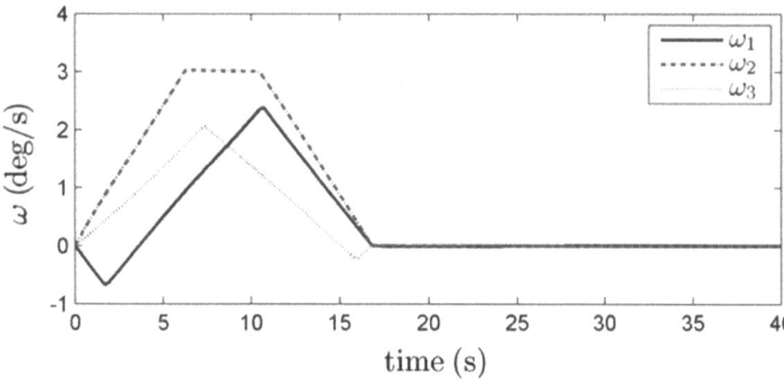

FIG. 10.3 Angular velocity of the spacecraft.

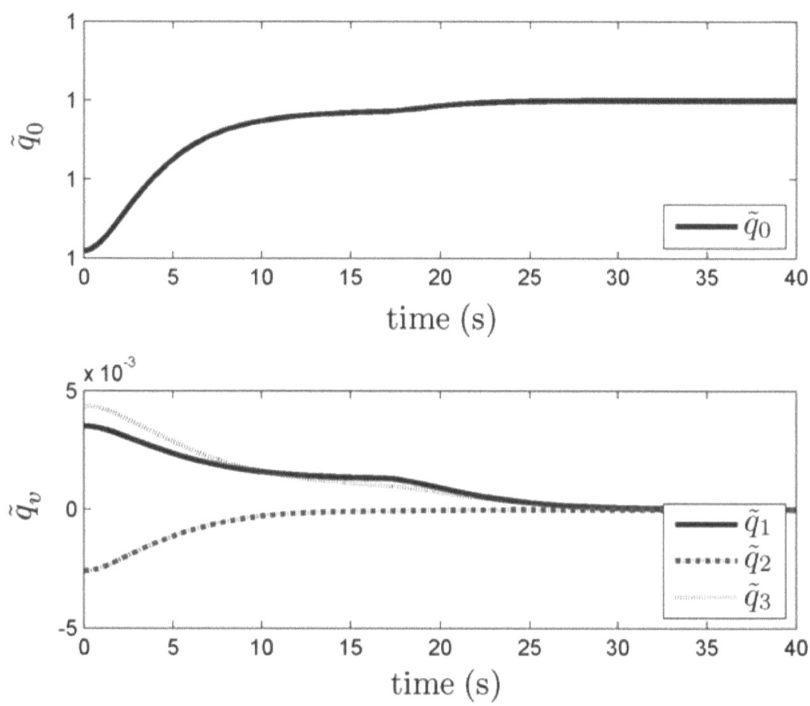

FIG. 10.4 Attitude tracking error in quaternion.

simulations in Figs. 10.9 and 10.10 to Figs. 10.4 and 10.5 respectively, the term $\text{sgn}\left(\tilde{q}_0\right)$ is useful to drive the system to the desired trajectory in the shortest distance.

Further, in order to validate the feasibility and effectiveness of the proposed control strategy under large initial condition deviations, we choose

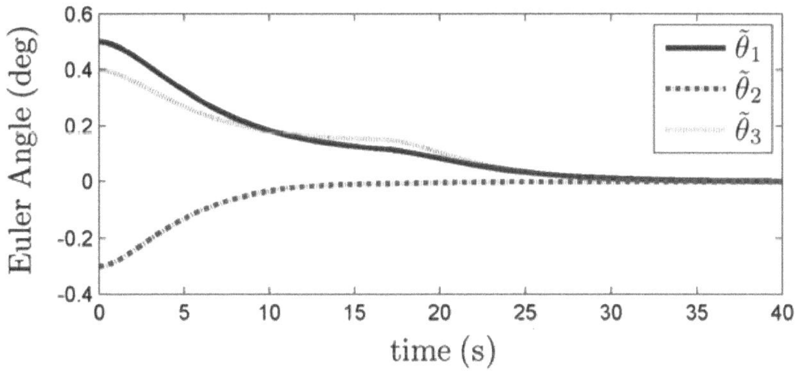

FIG. 10.5 Attitude tracking error in Euler angles.

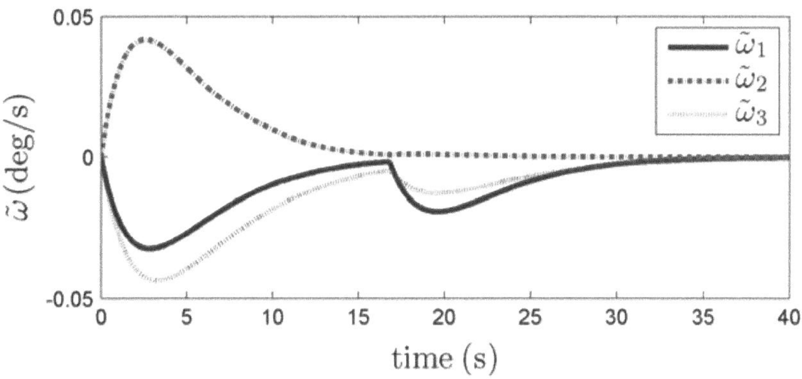

FIG. 10.6 Angular velocity tracking error.

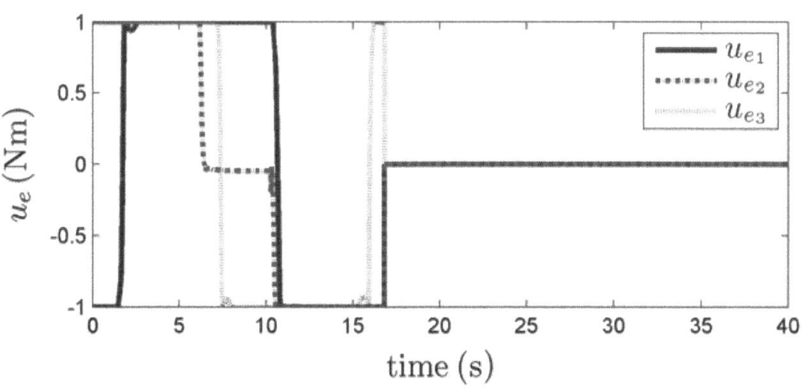

FIG. 10.7 Control torque performed by thrusters.

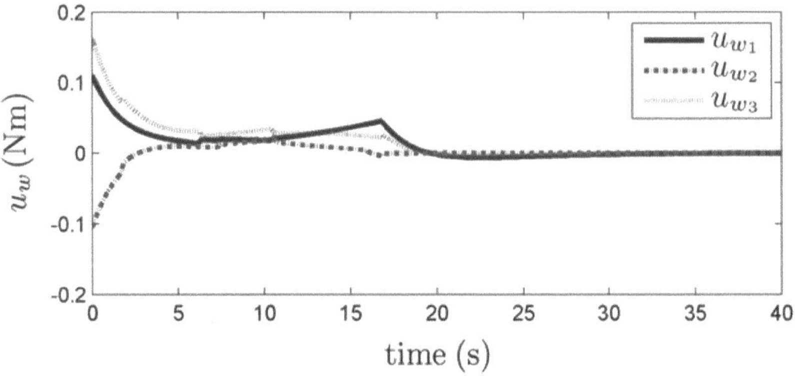

FIG. 10.8 Control torque performed by reaction wheels.

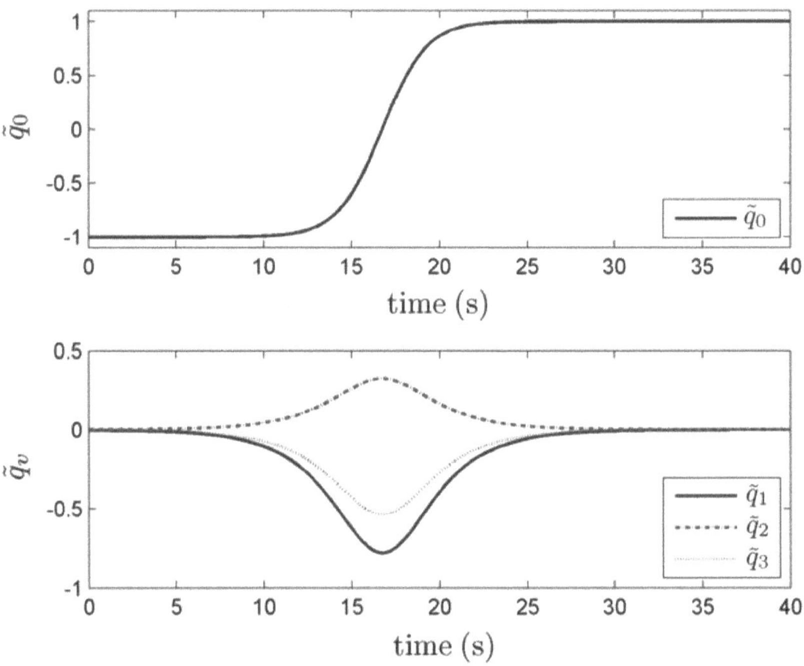

FIG. 10.9 Attitude tracking error in quaternion without signum function in the switching surface (10.21).

the initial actual attitude of the spacecraft in another simulation is $[0.9997\ \ 0.0132\ \ -0.0085\ \ 0.0176]$, which has deviation of $[2\ \ -11.5]^\circ$ in Euler angles from the nominal initial attitude. The simulation results presented in Figs. 10.11 and 10.12 show that the control strategy is still work well even though with large initial condition deviations.

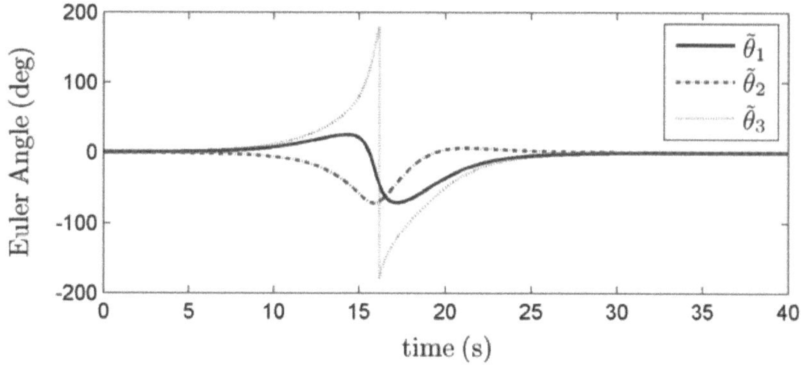

FIG. 10.10 Attitude tracking error in Euler angles without signum function in the switching surface (10.21).

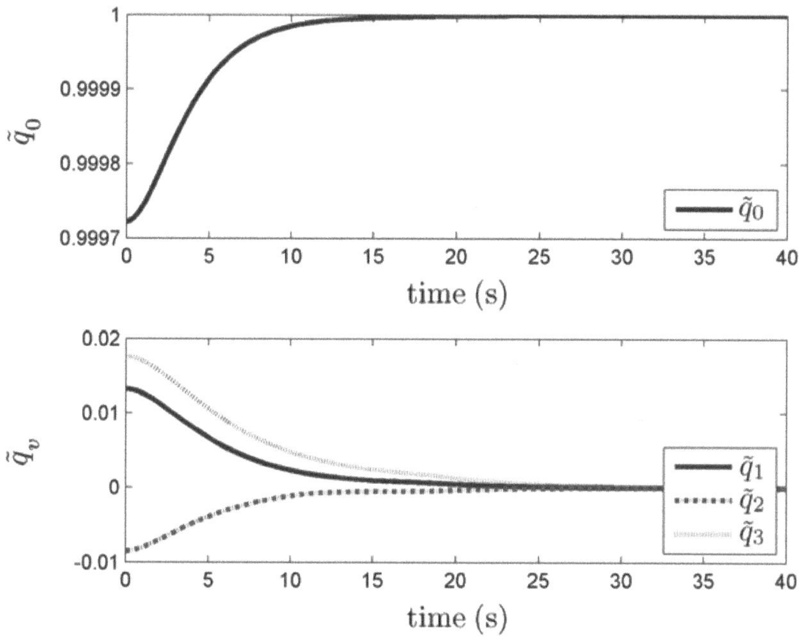

FIG. 10.11 Attitude tracking error in quaternion with large initial condition deviations.

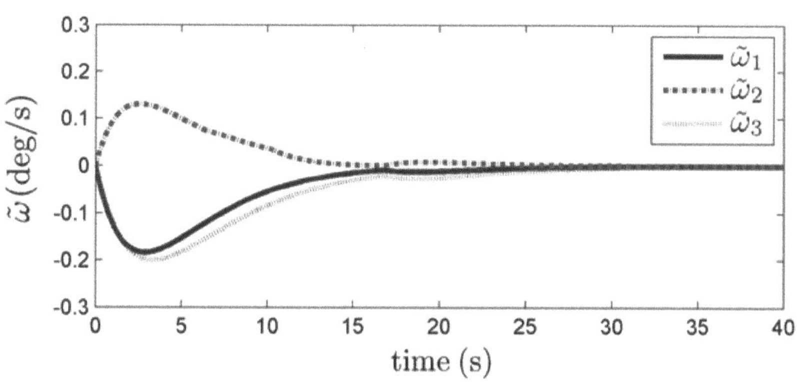

FIG. 10.12 Angular velocity tracking error with large initial condition deviations.

10.5 Conclusion

In this chapter, we present a control strategy that enables large-angle rapid reorientation of rigid spacecraft with high precision. The control strategy uses thrusters in open-loop control to drive the spacecraft to a reference path, with reaction wheels implementing a closed-loop variable structure control law to correct for tracking errors due to the initial condition errors and the torque bias of thrusters. Additionally, the term $\operatorname{sgn}(\widetilde{q}_0)$ is used in the switching surface to drive the spacecraft to the desired trajectory in the shortest distance. Theoretical proofs of the convergence of the control systems are provided. Numerical simulations are carried out to support the analysis of the control law presented. The results demonstrate that the effectiveness of the control strategy and the convergence of the attitude and angular velocity error can be obtained by the proposed control law.

References

[1] J.D. Boskovic, S.M. Li, R.K. Mehra, Robust tracking control design for spacecraft under control input saturation, J. Guid. Control Dynam. 27 (4) (2004) 627–633.

[2] J.D. Boskovic, S.M. Li, R.K. Mehra, Robust adaptive variable structure control of spacecraft under control input saturation, J. Guid. Control Dynam. 24 (2001) 14–22.

[3] M.D. Tandale, K. Subbarao, J. Valasek, M.R. Akella, Structured adaptive model inversion control with actuator saturation constraints applied to tracking spacecraft maneuvers, in: Proceedings of the 2004 American Control Conference, (2004) 1031–1036.

[4] R.J. Wallsgrove, M.R. Akella, Globally stabilizing saturated attitude control in the presence of bounded unknown disturbances, J. Guid. Control Dynam. 28 (2005) 957–963.

[5] B.H. Lee, B.U. Lee, H.S. Oh, et al., Time optimal attitude maneuver strategies for the agile spacecraft with reaction wheels and thrusters, J. Mech. Sci. Technol. 19 (9) (2005) 1695–1705.

[6] C.D. Hall, P. Tsiotras, H.J. Shen, Tracking rigid body motion using thrusters and momentum wheels, J. Astronaut. Sci. 50 (3) (2002) 311–323.

[7] Z.W. Sun, Y.H. Geng, G.D. Xu, et al., The combined control algorithm for large-angle maneuver of HITSAT-1 small satellite, Acta Astronaut. 54 (7) (2004) 463–469.

[8] V.J. Lappas, A Control Moment Gyro (CMG) Based Attitude Control System (ACS) for Agile Small Satellites, University of Surrey, 2002.

[9] M.J. Sidi, Spacecraft Dynamics and Control: A Practical Engineering Approach, Cambridge Univ Pr, 2000.

[10] V. Lappas, B. Wie, Robust control moment gyroscope steering logic with gimbal angle constraints, J. Guid. Control Dynam. 32 (5) (2009) 1662–1666.

[11] A.V. Rao, D.A. Benson, C. Darby, et al., Algorithm 902: GPOPS, a matlab software for solving multiple-phase optimal control problems using the gauss pseudospectral method, ACM Trans. Math. Softw. 37 (2010).

[12] M.D. Shuster, Survey of attitude representations, J. Astronaut. Sci. 41 (1993) 439–517.

[13] D. Seo, M.R. Akella, Separation property for the rigid-body attitude tracking control problem, J. Guid. Control Dynam. 30 (6) (2007) 1569–1576.

[14] J. Crassidis, S. Vadali, F. Markley, Optimal variable-structure control tracking of spacecraft maneuvers, J. Guid. Control Dynam. 23 (3) (2000) 564–565.

[15] S.R. Vadali, Variable-structure control of spacecraft large-angle maneuvers, J. Guid. Control Dynam. 9 (2) (1986) 235–239.

[16] Z.X. Li, B.L. Wang, Robust attitude tracking control of spacecraft in the presence of disturbances, J. Guid. Control Dynam. 30 (4) (2007) 1156–1159.

Chapter 11

Finite-time attitude control with hybrid actuators

11.1 Introduction

Satellite attitude control plays an important role in spacecraft technology, and it has been widely used in space missions such as rendezvous and docking, space target detection and earth observation. These missions require the satellite to maintain high-precision pointing during rapid maneuver. Compared with the conventional asymptotic stability and exponential stability methods that render system states converge to the equilibrium point in infinite time, the new finite-time stability algorithm makes the system converge to the equilibrium point in the finite time. Also, the technology is more practical in advanced modern satellite attitude control technology. Therefore, control methods with features such as strong robustness and finite-time convergence are widely used in satellite attitude control tasks.

A finite-time controller [1] was designed based on homogeneous principle for attitude tracking of satellite. Also, another kind of finite-time attitude controller [2] was designed with back-stepping methods. Continuous TSM is applied to the attitude control of spacecraft [3]. Xiao and Yin [4] present a theoretical framework for designing attitude control with simple structures to perform satellite attitude-tracking maneuver. A finite-time attitude-tracking control scheme [5] is proposed for spacecraft using TSM and Chebyshev neural network. However, the said research works are mainly aimed at a single actuator, which cannot guarantee the requirements of high mobility and precision simultaneously during attitude maneuver. New concept of hybrid actuators [6] was first proposed to solve these problems, but the question of imprecise output torque was left over. A two-steps strategy [7] was adopted to solve this imprecision problem. In this strategy, control moment gyros (CMGs) generate rough control torque, and reaction wheels produce precise attitude adjustment for error correction. However, this strategy did not solve the singularity problem of the CMGs. Moreover, the mechanical complexity of the CMGs limits its use on satellite. Thus, thrusters and flywheels are still the most commonly used attitude-tracking controllers by now.

In practical situations, a rational control law is necessary to coordinate thrusters and flywheels effectively for attitude maneuver. Currently, in the

Fast Satellite Attitude Maneuver and Control. https://doi.org/10.1016/B978-0-323-95455-6.00013-5

operation of hybrid actuators, thrusters generate large torque to implement rapid attitude maneuver for the satellite. When the satellite approaches the target attitude, the control system would turn off thrusters and switch to the flywheel working mode. Through the precise control torque produced by flywheels, the satellite can steadily reach the target attitude. However, this switching strategy does not guarantee the high precise control during attitude maneuver, and it does not have the ability of real-time attitude correction. Ye et al. [8] solved the problem of real-time attitude correction, but he did not give solutions regarding finite-time stability during attitude maneuver. Terminal sliding mode control (TSMC) is a finite-time control method and it can be adopted to achieve finite-time ability. It has already been widely used in many practical applications in spacecraft control tasks. Some novel nonsingular saturated TSM [9,10] based control laws are investigated for a rigid spacecraft, in which the actuator control input saturation and external disturbances are explicitly considered simultaneously. A continuous finite-time scheme [11] for rigid robotic manipulators is proposed suing a new form of TSM. Considering that the full-state measurements are not available, Xiao and Yin [12] proposed a velocity-free fault-tolerant and uncertainty attenuation control algorithm. Besides, several works [13,14] have described the robust control or filtering method to deal with the communication problem.

In order to solve the finite-time tracking control issue during satellite attitude maneuvering process, this chapter proposes a novel nonsingular terminal sliding mode (NTSM) attitude-tracking controller based on the hybrid thruster-flywheel actuator. The main contributions of this work are as follows: (1) in the design of attitude maneuver reference trajectory, the time optimal performance index is used in this chapter; (2) adopt two actuators simultaneously in attitude-tracking control; and (3) solve the singularity problem in finite-time attitude-tracking control.

The reminder of this chapter is organized as follows. First, the control strategy of thruster-flywheel hybrid actuators is introduced. Second, time optimal trajectory of the attitude maneuver task is developed, and it is used as the reference attitude trajectory for the subsequent attitude-tracking controller design. Third, the motion equations are developed based on thruster-flywheel hybrid actuator, and the relative dynamics and relative kinematics models are built for designing the attitude-tracking controllers. Then, satellite attitude-tracking control law is proposed based on Lyapunov principle and TSM control theory. To avoid the singularity of the algorithm, a NTSM attitude-tracking control law is developed in the latter part of this section. Finally, the performance of the proposed control algorithm is demonstrated by numerical simulation.

11.2 Problem formulation

11.2.1 Strategy for hybrid actuators

In this section, a rapid maneuver strategy is developed based on thruster-flywheel hybrid actuators. The detailed block diagram of the entire control system is shown in Fig. 11.1, and the dashed box represents the offline trajectory

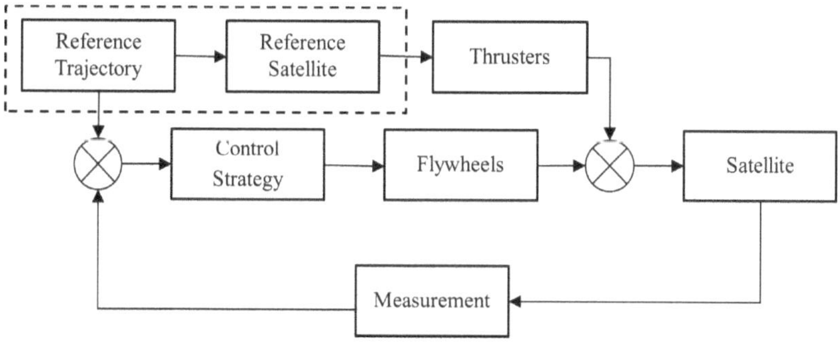

FIG. 11.1 Control organization for hybrid actuators.

planning process. In this section, the time-optimal attitude maneuver trajectory is adopted as the reference trajectory. Then the reference torque is obtained from the reference trajectory and implemented by thrusters. It works out to achieve rapid maneuver for satellite in the mode of feedforward compensation. In practical operation of the thrusters, there would inevitably be deviations in the output of the thrusters. The adverse effect such as mechanical error and measurement error would cause relatively large error between the actual maneuver trajectory and the reference trajectory of the satellite. The "control algorithm" module shown in Fig. 11.1 calculates the required correction torque according to the deviation between the reference trajectory and measured outputs. The flywheel provides the calculated control torque in the form of feedback compensation for the adverse effect above and to improve the attitude-tracking control precision during satellite maneuver.

11.2.2 Time-optimal trajectory planning

The spacecraft is modeled as a rigid body with hybrid thrusters-flywheel actuator that provide torque along three orthogonal axes of the body-fixed frame B. In order to make hybrid thruster-flywheel actuators work effectively, it is essential to design a reasonable control strategy to obtain the time-optimal maneuver trajectory. This is an optimal control problem with bounded thruster's output torque, constrained initial values, and constrained final values. Adopt quaternion as the attitude description then the attitude control issue in this chapter can be described as follows:

For three-axis stabilized satellite system equations,

$$\begin{cases} \boldsymbol{J}\dot{\boldsymbol{\omega}} = -\boldsymbol{\omega}^{\times}\boldsymbol{J}\boldsymbol{\omega} + \boldsymbol{T} + \boldsymbol{d} \\ \dot{\boldsymbol{q}} = \dfrac{1}{2}\begin{bmatrix} -\boldsymbol{q}_{\mathrm{v}}^{\mathrm{T}} \\ \boldsymbol{q}_{\mathrm{v}}^{\times} + q_0\boldsymbol{I}_{3\times 3} \end{bmatrix}\boldsymbol{\omega} = \dfrac{1}{2}\boldsymbol{E}(q)\boldsymbol{\omega} \end{cases} \tag{11.1}$$

where $\boldsymbol{\omega} = [\omega_1, \omega_2, \omega_3]^T$ is the angular rate of the spacecraft expressed in the body frame B with respect to the inertial frame I; $J \in R^{3\times 3}$ is the inertia matrix of the spacecraft; $T \in R^3$ is the torque control; $d \in R^3$ is the external disturbance. And the unit quaternion $q = (q_0, q_v)^T = (q_0, q_1, q_2, q_3)^T$.

Initial constraints:

$$\begin{cases} (q_0(0), q_1(0), q_2(0), q_3(0)) = (q_{0i}, q_{1i}, q_{2i}, q_{3i}) \\ (\omega_x(0), \omega_y(0), \omega_z(0)) = (0, 0, 0) \end{cases} \tag{11.2}$$

Final constraints:

$$\begin{cases} q_0(t_f), q_1(t_f), q_2(t_f), q_3(t_f)) = (q_{0f}, q_{1f}, q_{2f}, q_{3f}) \\ (\omega_x(t_f), \omega_y(t_f), \omega_z(t_f)) = (0, 0, 0) \end{cases} \tag{11.3}$$

Performance index:

$$J = t_f \tag{11.4}$$

Control constraints:

$$\boldsymbol{T} \in \boldsymbol{U} \tag{11.5}$$

Obviously in this time-optimal control problem, when relevant parameters of the satellite are determined, the reference satellite attitude and reference torque can be derived correspondingly. Subsequently, this torque can be used as the reference input to design control law of the thrusters.

11.2.3 Dynamics and kinematics description of thruster-flywheel hybrid actuators

In this section, kinematics and dynamics models of the hybrid thruster-flywheel actuators are built to satisfy needs in designing the finite-time sliding mode control laws.

In the dynamics model, define $\boldsymbol{J}_b$ as inertia matrix of the satellite body and $\boldsymbol{J}_{\mathrm{w}}$ as inertial matrix of reaction wheels which can be expanded as $\boldsymbol{J}_{\mathrm{w}} = \mathrm{diag}([\boldsymbol{J}_{\mathrm{w}1}, \boldsymbol{J}_{\mathrm{w}2}, \boldsymbol{J}_{\mathrm{w}3}])$.

The total angular momentum of the flywheels can be expressed as

$$\boldsymbol{h}_{\mathrm{w}} = \boldsymbol{J}_{\mathrm{w}} \left(\boldsymbol{A}^{\mathrm{T}} \boldsymbol{\omega}_{\mathrm{b}} + \boldsymbol{\omega}_{\mathrm{w}} \right) \tag{11.6}$$

where $\boldsymbol{\omega}_{\mathrm{b}}$ denotes the angular rate of the satellite expressed in the body frame $B$with respect to the inertia frame I; $\boldsymbol{A}$ represents the coefficient matrix of the flywheel with respect to the body frame B; $\boldsymbol{\omega}_{\mathrm{w}}$ represents the angular rates of the flywheels with respect to the body frame.

Therefore, the total angular momentum of the satellite can be expressed as

$$\boldsymbol{h} = \boldsymbol{h}_{\mathrm{b}} + \boldsymbol{A}\boldsymbol{h}_{\mathrm{w}} = \boldsymbol{J}_{\mathrm{b}}\boldsymbol{\omega}_{\mathrm{b}} + \boldsymbol{A}\boldsymbol{J}_{\mathrm{w}} \left(\boldsymbol{A}^{\mathrm{T}} \boldsymbol{\omega}_{\mathrm{b}} + \boldsymbol{\omega}_{\mathrm{w}} \right). \tag{11.7}$$

Letting $\boldsymbol{J} = \boldsymbol{J}_{\mathrm{b}} + \boldsymbol{A}\boldsymbol{J}_{\mathrm{w}}\boldsymbol{A}^{\mathrm{T}}$, the above formula can be rewritten as

$$\boldsymbol{h} = \boldsymbol{J}\boldsymbol{\omega}_{\mathrm{b}} + \boldsymbol{A}\boldsymbol{J}_{\mathrm{w}}\boldsymbol{\omega}_{\mathrm{w}}. \tag{11.8}$$

Then the satellite dynamics equations can be expressed as

$$\begin{cases} \dot{\boldsymbol{h}} = \boldsymbol{h}^{\times}\boldsymbol{J}_{\mathrm{b}}^{-1}(\boldsymbol{h} - \boldsymbol{A}\boldsymbol{h}_{\mathrm{w}}) + \boldsymbol{u}_{\mathrm{e}} \\ \dot{\boldsymbol{h}}_{\mathrm{w}} = \boldsymbol{u}_{\mathrm{w}} \end{cases} \tag{11.9}$$

where $\boldsymbol{u}_{\mathrm{e}}$ represents the control torque generated by thrusters and $\boldsymbol{u}_{\mathrm{w}}$ represents the control torque generated by flywheels.

The kinematics equation of the satellite can be expressed using quaternions as

$$\dot{\boldsymbol{q}} = 0.5\boldsymbol{E}(\boldsymbol{q})\boldsymbol{\omega}_{\mathrm{b}} \tag{11.10}$$

where operator $\boldsymbol{E}(\cdot):\boldsymbol{R}^4 \to \boldsymbol{R}^{4\times 3}$ is defined as

$$\boldsymbol{E}(\boldsymbol{q}) = \begin{bmatrix} -\boldsymbol{q}_{\mathrm{v}}^{\mathrm{T}} \\ \boldsymbol{q}_{\mathrm{v}}^{\times} + q_0\boldsymbol{I}_{3\times 3} \end{bmatrix}. \tag{11.11}$$

When designing the reference satellite attitude, we make the following assumption: only thrusters work and flywheels do not. Therefore, in the reference coordinate system, the attitude kinematics equation can be written as

$$\dot{\boldsymbol{h}}_{\mathrm{R}} = \boldsymbol{h}_{\mathrm{R}}^{\times}\boldsymbol{J}^{-1}\boldsymbol{h}_{\mathrm{R}} + \boldsymbol{u}_{\mathrm{R}} \tag{11.12}$$

where $\boldsymbol{u}_{\mathrm{R}}$ is the reference torque generated by thrusters. The calculation result of the formula (11.12) is the reference satellite attitude.

Further deducing formula (11.12) yields:

$$\dot{\boldsymbol{w}}_{\mathrm{R}} = \boldsymbol{J}^{-1}\boldsymbol{h}_{\mathrm{R}}^{\times}\boldsymbol{J}^{-1}\boldsymbol{h}_{\mathrm{R}} + \boldsymbol{J}^{-1}\boldsymbol{u}_{\mathrm{R}}. \tag{11.13}$$

The angular rate tracking error is written as

$$\delta\boldsymbol{\omega} = \boldsymbol{\omega}_{\mathrm{b}} - \boldsymbol{C}_{\mathrm{R}}^{\mathrm{B}}(\delta\boldsymbol{q})\boldsymbol{\omega}_{\mathrm{R}} \tag{11.14}$$

where $\boldsymbol{\omega}_{\mathrm{R}}$ is the reference angular rate; $\boldsymbol{C}_{\boldsymbol{R}}^{\boldsymbol{B}}(\boldsymbol{\delta q})$ is the transformation matrix of reference coordinate frame with respect to body coordinate frame.

Angular tracking error is defined as follow:

$$\begin{aligned} \delta\boldsymbol{h} &= \boldsymbol{h} - \boldsymbol{J}_{\mathrm{b}}\boldsymbol{C}_{\mathrm{R}}^{\mathrm{B}}(\delta\boldsymbol{q})\boldsymbol{\omega}_{\mathrm{R}} \\ &= \boldsymbol{J}_{\mathrm{b}}\boldsymbol{\omega}_{\mathrm{b}} + \boldsymbol{A}\boldsymbol{J}_{\mathrm{w}}\left(\boldsymbol{A}^{\mathrm{T}}\boldsymbol{\omega}_{\mathrm{b}} + \boldsymbol{\omega}_{\mathrm{w}}\right) - \boldsymbol{J}_{\mathrm{b}}\boldsymbol{C}_{\mathrm{R}}^{\mathrm{B}}(\delta\boldsymbol{q})\boldsymbol{\omega}_{\mathrm{R}} \\ &= \boldsymbol{J}_{\mathrm{b}}\left(\boldsymbol{\omega}_{\mathrm{b}} - \boldsymbol{C}_{\mathrm{R}}^{\mathrm{B}}(\delta\boldsymbol{q})\boldsymbol{\omega}_{\mathrm{R}}\right) + \boldsymbol{A}\boldsymbol{J}_{\mathrm{w}}\left(\boldsymbol{A}^{\mathrm{T}}\boldsymbol{\omega}_{\mathrm{b}} + \boldsymbol{\omega}_{\mathrm{w}}\right) \\ &= \boldsymbol{J}_{\mathrm{b}}\delta\boldsymbol{\omega} + \boldsymbol{A}\boldsymbol{h}_{\mathrm{w}} \end{aligned} \tag{11.15}$$

Substituting Eqs. (11.13) and (11.14) into Eq. (11.15) yields:

$$\begin{aligned} \delta\dot{\boldsymbol{h}} &= \dot{\boldsymbol{h}} - \boldsymbol{J}_{\mathrm{b}}\frac{d\boldsymbol{C}_{\mathrm{R}}^{\mathrm{B}}(\delta\boldsymbol{q})}{dt}\boldsymbol{\omega}_{\mathrm{R}} - \boldsymbol{J}_{\mathrm{b}}\boldsymbol{C}_{\mathrm{R}}^{\mathrm{B}}(\delta\boldsymbol{q})\dot{\boldsymbol{\omega}}_{\mathrm{R}} \\ &= \boldsymbol{h}^{\times}\boldsymbol{J}_{\mathrm{b}}^{-1}(\boldsymbol{h} - \boldsymbol{A}\boldsymbol{h}_{\mathrm{w}}) + \boldsymbol{u}_{\mathrm{e}} - \boldsymbol{J}_{\mathrm{b}}\boldsymbol{\omega}_{\mathrm{b}}^{\times}\delta\boldsymbol{\omega} - \boldsymbol{J}_{\mathrm{b}}\boldsymbol{C}_{\mathrm{R}}^{\mathrm{B}}(\delta\boldsymbol{q})\boldsymbol{J}^{-1}\boldsymbol{h}_{\mathrm{R}}^{\times}\boldsymbol{J}^{-1}\boldsymbol{h}_{\mathrm{R}} - \boldsymbol{J}_{\mathrm{b}}\boldsymbol{C}_{\mathrm{R}}^{\mathrm{B}}(\delta\boldsymbol{q})\boldsymbol{J}^{-1}\boldsymbol{u}_{\mathrm{R}} \end{aligned} \tag{11.16}$$

11.3 Design of controllers

To construct fundamentals on which the design and proof of the subsequent controller can be based, the following lemma and assumption are given first and the mathematical expression of the upper bound of the output torque of the thrusters is given at the same time.

Consider the following nonlinear system:

$$\dot{\boldsymbol{x}} = f(\boldsymbol{x}), \quad \boldsymbol{x} \in \boldsymbol{R}^n \tag{11.17}$$

where $f(\boldsymbol{x})$ denotes a continuous function and $f(0)=0$. If system (11.17) satisfies Lyapunov stability condition, and there exists a time function $T(\boldsymbol{x})$ such that for all $t \geq T(\boldsymbol{x})$ we have $\boldsymbol{x}(t)=0$, then system (11.17) would be able to converge to zero in finite time.

Lemma 11.1 [15]: Considering system (11.17), if there exists a continuous positive function $\boldsymbol{V}$ on the open set Φ which contains the origin such that $\dot{\boldsymbol{V}} + k\boldsymbol{V}^{\alpha} \leq 0$, where $k>0$ and $\alpha \in (0,1)$. Then the origin of (17) is finite-time stable, and an estimation of the consumed time is $T(x) \leq (k(1-\alpha))^{-1}\boldsymbol{V}(x_0)^{1-\alpha}$.

Lemma 11.2 [16]: If $a \in (0,1)$, then there is

$$\sum_{i=1}^{3} |x_i|^{1+a} \geq \left(\sum_{i=1}^{3} |x_i|^2\right)^{\frac{1+a}{2}} \tag{11.18}$$

Assumption 11.1 There exists a positive constant $d_i^{\max} > 0$, which makes each component of the external disturbance torque vector d_i satisfies the following inequality:

$$|d_i| \leq d_i^{\max}, \quad i = 1,2,3 \tag{11.19}$$

From Eq. (11.19), the external disturbance can be obtained as

$$\|\boldsymbol{d}\| \leq \sqrt{\sum_{i=1}^{3} \left(d_i^{\max}\right)^2} \triangleq d_{\max} \tag{11.20}$$

Considering practical situation, there also exist errors in the output torque of the thrusters. The relationship between the thrusters output torque and the action time satisfy:

$$t\overline{\boldsymbol{u}}_{\mathrm{e}} = \boldsymbol{u}_{\mathrm{e}i} \tag{11.21}$$

where $\overline{\boldsymbol{u}}_{\mathrm{e}}$ is the constant torque of pulsed thrusters; t is the periodic working time of the thruster; and $\boldsymbol{u}_{\mathrm{e}i}$ is the equivalent output torque within one period. $\Delta\overline{u}_{\mathrm{e}i}$ is defined as the constant value of the output torque deviation, then thruster deviations can be described as $d_i = t\Delta\overline{u}_{\mathrm{e}i} = u_{\mathrm{e}i}/\overline{\boldsymbol{u}}_{\mathrm{e}} \cdot \Delta\overline{u}_{\mathrm{e}i} = v_i\Delta\overline{u}_{\mathrm{e}i}$, where v_i is a positive number with a upper bound.

11.3.1 Design of terminal sliding mode control law

According to the dynamics and kinematics model of the thruster-flywheel hybrid actuators in Section 11.2.3, sliding surface of the variable structure control can be chosen as

$$s = \delta\omega + \beta sig(\delta q_v)^r, \quad 0 < r < 1 \tag{11.22}$$

where $\mathrm{sig}(\delta q_v)^r = \left[|\delta q_1|^r \mathrm{sign}(\delta q_1) \; |\delta q_2|^r \mathrm{sign}(\delta q_2) \; |\delta q_3|^r \mathrm{sign}(\delta q_3)\right]^T$. And $\lambda_{\max}$ is the maximum eigenvalue of s.

Consider the following Lyapunov function candidate:

$$V(s) = 0.5 s^{\mathrm{T}} J_{\mathrm{b}} s \tag{11.23}$$

Taking-time derivative of function (11.23) yields:

$$\dot{V}(s) = s^T J_{\mathrm{b}} \dot{s} \tag{11.24}$$

$$\begin{aligned} J_{\mathrm{b}}\dot{s} &= J_{\mathrm{b}}\left[J_{\mathrm{b}}^{-1}\left(\delta\dot{h} - A\dot{h}_{\mathrm{w}}\right) + \beta r|\delta q_{\mathrm{v}}|^{r-1}\delta\dot{q}_{\mathrm{v}}\right] \\ &= \delta\dot{h} - A\dot{h}_{\mathrm{w}} + \beta r J_{\mathrm{b}}|\delta q_{\mathrm{v}}|^{r-1}\delta\dot{q}_{\mathrm{v}} \end{aligned} \tag{11.25}$$

Thus

$$\begin{aligned} \dot{V}(s) &= s^{\mathrm{T}}\left(\delta\dot{h} - A\dot{h}_{\mathrm{w}} + \beta r J_{\mathrm{b}}|\delta q_{\mathrm{v}}|^{r-1}\delta\dot{q}_{\mathrm{v}}\right) \\ &= s^{\mathrm{T}}\left(h^{\times}J_{\mathrm{b}}^{-1}\left(\delta\dot{h} - A\dot{h}_{\mathrm{w}}\right) + u_{\mathrm{e}} + d - J_{\mathrm{b}}\omega_{\mathrm{b}}^{\times}\delta\omega - \right. \\ & J_{\mathrm{b}}C_{\mathrm{R}}^{\mathrm{B}}(\delta q_{\mathrm{v}})J^{-1}h_{\mathrm{R}}^{\times}J^{-1}h_{\mathrm{R}} - \\ & J_{\mathrm{b}}C_{\mathrm{R}}^{\mathrm{B}}(\delta q_{\mathrm{v}})J^{-1}u_{\mathrm{R}}) - \\ & A u_{\mathrm{w}} + \beta r J_{\mathrm{b}}|\delta q_{\mathrm{v}}|^{r-1}\delta\dot{q} \end{aligned} \tag{11.26}$$

Remark 11.1 In order to make the entire control system stable, the design of the output torque u_{e} and u_{w} must satisfy $\dot{V} < 0$.

In accordance with the design of the reference satellite attitude where flywheels do not work, the reference torque is derived from the reference trajectory, and the specific form of the output torque of the thrusters should be composed by h_{R} and u_{R}, which can be written as:

$$u_{\mathrm{e}} = J_{\mathrm{b}}C_{\mathrm{R}}^{\mathrm{B}}(\delta q_{\mathrm{v}})J^{-1}h_{\mathrm{R}}^{\times}J^{-1}h_{\mathrm{R}} + J_{\mathrm{b}}C_{\mathrm{R}}^{\mathrm{B}}(\delta q_{\mathrm{v}})J^{-1}u_{\mathrm{R}} \tag{11.27}$$

Considering the effect of initial measurement error, thrusters output deviation, and external disturbances, the actual attitude of the satellite will deviate from the reference attitude trajectory, which requires flywheels to help correct the attitude-tracking errors. The control law of flywheels is designed as follows:

$$A u_{w} = h^{\times}J_{\mathrm{b}}^{-1}\left(\delta\dot{h} - A\dot{h}_{\mathrm{w}}\right) - J_{\mathrm{b}}\omega_{\mathrm{b}}^{\times}\delta\omega + \beta r J_{\mathrm{b}}|\delta q_{\mathrm{v}}|^{r-1}\delta\dot{q}_{\mathrm{v}} + K_1 s + \lambda\,\mathrm{sgn}\,(s) \tag{11.28}$$

Proof: Substituting (11.27) and (11.28) into (11.26) yields:

$$\begin{aligned}\dot{V}(\boldsymbol{s}) &= \boldsymbol{s}^{\mathrm{T}}[-\boldsymbol{K}_1\boldsymbol{s} - \lambda\,\mathrm{sgn}\,(\boldsymbol{s}) + \boldsymbol{d}] \\ &= -\boldsymbol{s}^{\mathrm{T}}\boldsymbol{K}_1\boldsymbol{s} - \lambda\boldsymbol{s}^{\mathrm{T}}\,\mathrm{sgn}\,(\boldsymbol{s}) + \boldsymbol{s}^{T}\boldsymbol{d} \\ &< -\lambda\boldsymbol{s}^{\mathrm{T}}\,\mathrm{sgn}\,(\boldsymbol{s}) + \boldsymbol{s}^{\mathrm{T}}\boldsymbol{d} \\ &\le -\lambda\|\boldsymbol{s}\| + \|\boldsymbol{s}\|\|\boldsymbol{d}\| \\ &\le -\sqrt{2}(\lambda - d_{\max})\lambda_{\max}^{-1/2}(\boldsymbol{J}_{\mathrm{b}})V^{1/2}(\boldsymbol{s})\end{aligned} \tag{11.29}$$

According to Lemma 11.1, the state variables of system would converge to the sliding surface in finite time.

On the sliding surface $s=0$, there exists

$$\boldsymbol{\delta\omega} + \rho sig(\boldsymbol{\delta q}_{\mathrm{v}})^{r} = 0 \tag{11.30}$$

Then select the Lyapunov function candidate as

$$V_1 = \boldsymbol{\delta q}_{\mathrm{v}}^{\mathrm{T}}\boldsymbol{\delta q}_{\mathrm{v}} + (1 - \delta q_0)^2 \tag{11.31}$$

Taking-time derivative of function (11.31) yields

$$\dot{V}_1 = -2\delta\dot{q}_0 = \boldsymbol{\delta q}_{\mathrm{v}}^{\mathrm{T}}\boldsymbol{\delta\omega} \tag{11.32}$$

Substituting (11.30) into (11.32) yields:

$$\begin{aligned}\dot{V}_1 &= -\boldsymbol{\beta}\boldsymbol{\delta q}_{\mathrm{v}}^{\mathrm{T}}|\boldsymbol{\delta q}_{\mathrm{v}}|^{r}\,\mathrm{sgn}\,(\boldsymbol{\delta q}_{\mathrm{v}}) \\ &= -\boldsymbol{\beta}\sum_{i=1}^{3}|\boldsymbol{\delta q}_i|^{r+1} \\ &\le -\boldsymbol{\beta}\left(\sum_{i=1}^{3}\boldsymbol{\delta q}_i^2\right)^{\frac{r+1}{2}} \\ &= -\boldsymbol{\beta}\left(\boldsymbol{\delta q}_{\mathrm{v}}^{\mathrm{T}}\boldsymbol{\delta q}_{\mathrm{v}}\right)^{\frac{r+1}{2}}\end{aligned} \tag{11.33}$$

And noting that

$$\begin{aligned}V_1 &= \boldsymbol{\delta q}_{\mathrm{v}}^{\mathrm{T}}\boldsymbol{\delta q}_{\mathrm{v}} + (1 - \delta q_0)^2 \\ &\le \boldsymbol{\delta q}_{\mathrm{v}}^{\mathrm{T}}\boldsymbol{\delta q}_{\mathrm{v}} + (1 - \delta q_0)(1 + \delta q_0) \\ &= \boldsymbol{\delta q}_{\mathrm{v}}^{\mathrm{T}}\boldsymbol{\delta q}_{\mathrm{v}} + 1 - \delta q_0^2 \\ &= 2\boldsymbol{\delta q}_{\mathrm{v}}^{\mathrm{T}}\boldsymbol{\delta q}_{\mathrm{v}}\end{aligned} \tag{11.34}$$

It can be easily obtained that:

$$\boldsymbol{\delta q}_{\mathrm{v}}^{\mathrm{T}}\boldsymbol{\delta q}_{\mathrm{v}} \ge \frac{1}{2}V_1 \tag{11.35}$$

Substituting (11.35) into (11.33) yields

$$\dot{V}_1 \le -\boldsymbol{\beta}\left(\frac{1}{2}V\right)^{\frac{r+1}{2}} = -\boldsymbol{\beta}\left(\frac{1}{2}\right)^{\frac{r+1}{2}}V_1^{\frac{r+1}{2}} \tag{11.36}$$

$$\dot{V}_1 + \beta\left(\frac{1}{2}\right)^{\frac{r+1}{2}} V_1^{\frac{r+1}{2}} \leq 0 \tag{11.37}$$

which satisfies the finite-time convergence condition in Lemma 11.1. The hybrid actuators control strategy consist of (11.27) and (11.28) can help satellite control system track reference attitude trajectory in finite time. However, due to the existence of $|\delta \boldsymbol{q}_{\mathrm{v}}|^{r-1}$ in the control law $\boldsymbol{A}\boldsymbol{u}_{\mathrm{w}}$, there will be singular at zero crossing. This problem will be investigated in Section 11.3.2.

11.3.2 Design of non-singular terminal sliding-mode control law

To address the singularity problem in the control process, a nonsingular terminal sliding-mode control law is proposed under the condition that there exist an invertible matrix $\boldsymbol{F}=\delta q_0 \boldsymbol{E}_{3\times 3}+\delta \boldsymbol{q}_{\mathrm{v}}^{\times}$.

The proposed nonsingular sliding-mode surface of variable structure control is chosen as [3]

$$s_1 = \delta \boldsymbol{q}_{\mathrm{v}} + a sig(\delta \dot{\boldsymbol{q}}_{\mathrm{v}})^{r_1}, \quad 1 < r_1 < 2 \tag{11.38}$$

Consider the following Lyapunov function candidate:

$$V_2(\boldsymbol{s}) = \frac{1}{2} \boldsymbol{s}_1{}^{\mathrm{T}} \boldsymbol{s}_1 \tag{11.39}$$

Taking time derivative of function V_2 yields

$$\begin{aligned}
\dot{V}_2(\boldsymbol{s}) &= \boldsymbol{s}_1{}^{\mathrm{T}} \dot{\boldsymbol{s}}_1 = \boldsymbol{s}_1{}^{\mathrm{T}}\left(\delta \dot{\boldsymbol{q}}_{\mathrm{v}} + ar|\delta \dot{\boldsymbol{q}}_{\mathrm{v}}|^{r-1} \delta \ddot{\boldsymbol{q}}_{\mathrm{v}}\right) \\
&= \boldsymbol{s}_1{}^{\mathrm{T}}\left[\delta \dot{\boldsymbol{q}}_{\mathrm{v}} + \frac{1}{2} ar|\delta \dot{\boldsymbol{q}}_{\mathrm{v}}|^{r-1}(\dot{\boldsymbol{F}}\delta\boldsymbol{\omega} + \boldsymbol{F}\delta\dot{\boldsymbol{\omega}})\right] \\
&= \boldsymbol{s}_1{}^{\mathrm{T}}\left[\delta \dot{\boldsymbol{q}}_{\mathrm{v}} + \frac{1}{2} ar|\delta \dot{\boldsymbol{q}}_{\mathrm{v}}|^{r-1}(\dot{\boldsymbol{F}}\delta\boldsymbol{\omega} + \right. \\
&\quad \boldsymbol{F}\boldsymbol{J}_{\mathrm{b}}^{-1}(\boldsymbol{h}^{\times}\boldsymbol{J}_{\mathrm{b}}^{-1}(\delta\boldsymbol{h} - \boldsymbol{A}\boldsymbol{h}_{\mathrm{w}}) + \boldsymbol{u}_{\mathrm{e}1} + \boldsymbol{d} - \boldsymbol{J}_{\mathrm{b}}\boldsymbol{\omega}_{\mathrm{b}}^{\times}\delta\boldsymbol{\omega} - \\
&\quad \left. \boldsymbol{J}_{\mathrm{b}}\boldsymbol{C}_{\mathrm{R}}^{\mathrm{B}}(\delta\boldsymbol{q}_{\mathrm{v}})\boldsymbol{J}^{-1}\boldsymbol{h}_{\mathrm{R}}^{\times}\boldsymbol{J}^{-1}\boldsymbol{h}_{\mathrm{R}} - \boldsymbol{J}_{\mathrm{b}}\boldsymbol{C}_{\mathrm{R}}^{\mathrm{B}}(\delta\boldsymbol{q}_{\mathrm{v}})\boldsymbol{J}^{-1}\boldsymbol{u}_{\mathrm{R}} - \boldsymbol{A}\boldsymbol{u}_{\mathrm{w}1}))\right]
\end{aligned} \tag{11.40}$$

Remark 11.2 In order to make the entire control system stable, the design of the output torque $\boldsymbol{u}_{\mathrm{e}1}$ and $\boldsymbol{A}\boldsymbol{u}_{\mathrm{w}1}$ must satisfy $\dot{V}_2 < 0$.

In accordance with the design of the reference satellite attitude where flywheels do not work, the reference torque is derived from the reference trajectory, and the specific form of the output torque of the thrusters should be composed by $\boldsymbol{h}_{\mathrm{R}}$ and $\boldsymbol{u}_{\mathrm{R}}$, which is

$$\boldsymbol{u}_{\mathrm{e}1} = \boldsymbol{J}_{\mathrm{b}}\boldsymbol{C}_{\mathrm{R}}^{\mathrm{B}}(\delta\boldsymbol{q}_{\mathrm{v}})\boldsymbol{J}^{-1}\boldsymbol{h}_{\mathrm{R}}^{\times}\boldsymbol{J}^{-1}\boldsymbol{h}_{\mathrm{R}} + \boldsymbol{J}_{\mathrm{b}}\boldsymbol{C}_{\mathrm{R}}^{\mathrm{B}}(\delta\boldsymbol{q}_{\mathrm{v}})\boldsymbol{J}^{-1}\boldsymbol{u}_{\mathrm{R}} \tag{11.41}$$

Due to initial measurement error, thrusters output deviations and external disturbance, the practical attitude of the satellite will deviate from the reference attitude trajectory, which requires the flywheels to exert precise control to correct the attitude-tracking errors. The control law of flywheels is chosen as follow:

$$\begin{aligned} \boldsymbol{A}\boldsymbol{u}_{\mathrm{w}1} &= \boldsymbol{h}^{\times}\boldsymbol{J}_{\mathrm{b}}^{-1}(\boldsymbol{\delta h} - \boldsymbol{A}\boldsymbol{h}_{\mathrm{w}}) - \boldsymbol{J}_{\mathrm{b}}\boldsymbol{\omega}_{\mathrm{b}}^{\times}\boldsymbol{\delta\omega} + \boldsymbol{J}_{\mathrm{b}}\boldsymbol{F}^{-1}\dot{\boldsymbol{F}}\boldsymbol{\delta\omega} \\ &+ \frac{2}{\alpha r}\boldsymbol{J}_{\mathrm{b}}\boldsymbol{F}^{-1}\mathrm{sig}(\boldsymbol{\delta\dot{q}}_{\mathrm{v}})^{2-r} + \rho\boldsymbol{J}_{\mathrm{b}}\boldsymbol{F}^{-1}\,\mathrm{sgn}\,(\boldsymbol{s}) \end{aligned} \tag{11.42}$$

Proof: Substituting (11.41) and (11.42) into (11.40) yields

$$\begin{aligned} \dot{V}_{2}(s) &= \boldsymbol{s}_1{}^{\mathrm{T}}\dot{\boldsymbol{s}}_1 = \boldsymbol{s}_1{}^{\mathrm{T}}\left(\boldsymbol{\delta\dot{q}}_{\mathrm{v}} + ar|\boldsymbol{\delta\dot{q}}_{\mathrm{v}}|^{r-1}\boldsymbol{\delta\ddot{q}}_{\mathrm{v}}\right) \\ &= \boldsymbol{s}_1{}^{\mathrm{T}}\left[\boldsymbol{\delta\dot{q}}_{\mathrm{v}} + \frac{1}{2}ar|\boldsymbol{\delta\dot{q}}_{\mathrm{v}}|^{r-1}\left(\dot{\boldsymbol{F}}\boldsymbol{\delta\omega} + \boldsymbol{F}\boldsymbol{\delta\dot{\omega}}\right)\right] \\ &= \boldsymbol{s}_1{}^{\mathrm{T}}\Big[\boldsymbol{\delta\dot{q}}_{\mathrm{v}} + \frac{1}{2}ar|\boldsymbol{\delta\dot{q}}_{\mathrm{v}}|^{r-1}(\dot{\boldsymbol{F}}\boldsymbol{\delta\omega} + \\ &\quad \boldsymbol{F}\boldsymbol{J}_{\mathrm{b}}^{-1}(\boldsymbol{h}^{\times}\boldsymbol{J}_{\mathrm{b}}^{-1}(\boldsymbol{\delta h} - \boldsymbol{A}\boldsymbol{h}_{\mathrm{w}}) + \boldsymbol{u}_{\mathrm{e}1} + \boldsymbol{d} - \boldsymbol{J}_{\mathrm{b}}\boldsymbol{\omega}_{\mathrm{b}}^{\times}\boldsymbol{\delta\omega} - \\ &\quad \boldsymbol{J}_{\mathrm{b}}\boldsymbol{C}_{\mathrm{R}}^{\mathrm{B}}(\boldsymbol{\delta q}_{\mathrm{v}})\boldsymbol{J}^{-1}\boldsymbol{h}_{\mathrm{R}}^{\times}\boldsymbol{J}^{-1}\boldsymbol{h}_{\mathrm{R}} - \boldsymbol{J}_{\mathrm{b}}\boldsymbol{C}_{\mathrm{R}}^{\mathrm{B}}(\boldsymbol{\delta q}_{\mathrm{v}})\boldsymbol{J}^{-1}\boldsymbol{u}_{\mathrm{R}} - \boldsymbol{A}\boldsymbol{u}_{\mathrm{w}1}))\Big] \end{aligned} \tag{11.43}$$

where $\rho - J_{\max}d_{\max} > 0$.

When $\delta\dot{q}_i \neq 0 (i = 1, 2, 3)$, $\dot{V} < 0$. Then the state variables of system can converge to the sliding mode in finite time.

When $\delta\dot{q}_i = 0 (i = 1, 2, 3)$, substituting (11.41) and (11.42) into (11.40) yields

$$\boldsymbol{\delta\ddot{q}}_{\mathrm{v}} = -\frac{2}{\alpha r}|\boldsymbol{\delta\dot{q}}_{\mathrm{v}}|^{2-r} - \rho\,\mathrm{sgn}\,(\boldsymbol{s}) + \boldsymbol{F}\boldsymbol{J}^{-1}\boldsymbol{d}(\text{not a fixed value}) \tag{11.44}$$

Thus $\delta\dot{q}_v = 0$ cannot be established all the time and the state variables of system can converge to the sliding mode in finite time.

On the sliding surface, $s_1 = 0$ can be written as

$$\boldsymbol{\delta q}_{\mathrm{v}} + \alpha sig(\boldsymbol{\delta\dot{q}}_{\mathrm{v}})^{r} = 0 \tag{11.45}$$

Then,

$$\boldsymbol{\delta\dot{q}}_{\mathrm{v}} = -\left(\frac{1}{\alpha}\right)^{\frac{1}{r}} sig(\boldsymbol{\delta q}_{\mathrm{v}})^{\frac{1}{r}} \tag{11.46}$$

Consider the following Lyapunov function candidate:

$$V_3 = \frac{1}{2}\boldsymbol{\delta q}_{\mathrm{v}}^{\mathrm{T}}\boldsymbol{\delta q}_{\mathrm{v}} \tag{11.47}$$

Taking time derivative of function V_3 yields

$$
\begin{aligned}
\dot{V}_3 &= \delta \boldsymbol{q}_{\mathrm{v}}^{\mathrm{T}} \delta \dot{\boldsymbol{q}}_{\mathrm{v}} \\
&= -\left(\frac{1}{\alpha}\right)^{\frac{1}{r}} \sum_{i=1}^{3} |\delta \boldsymbol{q}_i|^{1+\frac{1}{r}} \\
&\le -\left(\frac{1}{\alpha}\right)^{\frac{1}{r}} \left(\sum_{i=1}^{3} \delta \boldsymbol{q}_i^2\right)^{\frac{1+\frac{1}{r}}{2}} \\
&= -\left(\frac{1}{\alpha}\right)^{\frac{1}{r}} 2^{\frac{1+\frac{1}{r}}{2}} V^{\frac{1+\frac{1}{r}}{2}}
\end{aligned}
\tag{11.48}
$$

Then,

$$
\dot{V}_3 + \left(\frac{1}{\alpha}\right)^{\frac{1}{r}} 2^{\frac{1+\frac{1}{r}}{2}} V_3^{\frac{1+\frac{1}{r}}{2}} \le 0 \tag{11.49}
$$

From (11.49), it can be obtained that this control method satisfies the finite-time convergence condition in Lemma 11.1. Therefore, the attitude-tracking error would converge to zero and be stabilized within finite time after applying this new hybrid-actuator control strategy.

Remark 11.3 In addition, both controllers in Eqs. (11.41) and (11.42) have simple structures, which consist of feed forward torque and feedback torque, so it has low on-line computational burden and can be directly applied in space.

11.4 Simulation results

In this section, numerical simulations of the proposed control laws under MALAB/SIMULINK environment are implemented to demonstrate its performance in practical situation.

The moment of inertia of the satellite body is chosen as $\boldsymbol{J}_{\mathrm{b}} = \mathrm{diag}(150, 120, 200)\mathrm{kg} \cdot \mathrm{m}^2$. Three flywheels are installed orthogonally, and the moment of inertia of each flywheel is $0.06\ \mathrm{kg} \cdot \mathrm{m}^2$. The initial speed of the flywheel is $0r/\mathrm{min}$.

In this control mission, the satellite is required to implement a rapid rotation from initial attitude to the desired attitude. The initial attitude quaternion is given by $[1\ \ 0\ \ 0\ \ 0]$, the target attitude quaternion is $[0.9305\ \ 0.2028\ \ 0.2280\ \ 0.2028]$, and its corresponding Euler angle is $[30^\circ\ \ 20^\circ\ \ 30^\circ]$. In order to verify the performance of the control method when the initial attitude of the controller deviates from the normal value, it is assumed that the Euler angle of the initial attitude of the satellite is $[0.5^\circ\ \ -0.3^\circ\ \ 0.4^\circ]$, and the initial angular rate is $[-0.01 - 0.010.01](^\circ)/\mathrm{s}$. $\overline{v} = 0.05$ is defined as the maximum value of the ratio between the disturbance torque and the constant torque generated by thrusters. At the beginning of each simulation, v should be derived as a random vector from function $0.1\ \mathrm{rand}(3,1) - 0.05$. The maximum control torque of a flywheel is chosen as $[0.06\quad 0.06\quad 0.06]\mathrm{N}\cdot\mathrm{m}$.

In the TSM control law, controller parameters are chosen as $\boldsymbol{K}_1 = \mathrm{diag}(40,40,40)$, $\boldsymbol{\beta} = \mathrm{diag}(0.8,0.8,0.8)$, $r = 0.5$, and $\lambda = 0.05$.

In the nonsingular NSTSM control law, controller parameters are chosen as $a = 8$, $\rho = 0.05$, and $r = 1.5$.

The time-optimal satellite reference attitude trajectory and reference torque are derived from the GPOPS program. GPOPS is an open source program on the MATLAB platform aiming at solving the optimal control problem, and it uses Radau pseudo-spectral method and Lagrange polynomial approximation method to transform the optimal problem of continuous time into nonlinear programming problem of finite dimension [17].

The simulation results of applying sliding mode control law (11.27) and (11.28) to the attitude-tracking task are shown in Figs. 11.2–11.7. Fig. 11.7 shows the torque performed by flywheels. A jumping change can be seen in the flywheel control torque as predicted in the theoretical analysis. Therefore, the TSM controller is not suitable for practical engineering applications.

Figs. 11.8–11.13 show the simulation results of the proposed NSTSM control law. From Figs. 11.8 and 11.9, it can be seen that the controller smoothly complete the time-optimal attitude maneuver in finite time. Figs. 11.10 and

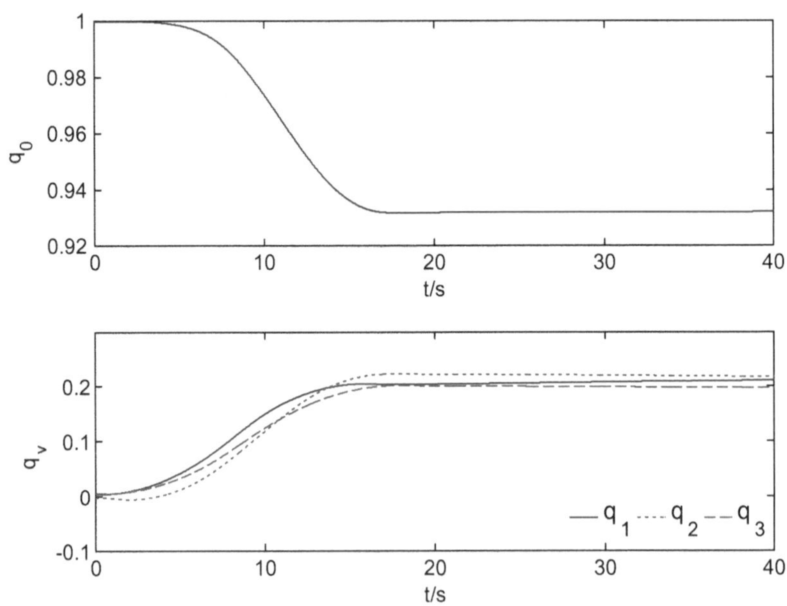

FIG. 11.2 Attitude of the satellite.

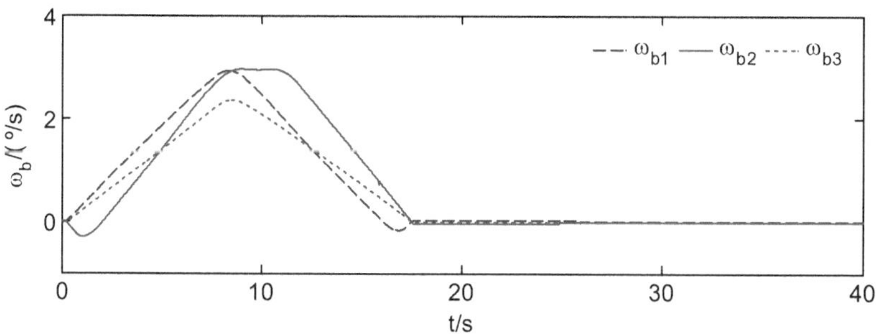

FIG. 11.3 Angular rate of the satellite.

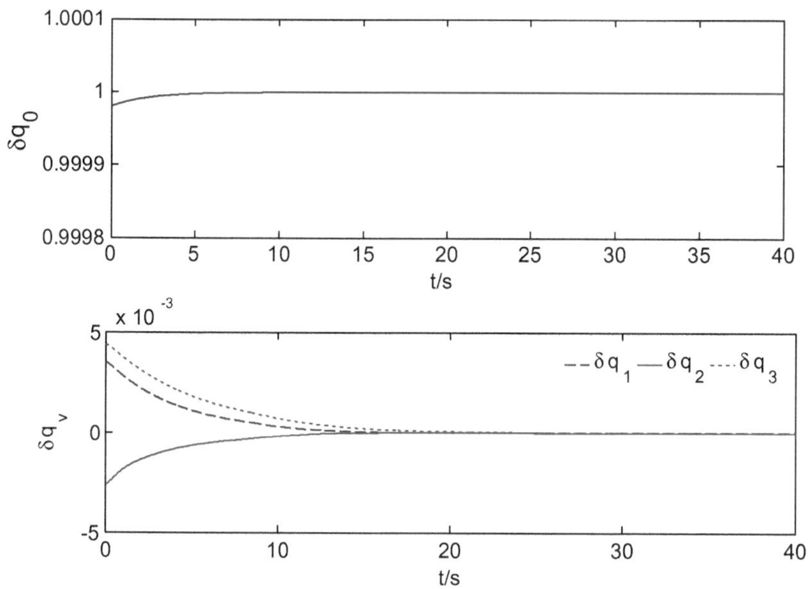

FIG. 11.4 Attitude-tracking error.

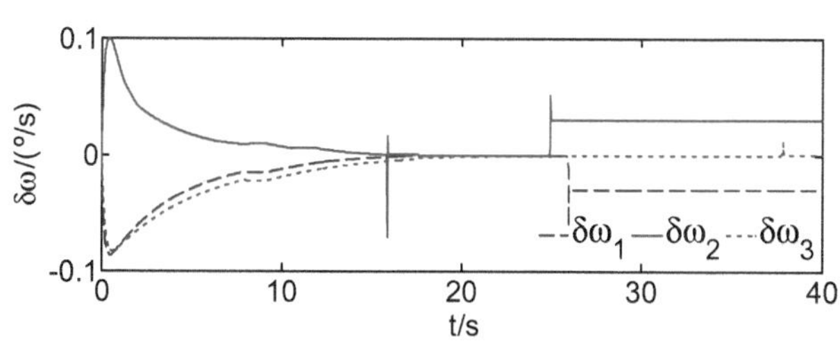

FIG. 11.5 Angular rate tracking error.

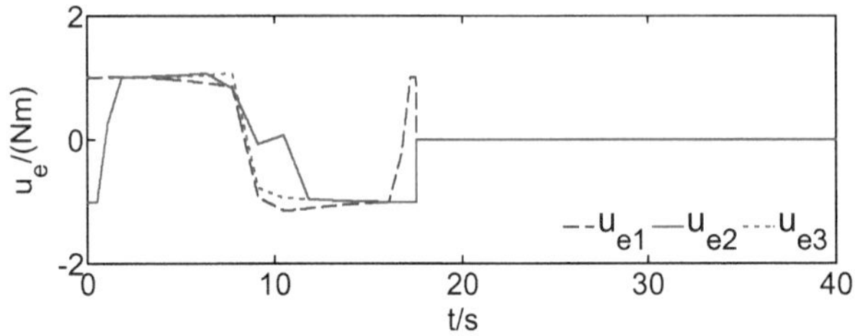

FIG. 11.6 Torque performed by thrusters.

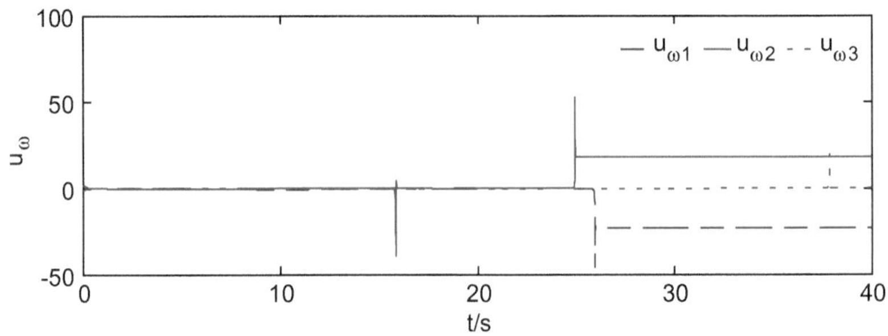

FIG. 11.7 Torque performed by flywheels.

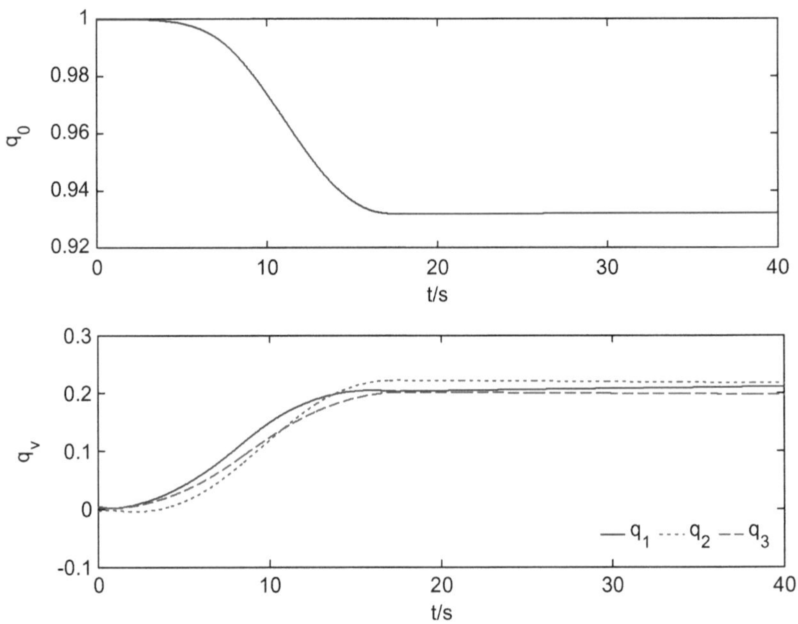

FIG. 11.8 Attitude of the satellite.

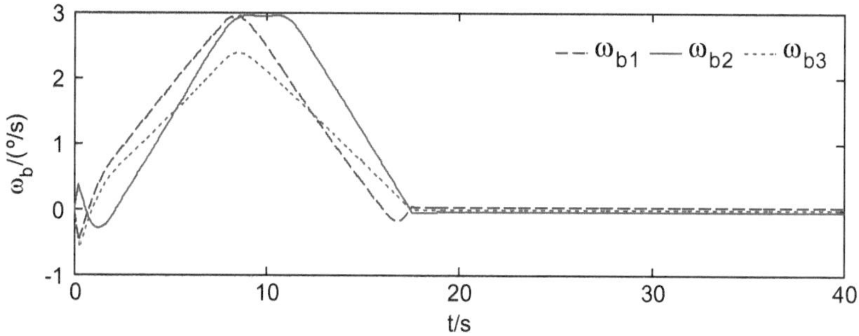

FIG. 11.9 Angular rate of the satellite.

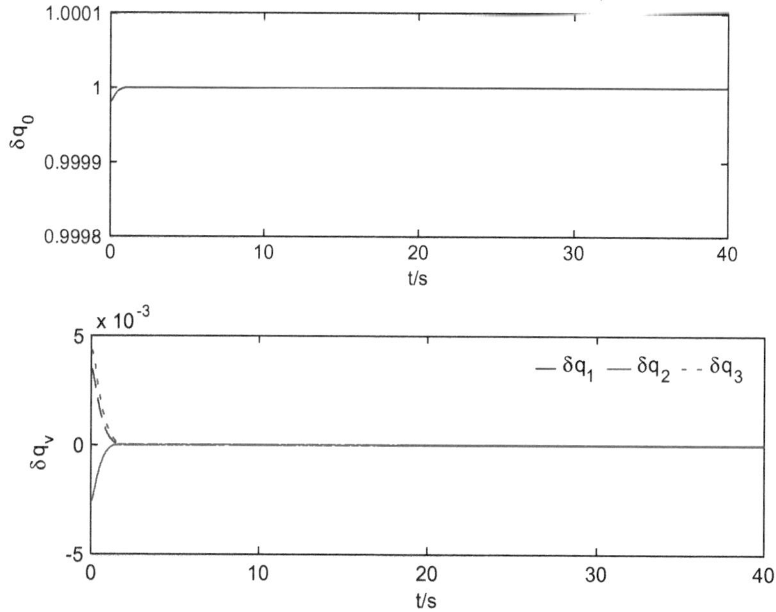

FIG. 11.10 Attitude-tracking error.

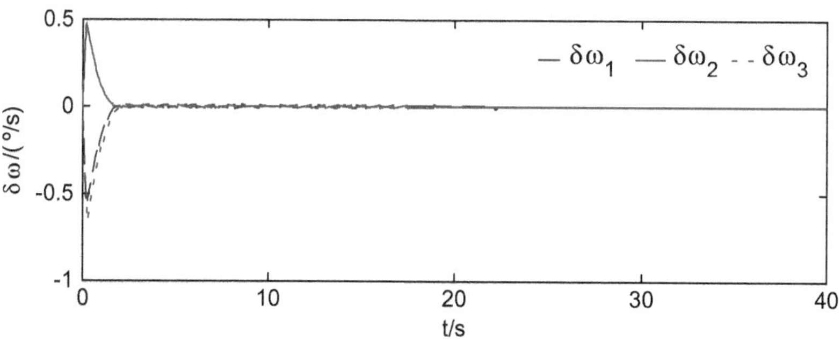

FIG. 11.11 Angular rate tracking error.

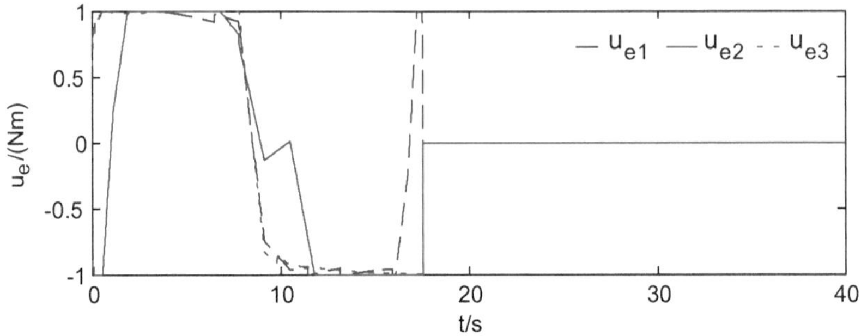

FIG. 11.12 Torque performed by thrusters.

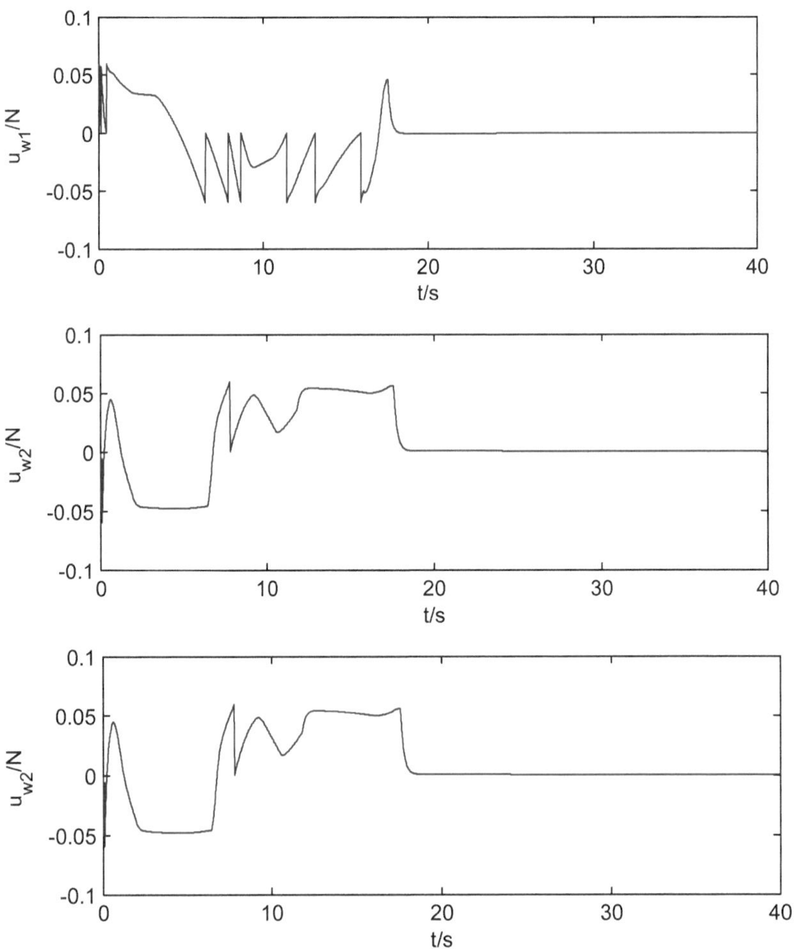

FIG. 11.13 Torque performed by flywheels.

11.11 show that the system states can track the satellite reference attitude in finite time with high precision. As shown in Fig. 11.13, the control torque of each flywheel has been maintained at 0.06 or less, and the singularity problem is avoided at the same time, both of which indicate the practicability of the designed control law.

11.5 Conclusion

In this chapter, the finite-time satellite attitude maneuver problem is investigated based on hybrid thruster-flywheel actuators controlling strategy. The model of satellite dynamics and kinematics based on hybrid actuators are built under consideration of output torque deviation and initial measurement error. A new NSTSM control law is proposed based on this model. This control law guarantees the stability of the closed-loop system and realizes the finite-time attitude tracking of the desired optimal-time trajectory. Finally, simulations are implemented to demonstrate the performance of this control law. The results show that this algorithm occupies high convergence speed, high control precision, and it has practical engineering application value.

References

[1] G.M. Li, L.D. Liu, Finite time stabilization method for the rigid spacecraft attitude control, Aerosp. Control Appl. 37 (3) (2011) 1–8.

[2] Y. Guo, S.M. Song, Adaptive finite-time backstepping control for attitude tracking of spacecraft based on rotation matrix, Chin. J. Aeronaut. 27 (2) (2014) 375–382.

[3] S.P. Bhat, D.S. Bernstein, Finite-time stability of continuous autonomous systems, Siam J. Control Optim. 38 (3) (2006) 751–766.

[4] P.M. Tiwari, S. Janardhanan, M. Un Nabi, A finite-time convergent continuous time sliding mode controller for spacecraft attitude control, in: International Workshop on Variable Structure Systems, 2010, pp. 399–403.

[5] B. Xiao, S. Yin, L.G. Wu, A structure simple controller for satellite attitude tracking maneuver, IEEE Trans. Ind. Electron. 64 (2) (2016) 1436–1446.

[6] A.M. Zou, K.D. Kumar, Z.G. Hou, et al., Finite-time attitude tracking control for spacecraft using terminal sliding mode and Chebyshev neural network, IEEE Trans. Syst. Man Cybern. B Cybern. 41 (4) (2011) 950–963.

[7] X.J. Chen, W. Steyn, Robust combined eigenaxis slew manoeuvre, in: AIAA Guidance, Navigation, and Control Conference, AIAA, Portland, 1999, pp. 521–529.

[8] C. Eugene Skelton II, Mixed Control Moment Gyro and Momentum Wheel Attitude Control Strategies, Virginia Polytechnic Institute and State University, USA, 2013.

[9] D. Ye, Z.W. Sun, J.Y. Wang, Control strategy of hybrid actuator for agile satellites, Acta Aeronaut. Astronaut. Sin. 33 (6) (2012) 1108–1115.

[10] Q.L. Hu, H. Wang, Z. Shi, et al., Novel non-singular saturated terminal sliding mode based attitude controller for spacecraft, J. Astronaut. 36 (4) (2015) 430–437.

[11] S.M. Song, Y. Guo, X.H. Li, Finite-time attitude tracking control for spacecraft with input saturation, Control Decis. 30 (11) (2015) 2004–2008.

[12] S.H. Yu, X.H. Yu, B. Shirinzadeh, et al., Continuous finite-time control for robotic manipulators with terminal sliding mode, Automatica 41 (11) (2005) 1957–1964.

[13] B. Xiao, Y. Shen, Velocity-free fault tolerant and uncertainty attenuation control for a class of nonlinear systems, IEEE Trans. Ind. Electron. 63 (7) (2016) 4400–4411.

[14] Y.Z. Zhu, L.X. Zhang, W.X. Zheng, Distributed H-infinity filtering for a class of discrete-time Markov Jump Lur'e systems with redundant channels, IEEE Trans. Ind. Electron. 63 (3) (2016) 1876–1885.

[15] M. Nazari, E.A. Butcher, T. Yucelen, et al., Decentralized consensus control of a rigid-body spacecraft formation with communication delay, J. Guid. Control Dynam. 39 (1) (2016) 1–14.

[16] E.D. Jin, Z.W. Sun, Robust controllers design with finite time convergence for rigid spacecraft attitude tracking control, Aerosp. Sci. Technol. 12 (4) (2008) 324–330.

[17] A.V. Rao, D.A. Benson, C. Darby, et al., Algorithm 902: GPOPS, A MATLAB software for solving multiple-phase optimal control problems using the gauss pseudospectral method, ACM Trans. Math. Softw. 37 (2) (2010) 163–172.

Chapter 12

A control allocation approach for hybrid actuators

12.1 Introduction

Nowadays, agile satellites are becoming a research hotspot in the current aerospace field. Agile satellites usually refer to small satellites that have fast maneuvering ability for attitude or orbit control, representing a higher requirement for the fast maneuver of spacecraft control systems [1,2]. Compared with traditional satellites, agile satellites can improve the application and efficiency, broaden the Earth observation capabilities of the satellite, ensure longer transmission time of payload data, and realize the missions that were difficult to complete before [3–5]. In addition, agile satellites provide technical support for tasks such as global communication, space monitoring, microinterstellar exploration, synthetic aperture radar remote sensing, land and resources observation, and rapid response in military reconnaissance [6,7].

For agile satellites, the core problem of the attitude fast maneuver is how to plan the process in order to shorten the maneuver time. In addition, the maximum output torque is limited under certain hardware conditions. Therefore, it is essential to adopt a method, which can maximize the ability of the existing actuators and realize the control technology of fast maneuver and high-precision pointing of the satellite attitude. At present, the satellite attitude control systems mostly use the hybrid actuators of thrusters and flywheels [8]. The thruster can output the required control torque in large-angle and fast maneuver process, but it is not precise enough. However, the flywheel that is commonly used as control actuator has the advantages of simple structure, mature manufacturing process, low cost, precisely control, and only electricity consumption [9]. Therefore, it will satisfy the requirement of the agile satellites if the advantages of these two actuators are combined.

However, the hybrid control strategy of thrusters and flywheels has some disadvantages in actual use: first, the thrust agent required by the thruster will increase the cost of satellites launching; second, the injection of the thrust agent may pollute the optical cameras, earth sensors, etc. If the CMG is adopted, the above problems can be effectively avoided. With the improvement of

Fast Satellite Attitude Maneuver and Control. https://doi.org/10.1016/B978-0-323-95455-6.00008-1

manufacturing technology, the CMG is bound to become the preferred actuator for agile satellites in the future because of the significant advantages of torque amplification and only electricity consumption [10]. For the CMG only providing the moment perpendicular to the frame angle, CMGs are usually combined used. In view of the singularity of the combined CMGs, although some scholars have proposed various methods to avoid or escape singularity [11–14], most of these methods are at the expense of control precision.

A method to ensure precision is to use variable speed control moment gyroscope (VSCMG) [15,16], but its mechanism is complex and the reduction of rotor speed is limited, and extremely low speed will destroy the characteristics of torque amplification. Considering the advantage of the flywheel that can be used as a functional supplement to the CMG, a hybrid actuators control strategy is proposed. So combining the advantages of the two actuators will effectively provide large output and high-precision control torque for the large-angle and fast maneuver of agile satellite [17].

It should be noted that using the hybrid control strategy of CMG and flywheel to realize satellite attitude control will make the actuators of the satellite stay in an overdrive state, so it is necessary to take reasonable control allocation. In order to solve this problem, this chapter will propose an optimization objective function that is related to the singularity of the CMG and the control energy, then obtain the allocation strategy on the basis of the optimization theory [18].

The rest of the chapter is outlined as follows. The dynamics model of the hybrid control algorithm of CMGs and flywheels is established in Section 12.2. Considering the singularity and energy cost problem of the hybrid control strategy, a control allocation method is designed based on the constrained optimization method in Section 12.3. Then the effectiveness of proposed control strategy is verified by mathematical simulation in Section 12.4. Finally, Section 12.5 concludes the work of this research.

Note: $\boldsymbol{I}_n$ is n rank unit matrix.

12.2 Problem formulation

In this section, the satellite dynamics equation with multiple single-frame VSCMG will be established first. On the basis of the model, the satellite dynamics equation with multiple VSCMGs, control moment gyroscopes, and reaction flywheels can be obtained through modification.

Single-frame VSCMG that contains a variable speed rotor can obtain the required control torque by rotating the frame or accelerating/decelerating the rotor. First, the frame coordinate system ($\boldsymbol{g}$, $\boldsymbol{s}$, $\boldsymbol{t}$) is defined in Fig. 12.1. The direction of the frame coordinate system relative to the satellite body coordinate system is represented by the following direction cosine matrix

$$\boldsymbol{C} = [\boldsymbol{g} \ \ \boldsymbol{s} \ \ \boldsymbol{t}] \tag{12.1}$$

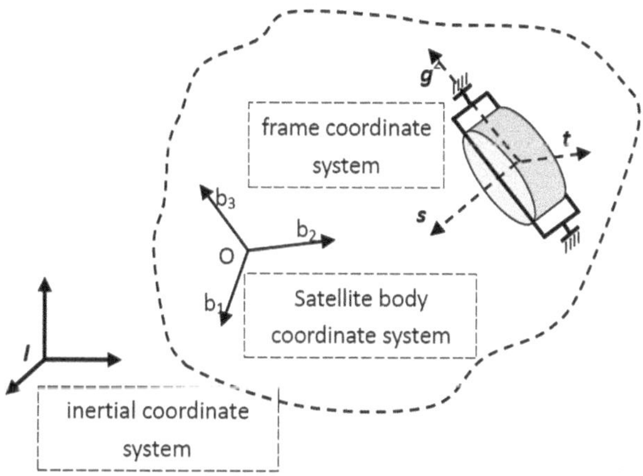

FIG. 12.1 Coordinate definition of spacecraft with single VSCMG.

The rotational inertia matrices of the frame and rotor are expressed as $\boldsymbol{I}_g = \mathrm{diag}[I_{g1},\ I_{g2},\ I_{g3}]$, $\boldsymbol{I}_w = \mathrm{diag}[I_{w1},\ I_{w2},\ I_{w3}]$ in the frame coordinate system, and $\boldsymbol{J}_g = \boldsymbol{C}\boldsymbol{I}_g\boldsymbol{C}^{\mathrm{T}}$, $\boldsymbol{J}_w = \boldsymbol{C}\boldsymbol{I}_w\boldsymbol{C}^{\mathrm{T}}$ in the body coordinate system.

Considering the satellite model with single VSCMG, the frame angle and rotor speed are expressed as $\dot{\gamma}_g$ and $\boldsymbol{\varpi}_g$, respectively. The angular velocity $\boldsymbol{\omega}$ can be denoted as $\boldsymbol{\omega} = \boldsymbol{C}\boldsymbol{\omega}_g$.

Then the angular momentum of the satellite can be expressed as

$$\begin{aligned} \boldsymbol{h} &= \boldsymbol{h}_s + \boldsymbol{h}_g + \boldsymbol{h}_w \\ &= \boldsymbol{J}_s\boldsymbol{\omega} + \boldsymbol{C}\boldsymbol{I}_g\left(\boldsymbol{\omega}_g + \dot{\boldsymbol{\gamma}}_g\right) + \boldsymbol{C}\boldsymbol{I}_w\left(\boldsymbol{\omega}_g + \dot{\boldsymbol{\gamma}}_g + \boldsymbol{\varpi}_g\right) \\ &= \left(\boldsymbol{J}_s + \boldsymbol{C}\left(\boldsymbol{I}_g + \boldsymbol{I}_w\right)\boldsymbol{C}^{\mathrm{T}}\right)\boldsymbol{\omega} + \boldsymbol{C}\left(\boldsymbol{I}_g + \boldsymbol{I}_w\right)\dot{\boldsymbol{\gamma}}_g + \boldsymbol{C}\boldsymbol{I}_w\boldsymbol{\varpi}_g \end{aligned} \tag{12.2}$$

where $\boldsymbol{h}_s$, $\boldsymbol{h}_g$, $\boldsymbol{h}_w$, and $\boldsymbol{J}_s$ denote angular momentum of the satellite, the VSCMG frame, the VSCMG rotor, and rotational inertia matrices of the satellite, respectively. In addition, $\boldsymbol{I}_c = \boldsymbol{I}_g + \boldsymbol{I}_w$, $\dot{\boldsymbol{\gamma}}_g = [\dot{\gamma},\ 0,\ 0]^{\mathrm{T}}$, and $\boldsymbol{\varpi}_g = [0,\ \boldsymbol{\varpi},\ 0]^{\mathrm{T}}$.

Thus, it can be deduced that the angular momentum of the satellite with N VSCMGs is

$$\boldsymbol{h} = \boldsymbol{J}\boldsymbol{\omega} + \boldsymbol{A}_{\mathrm{g}}\boldsymbol{I}_{\mathrm{cg}}\dot{\boldsymbol{\gamma}} + \boldsymbol{A}_{\mathrm{s}}\boldsymbol{I}_{\mathrm{ws}}\boldsymbol{\varpi} \tag{12.3}$$

$$\boldsymbol{J} = \boldsymbol{J}_{\mathrm{s}} + \boldsymbol{A}_{\mathrm{s}}\boldsymbol{I}_{\mathrm{cs}}\boldsymbol{A}_s{}^{\mathrm{T}} + \boldsymbol{A}_{\mathrm{t}}\boldsymbol{I}_{\mathrm{ct}}\boldsymbol{A}_{\mathrm{t}}{}^{\mathrm{T}} + \boldsymbol{A}_{\mathrm{g}}\boldsymbol{I}_{\mathrm{cg}}\boldsymbol{A}_{\mathrm{g}}{}^{\mathrm{T}} \tag{12.4}$$

where $\boldsymbol{\gamma} = [\gamma_1,\ \gamma_2,\ \cdots,\ \gamma_N]^{\mathrm{T}}$, $\boldsymbol{\varpi} = [\varpi_1,\ \varpi_2,\ \cdots,\ \varpi_N]^{\mathrm{T}}$, $\boldsymbol{A}_{\mathrm{s}} = [s_1,\ s_2,\ \cdots, s_N]$, $\boldsymbol{A}_{\mathrm{t}} = [\boldsymbol{t}_1,\ \boldsymbol{t}_2,\ \cdots,\ \boldsymbol{t}_N]$, and $\boldsymbol{A}_{\mathrm{g}} = [\boldsymbol{g}_1,\ \boldsymbol{g}_2,\ \cdots,\ \boldsymbol{g}_N]$. Furthermore, $\boldsymbol{I}_{\mathrm{ci}} = \boldsymbol{I}_{\mathrm{gi}} + \boldsymbol{I}_{\mathrm{wi}}$, $\boldsymbol{I}_{\mathrm{gi}} = \mathrm{diag}[\boldsymbol{I}_{\mathrm{gi1}},\ \boldsymbol{I}_{\mathrm{gi2}},\ \cdots,\ \boldsymbol{I}_{\mathrm{gi}N}]$, and $\boldsymbol{I}_{wi} = \mathrm{diag}[\boldsymbol{I}_{\mathrm{wi1}},\ \boldsymbol{I}_{\mathrm{wi2}},\ \cdots,\ \boldsymbol{I}_{\mathrm{wi}N}]$, here i denotes s, g, or t. $\boldsymbol{A}_{\mathrm{s}}$ and $\boldsymbol{A}_{\mathrm{t}}$ are both functions of the frame angle, $\boldsymbol{A}_{\mathrm{s}} = \boldsymbol{A}_{\mathrm{s0}}[\cos]^d + \boldsymbol{A}_{\mathrm{t0}}[\sin]^d, \boldsymbol{A}_{\mathrm{t}} = \boldsymbol{A}_{\mathrm{t0}}[\cos]^d - \boldsymbol{A}_{\mathrm{s0}}[\sin]^d$, $[\boldsymbol{x}]^d = \mathrm{diag}(x_1,\ x_2,\ \cdots,\ x_N)$.

According to the angular momentum theorem, the dynamics equation of the satellite system is

$$\begin{aligned}&\left(\boldsymbol{A}_{\mathrm{t}}[\dot{\gamma}]^{d}(\boldsymbol{I}_{\mathrm{cs}}-\boldsymbol{I}_{\mathrm{ct}})\boldsymbol{A}_{\mathrm{s}}{}^{\mathrm{T}}+\boldsymbol{A}_{\mathrm{s}}[\dot{\gamma}]^{d}(\boldsymbol{I}_{\mathrm{cs}}-\boldsymbol{I}_{\mathrm{ct}})\boldsymbol{A}_{\mathrm{t}}{}^{\mathrm{T}}\right)\boldsymbol{\omega}+\boldsymbol{J}\dot{\boldsymbol{\omega}}+\\&\boldsymbol{A}_{\mathrm{g}}\boldsymbol{I}_{\mathrm{cg}}\ddot{\gamma}+\boldsymbol{A}_{\mathrm{t}}\boldsymbol{I}_{\mathrm{ws}}[\boldsymbol{\varpi}]^{d}\dot{\gamma}+\boldsymbol{A}_{\mathrm{s}}\boldsymbol{I}_{\mathrm{ws}}\dot{\boldsymbol{\varpi}}+\boldsymbol{\omega}^{\times}\left(\boldsymbol{J}\boldsymbol{\omega}+\boldsymbol{A}_{\mathrm{g}}\boldsymbol{I}_{\mathrm{cg}}\dot{\gamma}+\boldsymbol{A}_{\mathrm{s}}\boldsymbol{I}_{\mathrm{ws}}\boldsymbol{\varpi}\right)=\boldsymbol{u}_{e}\end{aligned} \quad (12.5)$$

where $\boldsymbol{u}_e$ represents the sum of external moments. Here it is set as $\boldsymbol{u}_e=0$, and it is assumed that the satellite is completely controlled by the momentum exchange actuator.

For Eq. (12.5), let the parameters related to the angular rotation of the frame be 0, that is, $\dot{\gamma}=\ddot{\gamma}=0$, and then the equation can be applied to the reaction flywheel system; If $\boldsymbol{\varpi}$ is constant and $\dot{\boldsymbol{\varpi}}=0$, the equation can also be applied in traditional CMG system. To simply Eq. (12.5), it is assumed that some small quantities can be omitted [19]. As a result, the simplified dynamics equation of N VSCMGs satellite systems is denoted as

$$\boldsymbol{J}\dot{\boldsymbol{\omega}}+\boldsymbol{\omega}^{\times}\boldsymbol{J}\boldsymbol{\omega}+\boldsymbol{\omega}^{\times}\boldsymbol{A}_{\mathrm{s}}\boldsymbol{I}_{\mathrm{ws}}\boldsymbol{\varpi}=-\boldsymbol{A}_{\mathrm{t}}\boldsymbol{I}_{\mathrm{ws}}\boldsymbol{\varpi}^{\mathrm{d}}\dot{\gamma}-\boldsymbol{A}_{\mathrm{s}}\boldsymbol{I}_{\mathrm{ws}}\dot{\boldsymbol{\varpi}} \quad (12.6)$$

Then, based on Eq. (12.6), assuming that the satellite system carries M CMGs and N flywheels, the rotor of M VSCMGs be constant, $\dot{\boldsymbol{\varpi}}=0$, and the frame speed of the remaining N VSCMGs is $\dot{\gamma}=0$, the dynamics equation of the joint actuator system is

$$\begin{aligned}&\boldsymbol{J}\dot{\boldsymbol{\omega}}+\boldsymbol{\omega}^{\times}\boldsymbol{J}\boldsymbol{\omega}+\boldsymbol{\omega}^{\times}(\boldsymbol{A}_{\mathrm{cs}}\boldsymbol{I}_{\mathrm{cws}}\boldsymbol{\varpi}_{\mathrm{c}}+\boldsymbol{A}_{\mathrm{ws}}\boldsymbol{I}_{\mathrm{wws}}\boldsymbol{\varpi}_{\mathrm{w}})\\&=-\boldsymbol{A}_{\mathrm{ct}}\boldsymbol{I}_{\mathrm{cws}}\boldsymbol{\varpi}_{\mathrm{c}}^{\mathrm{d}}\dot{\gamma}_{\mathrm{c}}-\boldsymbol{A}_{\mathrm{ws}}\boldsymbol{I}_{\mathrm{wws}}\dot{\boldsymbol{\varpi}}_{\mathrm{w}}\end{aligned} \quad (12.7)$$

where the first letter c or w in the subscript indicates the matrix of CMG or flywheel, respectively. In addition, $\boldsymbol{h}_{\mathrm{cw}}=\boldsymbol{I}_{\mathrm{cws}}\boldsymbol{\varpi}_{\mathrm{c}}$, $\boldsymbol{h}_{\mathrm{ww}}=\boldsymbol{I}_{\mathrm{wws}}\boldsymbol{\varpi}_{\mathrm{w}}$, and $\boldsymbol{H}_{\mathrm{cw}}=\boldsymbol{I}_{\mathrm{cws}}\boldsymbol{\varpi}_{\mathrm{c}}^{\mathrm{d}}$.

12.3 Control torque allocation considering singularity of CMGs

12.3.1 Control torque allocation

This section proposes a reasonable control allocation method to deal with the overdrive of the system and the speed saturation of the flywheel and the singularity of the CMGs. In order to solve this problem, the objective function is selected as follows:

$$\min\left[\frac{1}{2}(\dot{\boldsymbol{\varpi}}-\dot{\boldsymbol{\varpi}}_{\mathrm{d}})^{\mathrm{T}}\boldsymbol{Z}_{d}(\dot{\boldsymbol{\varpi}}-\dot{\boldsymbol{\varpi}}_{\mathrm{d}})+\frac{1}{2}\dot{\gamma}^{\mathrm{T}}\boldsymbol{W}\dot{\gamma}+\frac{1}{2}\dot{\boldsymbol{\varpi}}^{\mathrm{T}}\boldsymbol{Z}_{\boldsymbol{\varpi}}\dot{\boldsymbol{\varpi}}+W_{\mathrm{V}}V(\gamma)\right] \quad (12.8)$$

In the above equation, the first term is used to weigh the speed saturation state of the flywheel, the second term and the third term are used to weigh the torque allocation proportion (energy) of the CMG and the flywheel, and the fourth term is used to weigh the singularity of the CMG, where $\boldsymbol{Z}_d$, $\boldsymbol{W}$, W_V, and $\boldsymbol{Z}_{\varpi}$ are weight matrices and diagonal matrices. $V(\boldsymbol{\gamma})$ is the singularity value index of the CMGs group. $\dot{}_{\mathrm{d}}$ is defined as $\dot{\boldsymbol{\varpi}}_{\mathrm{d}} = s\frac{\boldsymbol{\varpi}_{\mathrm{wd}} - \boldsymbol{\varpi}_{\mathrm{w}}}{\Delta t}$, where s is a positive constant, Δt is a constant of flywheel momentum unloading time, and $\boldsymbol{\varpi}_{\mathrm{wd}}$ is the reference speed of the flywheel which can be set to 0 under no special circumstances.

Since $V(\boldsymbol{\gamma})$ is a nonlinear function about the frame angle $\boldsymbol{\gamma}$, it is hard to obtain the analytical solution directly. Here we use $V(\boldsymbol{\gamma} + \dot{\boldsymbol{\gamma}}\boldsymbol{\Delta t})$ to replace $V(\boldsymbol{\gamma})$. The Taylor expansion of $V(\boldsymbol{\gamma})$ around $\boldsymbol{\gamma}$ can be obtained

$$W_{\mathrm{v}}V(\gamma+\dot{\gamma}\Delta t) = W_{\mathrm{v}}V(\gamma) + W_{\mathrm{v}}\boldsymbol{g}^{\mathrm{T}}\dot{\gamma}\Delta t + \frac{1}{2}W_{\mathrm{v}}\dot{\gamma}^{\mathrm{T}}\boldsymbol{H}\dot{\gamma}\Delta t^2 + \cdots \tag{12.9}$$

where

$$\boldsymbol{g} = V'(\gamma) = \left[\frac{\partial V(\gamma)}{\partial \gamma_1}, \frac{\partial V(\gamma)}{\partial \gamma_2}, \ldots, \frac{\partial V(\gamma)}{\partial \gamma_m}\right]^{\mathrm{T}} \in \mathbb{R}^m$$

$$\boldsymbol{H} = V''(\gamma) = \begin{bmatrix} \frac{\partial^2 V(\gamma)}{\partial \gamma_1^2} & \cdots & \frac{\partial^2 V(\gamma)}{\partial \gamma_m \partial \gamma_1} \\ \vdots & \ddots & \vdots \\ \frac{\partial^2 V(\gamma)}{\partial \gamma_1 \partial \gamma_m} & \cdots & \frac{\partial^2 V(\gamma)}{\partial \gamma_m^2} \end{bmatrix} \in \mathbb{R}^{m \times m}$$

Select $W_V = \frac{\kappa}{\Delta t}$, $\bar{\boldsymbol{g}} = \kappa V'(\gamma)$, and $\overline{\boldsymbol{H}} = \kappa \Delta t V''(\gamma)$, where κ is a positive constant.

Then,

$$W_{\mathrm{v}}V(\gamma + \dot{\gamma}\Delta t) \approx \frac{\kappa}{\Delta t}V(\gamma) + \bar{\boldsymbol{g}}^{\mathrm{T}}\dot{\gamma} + \frac{1}{2}\dot{\gamma}^{\mathrm{T}}\overline{\boldsymbol{H}}\dot{\gamma} \tag{12.10}$$

Eq. (12.8) should satisfy the following constraint

$$\boldsymbol{A}\dot{\gamma} + \boldsymbol{B}\dot{\boldsymbol{\varpi}} = \boldsymbol{u} \tag{12.11}$$

where $\boldsymbol{u}$ is the required control torque of the satellite, and $\boldsymbol{A} = -\boldsymbol{A}_{\mathrm{ct}}\boldsymbol{H}_{\mathrm{cw}}$, $\boldsymbol{B} = -\boldsymbol{A}_{\mathrm{ws}}\boldsymbol{I}_{\mathrm{wws}}$.

In order to obtain the control allocation strategy, construct the following Lagrange function:

$$\begin{aligned} L = & \frac{1}{2}(\dot{\boldsymbol{\varpi}} - \dot{\boldsymbol{\varpi}}_{\mathrm{d}})^{\mathrm{T}}\boldsymbol{Z}_d(\dot{\boldsymbol{\varpi}} - \dot{\boldsymbol{\varpi}}_{\mathrm{d}}) + \frac{1}{2}\dot{\gamma}^{\mathrm{T}}\boldsymbol{W}\dot{\gamma} + \frac{1}{2}\dot{\boldsymbol{\varpi}}^{\mathrm{T}}\boldsymbol{Z}_{\varpi}\dot{\boldsymbol{\varpi}} \\ & + W_V V(\gamma + \dot{\gamma}\Delta t) + \boldsymbol{\lambda}^{\mathrm{T}}(\boldsymbol{A}\dot{\gamma} + \boldsymbol{B}\dot{\boldsymbol{\varpi}} - \boldsymbol{u}) \end{aligned} \tag{12.12}$$

Substitute Eq. (12.10) into Eq. (12.12) and calculate the partial derivative of L for $\dot{\boldsymbol{\varpi}}$, $\dot{\gamma}$ and $\boldsymbol{\lambda}$ as

$$
\begin{aligned}
&\frac{\partial L}{\partial \dot{\boldsymbol{\varpi}}} = \boldsymbol{Z}_d(\dot{\boldsymbol{\varpi}} - \dot{\boldsymbol{\varpi}}_{\mathrm{d}}) + \boldsymbol{Z}_{\boldsymbol{\varpi}}\dot{\boldsymbol{\varpi}} + B^{\mathrm{T}}\boldsymbol{\lambda} = 0 \\
&\frac{\partial L}{\partial \dot{\gamma}} = \boldsymbol{W}\dot{\gamma} + \boldsymbol{A}^{\mathrm{T}}\boldsymbol{\lambda} + \bar{\boldsymbol{g}} + \overline{\boldsymbol{H}}\dot{\gamma} = 0 \\
&\boldsymbol{A}\dot{\gamma} + \boldsymbol{B}\dot{\boldsymbol{\varpi}} - \boldsymbol{u} = 0
\end{aligned} \tag{12.13}
$$

Solving the Eq. (12.13), we can get the controllers as

$$
\dot{\gamma} = \boldsymbol{P}\boldsymbol{C}^*\boldsymbol{u} + \boldsymbol{P}\boldsymbol{C}^*\boldsymbol{Q}^{\mathrm{T}}\boldsymbol{g}_{\mathrm{w}} + \left(\boldsymbol{P}\boldsymbol{C}^*\boldsymbol{P}^{\mathrm{T}} - \overline{\boldsymbol{W}}^{-1}\right)\bar{\boldsymbol{g}} \tag{12.14}
$$

$$
\dot{\boldsymbol{\varpi}} = \boldsymbol{Q}\boldsymbol{C}^*\boldsymbol{u} + \left(\boldsymbol{Q}\boldsymbol{C}^*\boldsymbol{Q}^{\mathrm{T}} - \boldsymbol{Z}^{-1}\right)\boldsymbol{g}_{\mathrm{w}} + \boldsymbol{Q}\boldsymbol{C}^*\boldsymbol{P}^{\mathrm{T}}\bar{\boldsymbol{g}} \tag{12.15}
$$

In the above equations, $\overline{\boldsymbol{W}} = \boldsymbol{W} + \overline{\boldsymbol{H}}$, $\boldsymbol{C}^* = \left(\boldsymbol{A}\overline{\boldsymbol{W}}^{-1}\boldsymbol{A}^{\mathrm{T}} + \boldsymbol{B}\boldsymbol{Z}^{-1}\boldsymbol{B}^{\mathrm{T}}\right)^{-1} = (\boldsymbol{A}\boldsymbol{P} + \boldsymbol{B}\boldsymbol{Q})^{-1}$, $\boldsymbol{P} = \overline{\boldsymbol{W}}^{-1}\boldsymbol{A}^{\mathrm{T}}$, $\boldsymbol{Q} = \boldsymbol{Z}^{-1}\boldsymbol{B}^{\mathrm{T}}$, $\boldsymbol{Z} = \boldsymbol{Z}_d + \boldsymbol{Z}_{\boldsymbol{\varpi}}$, and $\boldsymbol{g}_{\mathrm{w}} = -\boldsymbol{Z}_d\dot{\boldsymbol{\varpi}}_{\mathrm{d}}$.

The first terms in Eqs. (12.14) and (12.15) are related to the control torque and $\boldsymbol{A}\boldsymbol{P}\boldsymbol{C}^*\boldsymbol{u} + \boldsymbol{B}\boldsymbol{Q}\boldsymbol{C}^*\boldsymbol{u} = \boldsymbol{u}$, which indicate that the first terms of the controllers provide the control torque required by the control system. $\boldsymbol{g}_{\mathrm{w}}$ can be esteemed as the gradient vector related to flywheel momentum unloading. The second terms are to achieve the aim of flywheel momentum unloading and transfer the angular momentum of the flywheel to the CMG without affecting the system output control torque, with $\boldsymbol{A}\boldsymbol{P}\boldsymbol{C}^*\boldsymbol{Q}^{\mathrm{T}}\boldsymbol{g}_{\mathrm{w}} + \boldsymbol{B}(\boldsymbol{Q}\boldsymbol{C}^*\boldsymbol{Q}^{\mathrm{T}} - \boldsymbol{Z}^{-1})\boldsymbol{g}_{\mathrm{w}} = 0$. The third terms are to manipulate the angular rotation of the CMG frame without affecting the satellite control output. Then, it can reduce the singularity value in order to avoid saturation, which is similar to the "idling" of the CMG [20,21] and $\boldsymbol{A}\left(\boldsymbol{P}\boldsymbol{C}^*\boldsymbol{P}^{\mathrm{T}} - \overline{\boldsymbol{W}}^{-1}\right)\bar{\boldsymbol{g}} + \boldsymbol{B}\boldsymbol{Q}\boldsymbol{C}^*\boldsymbol{P}^{\mathrm{T}}\bar{\boldsymbol{g}} = 0$.

The control torque of CMGs and flywheels can be well allocated by adjusting the weight matrices. If the required control torque is large, we can increase $\boldsymbol{W}$ or decrease $\boldsymbol{Z}_{\boldsymbol{\varpi}}$. If the required control accuracy is high, we can reduce $\boldsymbol{W}$ or increase $\boldsymbol{Z}_{\boldsymbol{\varpi}}$. If the flywheels speed is close to saturation, we can increase $\boldsymbol{Z}_d$ to unload the flywheel momentum. In order to improve the control performance, a weight matrix scheme is proposed as

$$
\boldsymbol{W} = \alpha_0 \exp\left(-\alpha_1 \|\boldsymbol{u}\|\right)\boldsymbol{I}_{\mathrm{m}} \tag{12.16}
$$

$$
\boldsymbol{Z}_d = \beta_0 \exp\left(\beta_1 \|\boldsymbol{\varpi}_{\mathrm{wd}} - \boldsymbol{\varpi}_{\mathrm{w}}\|\right)\mathbf{I}_{\mathrm{n}} \tag{12.17}
$$

where α_0, α_1, β_0, and β_1 are positive constants.

The inverse of matrix $\overline{\boldsymbol{W}}$ is contained in Eqs. (12.14) and (12.15), and it should always be reversible in order to ensure the effectiveness of the control law. By selecting an appropriate singularity representation, each parameter in the Hessian matrix $V''(\gamma)$ can be bounded. Combined with $\bar{g} = \kappa V'(\gamma)$ and $\overline{\boldsymbol{H}} = \kappa \Delta t V''(\gamma)$, Eq. (12.18) is obtained as follows:

$$
\left\|\overline{\boldsymbol{H}}\right\| < \kappa \Delta t v \tag{12.18}
$$

where v is a positive constant. $\overline{W}$ can be proved to be reversible through the following theorem.

Theorem 12.1 *let* $\mathbf{X}$ *be a linear operator. If* $\|\mathbf{X}\|<1$, *then* $(\mathbf{I}-\mathbf{X})^{-1}$ *exists, and it is a constrained linear operator.*

For the convenience of analysis, select the weight matrix $\mathbf{W}$ *as* $\alpha\boldsymbol{I}_3$, *where* $\alpha=\kappa\Delta tv$, *and* $\overline{\mathbf{W}}=\mathbf{W}+\overline{\mathbf{H}}$ *can be rewritten as*

$$\overline{W}=W+\overline{H}=\kappa\Delta tv\left(\mathbf{I}+\frac{\overline{H}}{\kappa\Delta tv}\right) \tag{12.19}$$

According to the Theorem 12.1, the matrix $\left(1+\overline{H}/\kappa\Delta tv\right)^{-1}$ will always exist. Therefore, the reversibility of matrix $\overline{W}$ can be guaranteed by choosing $\alpha\geq\kappa\Delta tv$. If α is fixed, nonsingularity of $\overline{W}$ can be guaranteed by selecting the time interval

$$0\leq\Delta t\leq\frac{\alpha}{\kappa v} \tag{12.20}$$

12.3.2 Singularity indexes

The proposed control torque allocation strategy above needs to calculate the gradient vector and the Hessian matrix of the singularity index of the Jacobian matrix A_{ct}. The traditional method using determinant as the singularity index to solve the problem is very difficult and complex. This chapter will introduce two singularity indexes based on which it is easy to calculate the gradient matrix and Hessian matrix: nondimensional condition number and inner product index.

A. nondimensional condition number.

Condition number or nondimensional condition number is a common index to measure the singularity of the matrix, which is defined as

$$V_c=\frac{\sigma_1}{\sigma_3} \tag{12.21}$$

where σ_i is the singular value of the matrix A_{ct}. Junkins presents a fast gradient vector calculation method aimed to reduce the calculation load in reference [22]. The partial differential of V_c relative to the frame angle is denoted as

$$g_i=\frac{\partial V_c}{\partial\gamma_i}=\frac{1}{\sigma_3}\frac{\partial\sigma_1}{\partial\gamma_i}-\frac{\sigma_1}{\sigma_3^2}\frac{\partial\sigma_3}{\partial\gamma_i} \tag{12.22}$$

The partial differential of singularity relative to frame angle is

$$\frac{\partial\sigma_k}{\partial\gamma_i}=U_k^{\mathrm{T}}\frac{\partial A}{\partial\gamma_i}V_k \tag{12.23}$$

where $U^{\mathrm{T}}U=UU^{\mathrm{T}}=I_m$, $V^{\mathrm{T}}V=VV^{\mathrm{T}}=I_n$, σ_i represents the singularity value of A and

$$\sum=\begin{bmatrix}\sigma_1 & 0 & \cdots & 0 & \cdots & 0\\ 0 & \ddots & \cdots & 0 & \cdots & 0\\ 0 & 0 & \sigma_m & 0 & \cdots & 0\end{bmatrix}$$

Define $\overline{\boldsymbol{h}}=\boldsymbol{A}_{\mathrm{cs}}=\left[\overline{\boldsymbol{h}}_1\ \overline{\boldsymbol{h}}\ \overline{\boldsymbol{h}}_3\ \overline{\boldsymbol{h}}_4\right]$, $\boldsymbol{A}_{\mathrm{cs}}\equiv[\boldsymbol{f}_1\boldsymbol{f}_2\boldsymbol{f}_3\boldsymbol{f}_4]\in\boldsymbol{R}^{3\times4}$. It should be noted that $\overline{\boldsymbol{h}}$ is a function with period 2π [23], and $\frac{\partial^2\overline{\boldsymbol{h}}_i}{\partial\gamma_i^2}=\frac{\partial\boldsymbol{f}_i}{\partial\gamma_i}=-\overline{\boldsymbol{h}}_i$.

Then, Eq. (12.23) can be written in a simple form as

$$\frac{\partial\sigma_k}{\partial\gamma_i}=-\boldsymbol{V}_{ik}\boldsymbol{U}_k^{\mathrm{T}}\overline{\boldsymbol{h}}_i \tag{12.24}$$

According to the proposed control allocation method, the Hessian matrix needs to be calculated. Therefore, further partial differentiation of Eq. (12.22) relative to the frame angle can lead to each parameter in the Hessian matrix as

$$\begin{aligned}\frac{\partial^2 V_{\mathrm{c}}}{\partial\gamma_i\partial\gamma_j}=&\frac{1}{\sigma_3}\frac{\partial^2\sigma_1}{\partial\gamma_i\partial\gamma_j}-\frac{1}{\sigma_3^2}\times\left(\frac{\partial\sigma_1}{\partial\gamma_i}\frac{\partial\sigma_3}{\partial\gamma_j}+\frac{\partial\sigma_3}{\partial\gamma_i}\frac{\partial\sigma_1}{\partial\gamma_j}+\sigma_1\frac{\partial^2\sigma_3}{\partial\gamma_i\partial\gamma_j}\right)\\&+\frac{2\sigma_1}{\sigma_3^3}\frac{\partial\sigma_3}{\partial\gamma_i}\frac{\partial\sigma_3}{\partial\gamma_j}\end{aligned} \tag{12.25}$$

According to Eq. (12.24)

$$\frac{\partial^2\sigma_k}{\partial\gamma_i\partial\gamma_j}=\begin{cases}-y_{ik}\boldsymbol{x}_k^{\mathrm{T}}\boldsymbol{f}_i & if\ i=j\\ 0 & if\ i\neq j\end{cases} \tag{12.26}$$

Substituting Eq. (12.26) into Eq. (12.25), the Hessian matrix of the nondimensional condition number relative to the frame angle vector γ can be obtained.

There are some disadvantages when adopting the gradient vector and Hessian matrix obtained from the nondimensional condition number: if the CMGs group approaches the singular state, that is, $\sigma_3\approx0$, it will lead to the condition that the denominator approaches 0 when calculating the gradient and Hessian matrix according to Eq. (12.22) and Eq. (12.25). It means that a large angular velocity output command will be generated to the frame angle of the CMGs, which will even exceed the ability of the CMGs system. To avoid this possibility, another singularity representation will be introduced in the following part.

B. Inner product index [24].

If the inner product of any two different column vectors in a known matrix is 0, that is, any two different column vectors are orthogonal, then the matrix has full rank and the matrix is not in a singular state. Based on this case, the inner product singularity index is defined as

$$V_{\mathrm{ip}} = \frac{1}{2} \sum_{i=j=1, i \neq j}^{4} \langle \boldsymbol{f}_i, \boldsymbol{f}_j \rangle^2 \tag{12.27}$$

When the index value is 0 or very small, the matrix is full rank. At this time, it can be considered that the CMGs group is not in a singular state. It is worth noting that the norm of the column vector $\boldsymbol{f}_i$ should be nonzero. If this condition is not satisfied, the inner product index will not be suitable. The column vectors of the Jacobian matrix $\boldsymbol{A}_{\mathrm{ct}}$ are unit vectors, and it ensure that the inner product index is available. At the same time, if the column vectors of Jacobian matrix are zero vectors, it means that the angular velocity of the CMG rotor is 0, which does not match the characteristics of CMGs. In conclusion, the inner product index can quantify the singularity of Jacobian matrix.

Adopting the inner product index as the singularity performance to realize control allocation needs to calculate the gradient vector and Hessian matrix. First, the gradient vector can be obtained by

$$\boldsymbol{g} = \sum_{i=j=1, i \neq j}^{4} \langle \boldsymbol{f}_i^{\mathrm{T}} \boldsymbol{f}_j \rangle \frac{\partial(\boldsymbol{f}_i^{\mathrm{T}} \boldsymbol{f}_j)}{\partial \gamma} \tag{12.28}$$

The partial differential of the inner product relative to the frame angle in Eq. (12.28) is

$$\frac{\partial(\boldsymbol{f}_i^{\mathrm{T}} \boldsymbol{f}_j)}{\partial \gamma_k} = \begin{cases} -\overline{\boldsymbol{h}}_i^{\mathrm{T}} \boldsymbol{f}_j & \text{if } k = i \\ 0 & \text{if } k \neq i \end{cases} \tag{12.29}$$

Then the ith parameter of the gradient vector is denoted as

$$g_i = \sum_{j=1, j \neq i}^{4} (\boldsymbol{f}_i^{\mathrm{T}} \boldsymbol{f}_j) \left(\overline{\boldsymbol{h}}_i^{\mathrm{T}} \boldsymbol{f}_j \right) \tag{12.30}$$

Calculate the partial differential of the gradient vector for the frame angle, and then each parameter in the Hessian matrix will be obtained

$$\frac{\partial^2 V_{\mathrm{ip}}}{\partial \gamma_i^2} = \sum_{j=1, j \neq i}^{4} \left[\left(\overline{\boldsymbol{h}}_i^{\mathrm{T}} \boldsymbol{f}_j \right)^2 - (\boldsymbol{f}_i^{\mathrm{T}} \boldsymbol{f}_j)^2 \right] \tag{12.31}$$

$$\frac{\partial^2 V_{\mathrm{ip}}}{\partial \gamma_i \partial \gamma_j} = (\boldsymbol{f}_i^{\mathrm{T}} \overline{\boldsymbol{h}}_j) \left(\overline{\boldsymbol{h}}_i^{\mathrm{T}} \boldsymbol{f}_j \right) + (\boldsymbol{f}_i^{\mathrm{T}} \boldsymbol{f}_j) \left(\overline{\boldsymbol{h}}_i^{\mathrm{T}} \overline{\boldsymbol{h}}_j \right) \tag{12.32}$$

From Eqs. (12.30)–(12.32), it can be seen that the gradient vector and Hessian matrix are calculated only by $\overline{\boldsymbol{h}}_i$ and $\boldsymbol{f}_i$, and these two vectors can be easily obtained from the installation matrix $\boldsymbol{A}_{\mathrm{cs}}$ and $\boldsymbol{A}_{\mathrm{ct}}$ of the CMGs. Therefore, the gimbal rates generated by the proposed index are more reliable than that of the nondimensional condition number.

12.4 Simulation results

Here, we define the inertia matrix of the satellite as

$$\boldsymbol{J} = \begin{bmatrix} 100 & & \\ & 120 & \\ & & 150 \end{bmatrix} \mathrm{kg \cdot m^2} \tag{12.33}$$

In the simulation, the satellite platform carries four CMGs with the pyramid configuration, as shown in Fig. 12.2, and the incidence is $\beta = 53.13°$. Assuming that the frame angle is 0 at initial time, and the angular momentum matrix and control output matrix corresponding to the CMGs group are

$$\boldsymbol{A}_{cs} = \begin{bmatrix} -c\beta\sin\gamma_1 & -\cos\gamma_2 & c\beta\sin\gamma_3 & \cos\gamma_4 \\ \cos\gamma_1 & -c\beta\sin\gamma_2 & -\cos\gamma_3 & c\beta\sin\gamma_4 \\ s\beta\sin\gamma_1 & s\beta\sin\gamma_2 & s\beta\sin\gamma_3 & s\beta\sin\gamma_4 \end{bmatrix}$$

$$\boldsymbol{A}_{ct} = \begin{bmatrix} -c\beta\cos\gamma_1 & \sin\gamma_2 & c\beta\cos\gamma_3 & -\sin\gamma_4 \\ -\sin\gamma_1 & -c\beta\cos\gamma_2 & \sin\gamma_3 & c\beta\cos\gamma_4 \\ s\beta\cos\gamma_1 & s\beta\cos\gamma_2 & s\beta\cos\gamma_3 & s\beta\cos\gamma_4 \end{bmatrix}$$

where $c\beta = \cos(\beta)$, $s\beta = \sin(\beta)$. The rotor moment of inertia of the CMG is $0.05\,\mathrm{kg\,m^2}$ and the rotor angular velocity is 6000 rpm. The rotor moments of inertia of the three flywheels are $0.07\,\mathrm{kg\,m^2}$, and the initial rotor angular velocities are 0.

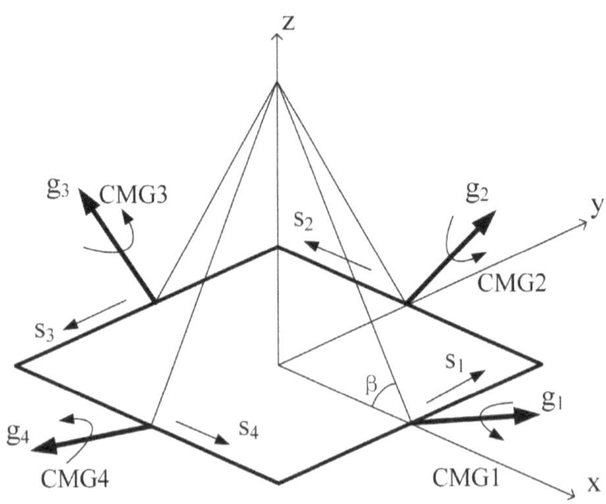

FIG. 12.2 A CMGs system with paramid configuration.

The task of large-angle fast maneuver of agile satellite is to maneuver the satellite attitude from the initial attitude $\boldsymbol{q}_0 = [0.6 \ \ 0.316 \ \ 0.2 \ \ 0.707]$ to the target attitude $\boldsymbol{q}_{\mathrm{f}} = [0001]$, and the initial and target angular velocities are 0.

A simple PD tracking control law can be adopted

$$\boldsymbol{u} = -\operatorname{sgn}(\tilde{q}_4)\boldsymbol{K}_{\mathrm{q}}\tilde{\boldsymbol{q}}_{\mathrm{v}} - \boldsymbol{K}_{\mathrm{w}}\tilde{\boldsymbol{\omega}} \tag{12.34}$$

where the control parameter is selected as $\mathbf{K}_{\mathrm{q}} = 2.28 * \boldsymbol{J}$, $\mathbf{K}_{\mathrm{w}} = 10.5 * \boldsymbol{J}$. The weight matrices are $\alpha_0 = 2 \times 10^7$, $\alpha_1 = 20$, $\boldsymbol{Z}_{\boldsymbol{\varpi}} = 110 * \mathbf{I}_3$, $\beta_0 = 100$, $\beta_1 = 1 \times 10^{-2}$, $W_{\mathrm{v}} = 100$.

In order to illustrate the effectiveness of the singular value index, a random torque is imposed to the satellite. Then the nondimensional condition number, inner product index, and the reciprocal of the determinant corresponding to the control torque matrix $\boldsymbol{A}_{\mathrm{ct}}$ of the CMGs can be obtained. The results are shown in Fig. 12.3, and the change trends of the three singularity indexes are completely consistent, which shows that the two kinds of singularity index mentioned in Section 12.3.2 can be used to measure the singularity of CMGs group.

To verify the effectiveness of the control allocation considering the singularity of the CMGs group, select the nondimensional condition number as the singularity index. The simulation results are shown in Figs. 12.4–12.10. It can be seen from Figs. 12.4 and 12.5 that the control allocation designed in

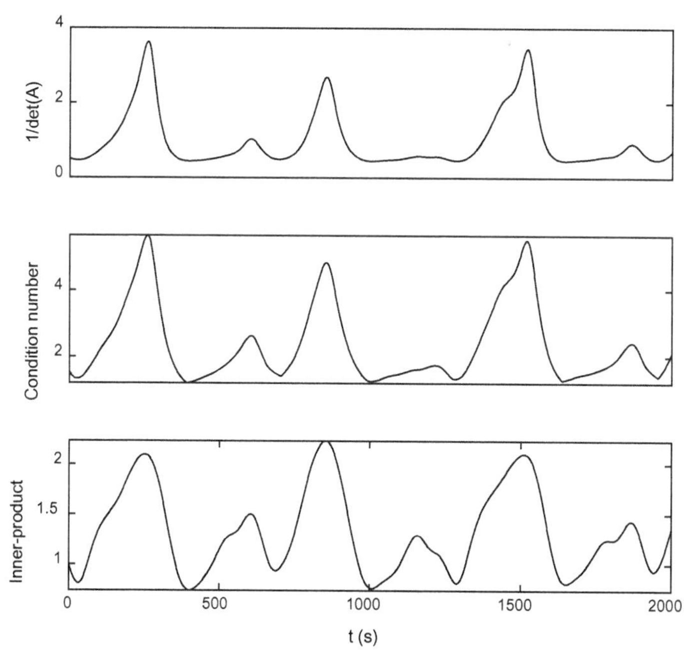

FIG. 12.3 The comparison of three singularity index.

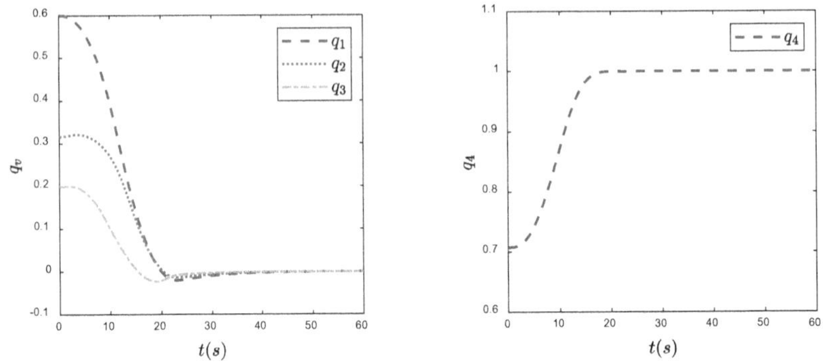

FIG. 12.4 Time histories of the attitude of the spacecraft.

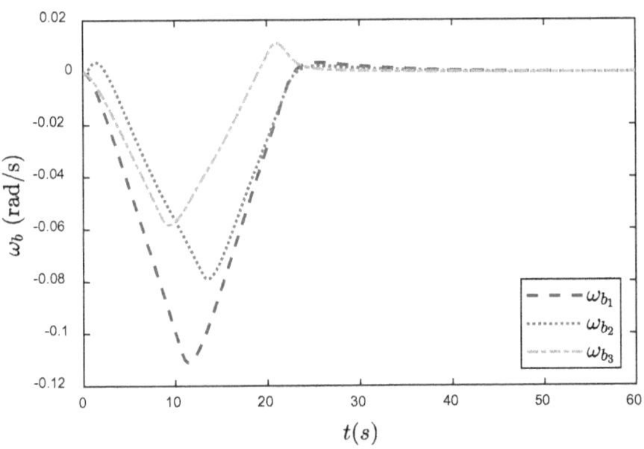

FIG. 12.5 Time histories of the angular velocity of the spacecraft.

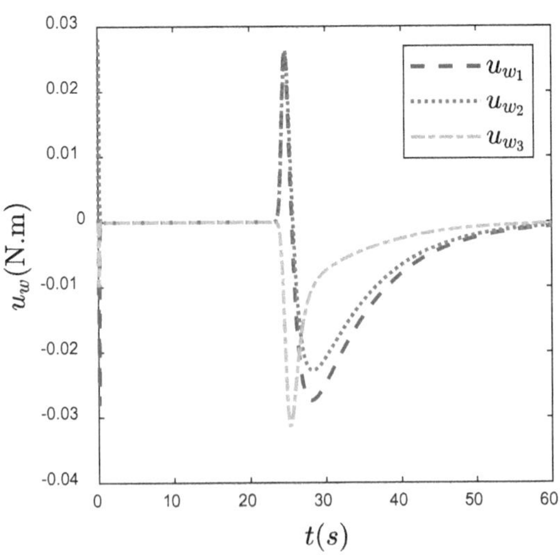

FIG. 12.6 The torque performed by flywheels.

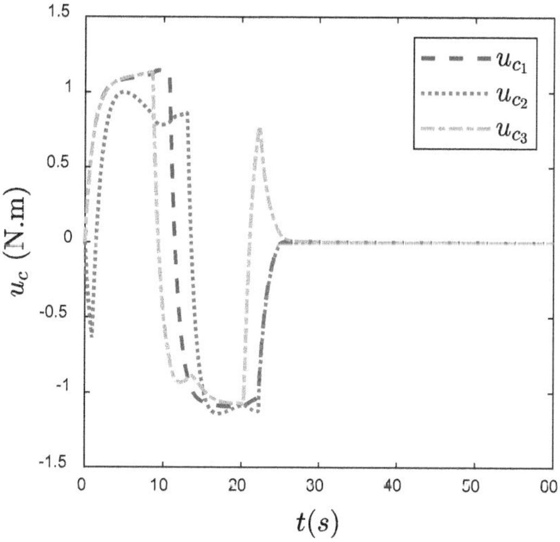

FIG. 12.7 The torque performed by CMGs.

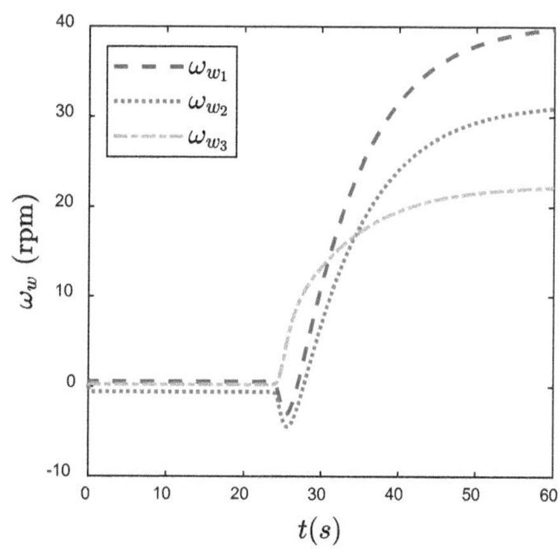

FIG. 12.8 The rotational speed of flywheels.

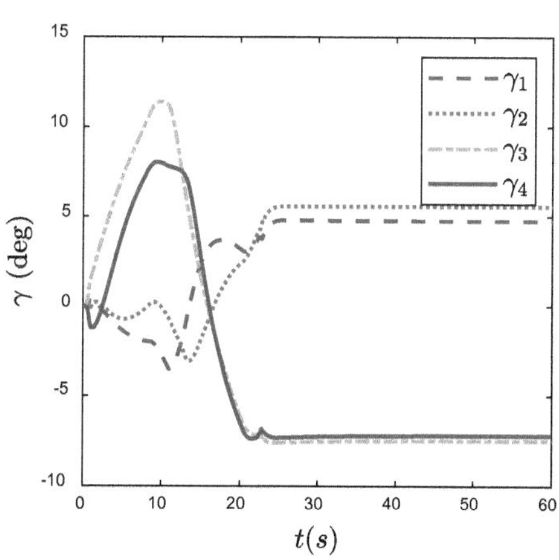

FIG. 12.9 The gimble angle speed of CMGs.

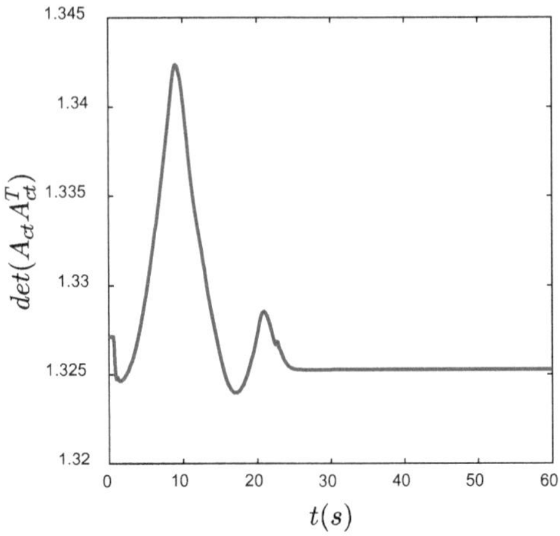

FIG. 12.10 Determinant history of control collocation.

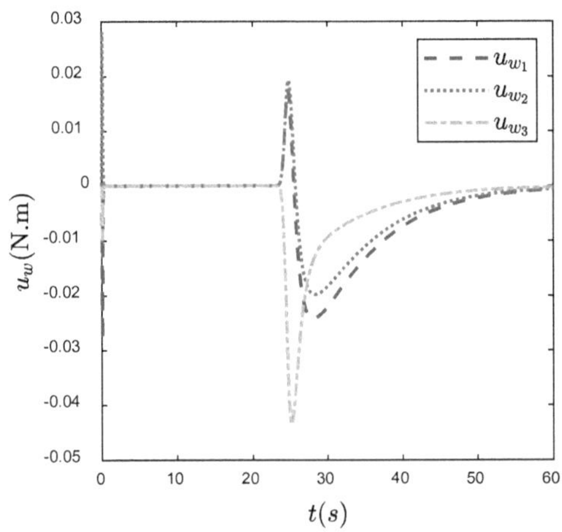

FIG. 12.11 The torque performed by flywheels.

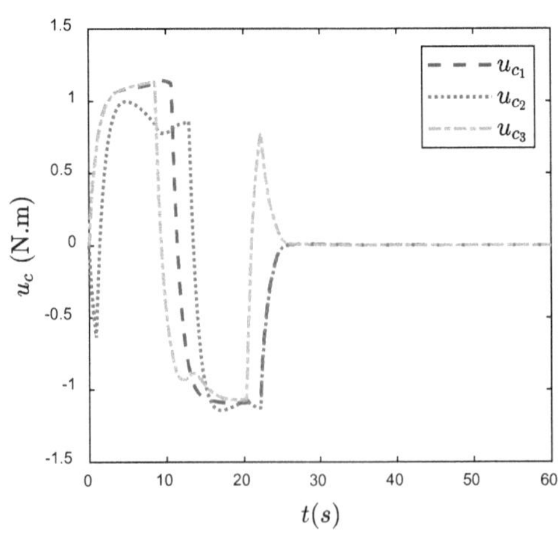

FIG. 12.12 The torque performed by CMGs.

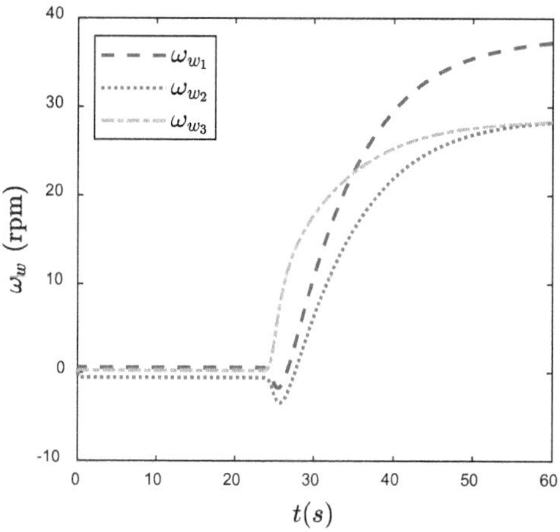

FIG. 12.13 The rotational speed of flywheels.

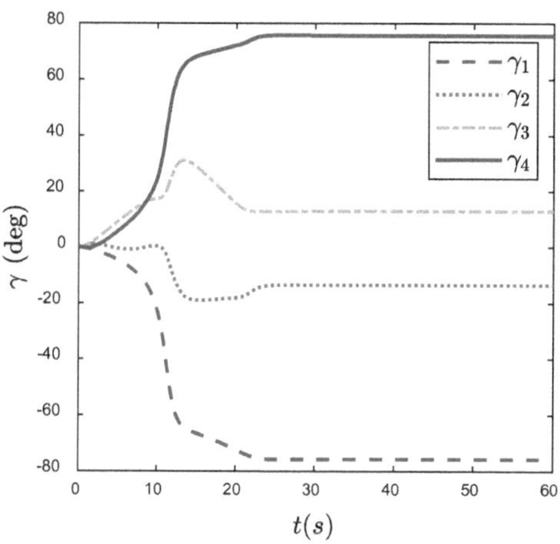

FIG. 12.14 The gimble angle speed of CMGs.

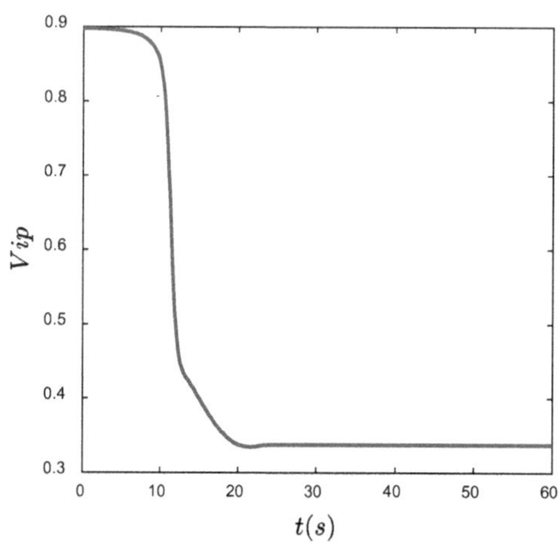

FIG. 12.15 Inner product index history.

Section 12.3 does not affect the control effect. In order to quantity the influence of the control allocation on the singularity index, the determinant of $\boldsymbol{A}_{\mathrm{ct}}\boldsymbol{A}_{\mathrm{ct}}^{\mathrm{T}}$ is taken as the singularity index, which is shown in Fig. 12.10. It can be concluded that the singularity index of the CMGs group has been kept within a small range of variation, and the proposed control strategy can prevent the CMGs group from entering the singular state. From Fig. 12.9, it can be seen that the variation of the CMGs group's frame angle is quite different. It shows that the CMGs group uses "idling," which can not only provide the control torque required for satellite maneuver but also reduce the singularity of the CMGs group through idling.

The simulation results of the inner product index are shown in Figs. 12.11–12.16. Comparing Fig. 12.16 with Fig. 12.10, when the inner product index is used as the singularity index, the determinant $\det(\boldsymbol{A}_{\mathrm{ct}}\boldsymbol{A}_{\mathrm{ct}}^{\mathrm{T}})$ related to the CMGs will be greatly improved under the control allocation, and the oscillation shown in Fig. 12.10 will be avoided, so that the CMGs are in a state of principle singularity, which can also be concluded from Fig. 12.14. Comparing Figs. 12.10 and 12.16, the inner product index can make the CMGs achieve better singularity performance in the process of satellite maneuver.

To analyze the role of the two singularity indexes in control allocation, the CMGs control law and the gradient vector norm of the singularity index in the above simulations are given. Fig. 12.17 shows that adopting the nondimensional conditions number as the singularity index, the steering law of the CMGs is about 0.7 rad/s in the process of satellite attitude maneuver, while in Fig. 12.19, when the inner product index is adopted as the singularity index, the steering law of the CMGs is about 0.05 rad/s. Therefore, the inner product index can reduce the frame angular velocity of the CMGs and make the motion

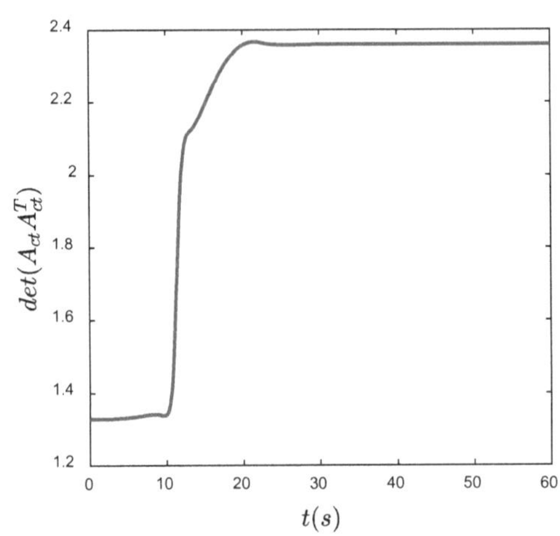

FIG. 12.16 Determinant history.

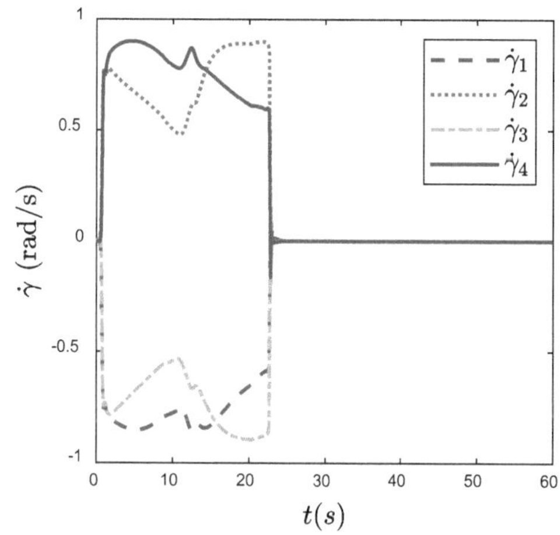

FIG. 12.17 The gimbal speed of CMGs corresponding to condition number.

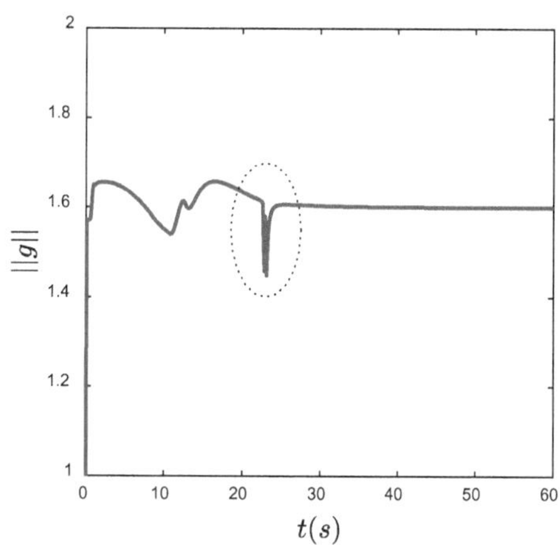

FIG. 12.18 The norm of gradient of condition number.

of the CMGs more effective. Flutter occurs in the gradient vector norm is shown as the circle in Fig. 12.18, which will definitely cause the flutter of the frame angular velocity of the CMGs, and it is unfavorable for the actual control. And the effect will be improved by adopting the inner product index which is shown in Fig. 12.20.

In conclusion, selecting the inner product index as the singularity performance can not only obtain better control effect but also get lower angular velocity, which makes the "idling" of the CMGs more effective and the angular velocity control smoother.

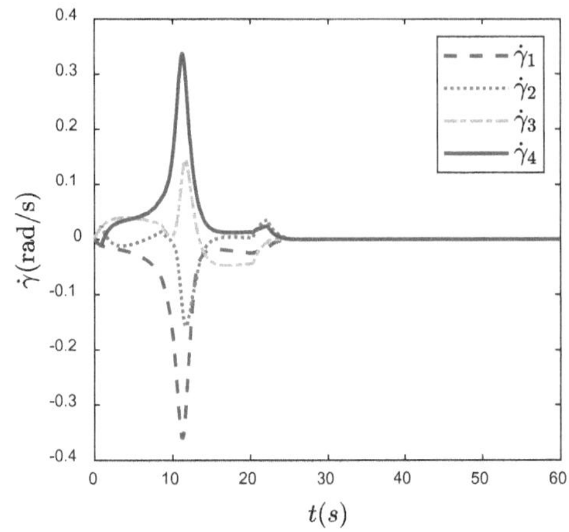

FIG. 12.19 The gimbal speed of CMGs corresponding to inner product.

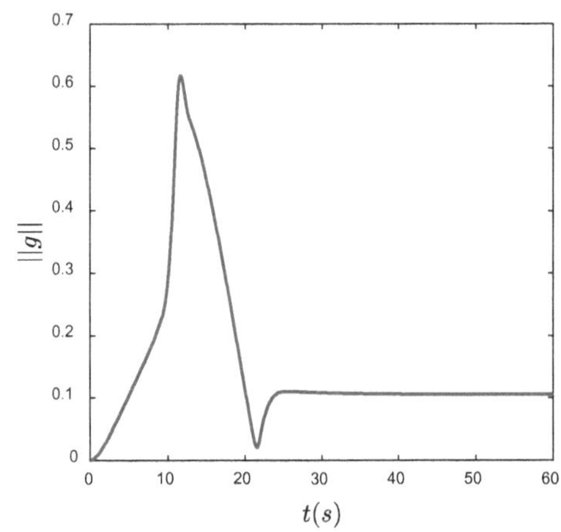

FIG. 12.20 The norm of gradient of inner product index.

12.5 Conclusion

Aiming at the control overdrive problem with the hybrid actuator of CMGs and flywheels, this chapter designs an optimization objective function including the singularity index of CMGs and the energy and proposes a control allocation strategy based on the constrained optimization method. By reasonably adjusting the weight matrices, the CMGs with large output torque are adopted for control in the process of rapid maneuver, and the flywheels are adopted in the stable control stage requiring high accuracy. The problems of flywheel speed saturation and singularity of CMGs can be effectively addressed. The simulation results show that using the inner product singularity index as the singularity index can make the CMG's frame angular velocity lower than that of the non-dimensional conditions number and make the "idling" of the control moment gyroscope more effective and the angular velocity control smoother.

References

[1] X.H. Wang, H. Zhang, S.Z Bai, et al. Design of agile satellite constellation based on hybrid-resampling particle swarm optimization method, Acta Astronaut. 178 (5) (2021) 595–605.

[2] Z.M. Song, X.Y. Chen, X. Luo, et al., Multi-objective optimization of agile satellite orbit design, Adv. Space Res. 62 (11) (2018) 3053–3064.

[3] X.W. Wang, Y. Gu, G.H. Wu, et al., Robust scheduling for multiple agile earth observation satellites under cloud coverage uncertainty, Comput. Ind. Eng. 156 (2021), 107292.

[4] X.G. Chu, Y.N. Chen, Y.J. Tan, An anytime branch and bound algorithm for agile earth observation satellite onboard scheduling, Adv. Space Res. 60 (9) (2017) 2077–2090.

[5] H.A. Lei, A. Xll, A. Ywc, et al., Hierarchical scheduling for real-time agile satellite task scheduling in a dynamic environment, Adv. Space Res. 63 (2) (2019) 897–912.

[6] J. Kim, J. Ahn, Integrated framework for task scheduling and attitude control of multiple agile satellites, J. Aerospace Inf. Syst. 18 (8) (2021) 539–552.

[7] F. Yao, J.T. Li, Y.N. Chen, et al., Task allocation strategies for cooperative task planning of multi-autonomous satellite constellation, Adv. Space Res. 63 (2) (2019) 1073–1084.

[8] J.C. Sanchez, F. Gavilan, R. Vazquez, et al., A flatness-based trajectory planning algorithm for rendezvous of single-thruster spacecraft, IFAC-PaperseOnLine 51 (12) (2018) 118–123.

[9] X.B. Cao, B.L. Wu, Robust spacecraft attitude tracking control using hybrid actuators with uncertainties, Acta Astronaut. 136 (2017) 1–8.

[10] Y.H. Wu, H. Feng, M.H. Zheng, et al., Attitude control for on-orbit servicing spacecraft using hybrid actuator, Adv. Space Res. 61 (6) (2018) 1600–1616.

[11] B. Wie, Singularity escape/avoidance steering logic for control moment gyro systems, J. Guid. Control. Dyn. 28 (5) (2005) 948–956.

[12] K.A. Ford, C.D. Hall, Singular direction avoidance steering for control-moment gyros, J. Guid. Control. Dyn. 23 (4) (2000) 648–656.

[13] F.A. Leve, N.G., Fitzcoy hybrid steering logic for single-gimbal control moment gyroscopes, J. Guid. Control. Dyn. 33 (4) (2010) 1202–1212.

[14] S.R. Vadali, H.S. Oh, S.R. Walker, Preferred gimbal angles for single gimbal control moment gyros, J. Guid. Control. Dyn. 13 (6) (1990) 1090–1095.

[15] H. Yoon, Spacecraft Attitude and Power Control Using Variable Speed Control Moment Gyros, Georgia Institute of Technology, Georgia, US, 2004, pp. 138–148.

[16] H. Yoon, P. Tsiotras, Singularity analysis of variable-speed control moment gyros, J. Guid. Control. Dyn. 27 (3) (2004) 374–386.

[17] Y.H. Wu, M.H. Zheng, C.Y. Li, et al., Hybrid actuator optimal angular momentum management with analytical solution for spacecraft attitude agile maneuvering mission, Aerosp. Sci. Technol. 97 (2020), 105597.

[18] K. Ishihara, J. Morimoto, An optimal control strategy for hybrid actuator systems: application to an artificial muscle with electric motor assist, Neural Netw. 99 (2018) 92–100.

[19] H. Leeghim, D. Lee, H. Bang, et al., Spacecraft attitude control by combination of various torquers, Int. J. Syst. Sci. 40 (10) (2009) 995–1008.

[20] H. Schaub, J.L. Junkins, Singularity avoidance using null motion and variable-speed control moment gyros, J. Guid. Control. Dyn. 23 (1) (2000) 11–16.

[21] D.E. Cornick, Singularity Avoidance Control Laws for Single Gimbal Control Moment, Guidance and Control Conference, American Institute of Aeronautics and Astronautics, 1979, pp. 1–21.

[22] J.L. Junkins, Y. Kim, Introduction to Dynamics and Control of Flexible Structures, American Institute of Aeronautics and Astronautics, Reston, VA, USA, 1993, pp. 9–68.

[23] B. Wie, Singularity analysis and visualization for single-gimbal control moment gyro systems, J. Guid. Control. Dyn. 27 (2) (2004) 271–282.

[24] H. Leeghim, H. Bang, J.O. Park, Singularity avoidance of control moment gyros by one-step ahead singularity index, Acta Astronaut. 64 (9–10) (2010) 935–945.

Chapter 13

Conclusions and future works

13.1 General conclusion

This book mainly discusses the attitude control strategy of spacecraft large-angle fast maneuvering and discusses the control methods such as PID+ controller, sliding mode controller, finite time controller, and path planning algorithm in satellite attitude control. The main contributions and innovations are as follows:

(1) Based on the idea of bang-bang control and by optimizing the total corner to give an analytical solution for a shorter path, the computational complexity of the path planning algorithm is greatly reduced and the system convergence rate is effectively improved, in order to solve a special situation of attitude maneuver, a kind of attitude maneuver path similar to "Horman maneuver" is designed, under the premise of large-angle attitude maneuvering, and the system convergence time is increased by more than 50% compared with the classic PD controller.

(2) By effectively combining PID control and sliding mode control and designing a PID+ controller to make the system state converge along the designed sliding mode surface, the convergence time of the system state under the premise of large-angle maneuvering is increased by more than 70% compared with the classic PID controller. The relationship between the parameters of the PID controller and the norm of the attitude angular velocity is proved, and the constraint of the attitude angular velocity on the controller parameters is determined;

(3) By combining the open-loop bang-bang control with the closed-loop sliding mode control, the system performance is improved to the utmost. The method of locking sliding mode parameters in classical sliding mode control is improved by designing the sliding mode parameter update law to increase the system convergence rate. Under the premise of large-angle attitude maneuver, the system convergence time of the two control algorithms is reduced by more than 60% compared with the classic sliding mode controller. The relationship between the control torque, the attitude angular velocity, and the control parameters is proved, and the strict constraints of the control parameters are determined.

Fast Satellite Attitude Maneuver and Control. https://doi.org/10.1016/B978-0-323-95455-6.00002-0

(4) A finite-time sliding surface with a three-stage structure is designed, and it uses a uniform speed process to maintain the convergence rate of the system when it is far from the equilibrium point, and it uses the Euler axis characteristics to eliminate the singularity of the controller, so that the system has finite time stability when approaching the equilibrium point. Under the premise of large-angle attitude maneuvering, the system convergence time is increased by more than 50% compared with the uniformly asymptotically stable sliding mode controller.

(5) A nonsingular fast terminal sliding mode attitude control scheme is proposed for the improved first-order approximate rigid-flexible coupling dynamic model of flexible satellites. The on-orbit condition when flexible vibration or angular velocity is unavailable is also considered in the controller design. Both theoretical analysis and numerical simulation show that the error angle and error angular velocity converge quickly in finite time with improved steady-state accuracy.

(6) With considering external disturbances, model uncertainties, and unmeasurable flexible coupling terms, an ESO-based continuous finite time attitude control is proposed for rigid–flexible coupling spacecraft in large-angle fast maneuver mode. The total perturbation of the system can be compensated without knowing the bound in prior by the ESO. The compensation effect of the ESO, which makes the control gain decrease, and the continuity contributed by the fast power reaching law together attenuate the chattering that exists in the sliding mode control. The effectiveness of this control scheme is further verified by the ground-based semi-physical simulation.

(7) Further considering actuator faults of rigid-flexible coupling spacecraft, a leNN-based finite time fault tolerant attitude control scheme is proposed based on the nonsingular fast terminal sliding mode. LeNN is incorporated to approximate the total disturbance of the fault mode system including the estimation error of the fault control torque, high-order coupling terms, external disturbances, and uncertain rotational inertia. A switch mechanism is introduced to generate a switching between the proposed fault-tolerant attitude controller and a robust controller, thus guarantees that the output of the Legendre polynomial-based neural network stays inside the bound of the lump perturbation,

(8) For large-angle rapid reorientation of rigid spacecraft, a hybrid actuators control strategy is proposed. Thrusters are applied with an open loop control to drive the spacecraft to a reference path, while reaction wheels are implemented with a closed-loop variable structure control law to correct for tracking errors due to the initial condition errors and the torque bias of thrusters.

(9) Under the consideration of output torque deviation and initial measurement error, a hybrid actuators finite time attitude control scheme for rigid spacecraft in rapid maneuver is presented. The finite-time attitude tracking of the desired optimal-time trajectory can be achieved based on a nonsingular terminal sliding mode-based hybrid thruster-flywheel control strategy.

(10) With considering the control overdrive problem with the hybrid actuators of CMGs and flywheels, a constrained optimization method-based control allocation strategy for rigid spacecraft is presented by designing an optimization objective function including the singularity index of CMGs and the rotational energy. The problems of flywheel speed saturation and CMG singularity can be effectively solved.

13.2 Future work

In general, this book discusses the design of the attitude control scheme for agile satellites in large-angle fast maneuver and obtains phased research results. Furthermore, the authors believe that it is still necessary to continue the follow-up research in the following aspects.

(1) In terms of sliding mode control, the control torque in the bang-bang control-based sliding mode controller proposed in this book does not deal with its amplitude, which means that there is still room for improvement in the convergence time.

(2) The path planning algorithm proposed in this research sacrifices part of the performance of the system in order to obtain the analytical solution, and the system needs to perform frequent acceleration and deceleration. How to obtain the analytical solution of the maneuvering path without sacrificing system performance still needs to be discussed.

(3) The simulations carried out in this research are mostly numerical simulations. Furthermore, more physical or semi-physical simulations can be carried out to verify the control algorithms proposed in this research if conditions permit.

(4) The rigid–flexible coupling effect in three-axis rotation should be further studied, with which a first-order approximation rigid-flexible three-axis dynamic model could be considered. Then attitude tracking for time-variant target attitude for flexible spacecraft with active vibration suppression can be addressed.

(5) Because the assumption of thruster disturbance torque in this book is relatively simple, it is necessary to deeply model and analyze it in order to achieve better engineering control performance.

Index

Note: Page numbers followed by *f* indicate figures and *t* indicate tables.

G

H

I

J

L

FSC
www.fsc.org
MIX
Papier | Fördert
gute Waldnutzung
FSC® C083411

Kolibri 360 GmbH
Ferdinand-Jühlke-Straße 7
99095 Erfurt, Deutschland
produktsicherheit@kolibri360.de